Environmental Economics

By the Same Author

Control of Air Pollution, 1963
The Prospects for Nuclear Power Generation in Australia, 1973
Dictionary of Environmental Terms, 1976
Air Pollution, 1978
The Australian Environment: Twelve Controversial Issues, 1979
Environment Policy in Australia, 1980
Dictionary of Energy Technology, 1982
The Human Environment: The World After Stockholm, 1975
Environmental Planning: A Condensed Encyclopaedia, 1986
Dictionary of Economics and Financial Markets, 1986
State of the Environment in Australia: Overview, 1986
An Australian Dictionary of Environment and Planning, 1990
Environmental Impact Assessment: Cutting Edge for the Twenty-first Century, 1995
Dictionary of Environment and Sustainable Development, 1996

Environmental Economics

A critical overview

ALAN GILPIN

JOHN WILEY & SONS, LTD
Chichester • New York • Weinheim • Brisbane • Singapore • Toronto

National 01243 779777
International (+44) 1243 779777
e-mail (for orders and customer service enquiries): cs-books@wiley.co.uk
Visit our Home Page on http://www.wiley.co.uk
or http://www.wiley.com

Other Wiley Editorial Offices

John Wiley & Sons, Inc., 605 Third Avenue,
New York, NY 10158-0012, USA

WILEY-VCH Verlag GmbH, Pappelallee 3,
D-69469 Weinheim, Germany

Jacaranda Wiley Ltd, 33 Park Road, Milton,
Queensland 4064, Australia

John Wiley & Sons (Asia) Pte Ltd, 2 Clementi Loop #02-01,
Jin Xing Distripark, Singapore 129809

John Wiley & Sons (Canada) Ltd, 22 Worcester Road,
Rexdale, Ontario M9W 1L1, Canada

Library of Congress Cataloging-in-Publication Data

Gilpin, Alan.
Environmental economics : a critical overview / Alan Gilpin.
p. cm.
Includes bibliographical references and indexes.
ISBN 0-471-98558-9. – ISBN 0-471-98559-7
1. Environmental economics. 2. Sustainable development.
I. Title.
HC79.E5G519 1999
333.7 – dc21 99-35168
CIP

British Library Cataloguing in Publication Data

A catalogue record for this book is available from the British Library

ISBN: 978-0-471-98559-4

To
Andrew and Benjamin
Lisa, Mimi and Nancy

Contents

About the Author

Alan Gilpin is a Visiting Fellow in the School of Civil and Environmental Engineering, University of New South Wales, Australia. He was previously, for ten years, a Commissioner of Inquiry conducting public inquiries into controversial development projects throughout New South Wales. A graduate of the University of London, Dr Gilpin gained his doctorate through the University of Queensland for a thesis on *The Economics of Nuclear Power*. He is also a Fellow of the Institution of Engineers Australia and the author of twelve books on economic and environmental terms, energy technology, pollution control, and environmental impact assessment. He has lectured widely in Britain, the USA, Australia and Asian countries.

He began his career in the British public health service, becoming the youngest chief environmental health officer in local government. He later joined the Central Electricity Generating Board, being involved in public inquiries into the design and location of major power stations. In 1965, he emigrated with his family to Australia to set up one of the earliest environment agencies, becoming chairman of the Environment Protection Authority, Victoria.

This book is based on extensive teaching at the University of New South Wales, mainly at fourth-year and Master's degree level, in environmental economics, environmental impact assessment, and environmental law, in engineering and social science faculties.

Preface

This book seeks to demonstrate that while the prophets of doom and gloom have been largely confounded, there are in the greater distance still limitations to growth. The key limitation is population and the rate of population growth, as it appears beyond doubt that further marginal increases in the production of food are being achieved with increasing rates of marginal environmental damage. World population needs to achieve a stable situation early in the 21st century, to ease the growing pressure on soil and agricultural resources, fisheries, and the availability of water. Population stability becomes particularly important when considering the need for a rising standard of living for most of the world's population and the urgent need for the alleviation of poverty experienced by too many. To achieve all this will impose demands on the entire range of resources within a finite world.

However, the tone of the book is essentially cornucopian; that a rising standard of living for all is within reach, that many of the Earth's resources are as yet unexploited, and that most environmental problems are amenable to solution. Further, this can be achieved without the imposition of restraints on consumption by the developed world, as urged by some. A declining standard of living in the West with the rationing of petrol and cars, with timed energy consumption, are not prerequisites for the material success of the developing world. Indeed, on the contrary.

In the longer term, as in the past, substitution will continue to play a key role in the sustainability of diverse activities. Exhaustible resources will gradually be replaced over the next hundred years by more economically acceptable substitutes, with much reduced adverse environmental effects.

The ten chapters of this book traverse a range of controversial issues and developments. Chapter 1 (Economics and the Environment) introduces the subject of environmental economics as a relatively new branch of welfare economics, a subject already well-established in mainstream neoclassical economics. The chapter describes briefly the relevant features of the market economy, now tending to dominate throughout the world. It describes the failures of this system in respect of adequate protection of the environment, examining some costs and benefits in the USA.

Chapter 2 (Luminaries) seeks to elaborate the history of environmental concerns through the lives of some of those who have brought attention to various crucial aspects of the subject. It begins with the contribution of John Evelyn and finishes with the flourishing of the Green Parties and NGOs. In Chapter 3 (Institutions) the stories of people blend with the development of international institutions, mainly as an outcome of a whole range of UN Conferences since 1949. Also embraced are the International Whaling Commission, the World Trade Organization, the International Institute for Environment and Development and the Beijer International Institute of Ecological Economics.

Chapter 4 (Sustainable Development) addresses a vexed question, for since the introduction of the concept of sustainable development by the World Conservation Strategy in 1980, no one has been quite sure what it means, or rather everyone has a different opinion as to what it means. Some definitions point to concern with the well-being of future generations, while this book argues that it is sufficient to concentrate on the well-being of those alive today. Those not yet born could well be many times better off than people today. Our main gift to the future is the technology of today, and hopefully a breath of civilization. Property rights are not seen as a simple solution to anything.

Chapter 5 (Direct Regulation) describes the history, scope and achievements of direct regulation, rarely acknowledged, ranging from the control of air and water pollution, environmental management, environmental impact assessment, and town planning, to international conventions.

Chapter 6 (Economic Instruments) is devoted to a review of a whole range of economic incentives towards proper and responsible environmental behaviour, including fines and ransoms. Such instruments will never replace direct regulation, but may in appropriate circumstances, prove a valuable augmentation. In this way, financial incentive and social responsibility may point in the same direction. The validity of the buying and selling of pollution and development rights is questioned.

Although the whole of environmental economics sees the activities of society within an overarching cost–benefit framework, Chapter 7 (Cost–Benefit Analysis and Valuation) is concerned with cost–benefit as an aid to the making of sound investment decisions for the community as well as private benefit. Its strengths and limitations are expounded, as well as the crucial problems of attaching economic values to intangible environmental assets.

Chapter 8 (Air, Land and Water Issues) addresses a somewhat arbitrary selection of issues, but which collectively strengthen the fabric of the book. Air pollution issues include urban air pollution, an increasingly damaging issue in developing countries, and acid rain, an issue of importance in North America and Europe. We then switch to the interrelated subjects of global biodiversity, forestry and the Amazon rain forest. In contrast these are followed by deserts, both sand and ice, with a special note on Egypt. Turning to water, the Baltic and the Caspian Seas are followed by the Murray-Darling Basin in Australia, and Asia's Mekong River. Two case studies follow: the Ok Tedi mine operation in

Papua New Guinea and the biggest dam in the world, the Three-Gorges water conservation and hydroelectric project, China.

Following the Kyoto conference, many nations are now committed to achieving reductions in emissions of carbon dioxide and other greenhouse gases. The basis for this decision is examined in Chapter 9 (Global Warming). The author's conclusion is that positive and expensive measures against greenhouse gases are probably not needed for thirty years. In the meanwhile, nations should accelerate current research into alternative sources of energy, with particular reference to solar energy and hot rock, the preferred front-runners for a pollution-free future. There is as yet no clear evidence that human activities have so far contributed in a substantial way to the warming of the atmosphere; severe and dramatic measures undertaken prematurely over a short period could result in energy disruption, privation, rationing and unemployment on a large scale. There is an easier way to cheap energy, without pollution.

Chapter 10 (Population and the Quality of Life) addresses several of the main issues in this book, world population and the rate of population growth, mortality and morbidity, the increasing rate of urbanization, world food production, fisheries, water quality and sanitation. Measures of progress are reviewed.

At this point, the controversial issue of the automobile is introduced. Suzuki envisages a carless society as an ultimate aim. On the other hand, everyone tries to own a car for its mobility and comfort, and many modern towns and cities require cars, public transport being scarce and inconvenient. This subject appears in this chapter for its links with population growth and rising standards of living. Can the world accommodate two billion cars?

Each chapter in this book has an introduction and, at the end, a summary which picks up the main conclusions. A list of further reading follows, while the book ends with a bibliography. The subject of environmental economics has attracted a considerable literature: selection has been difficult.

Abbreviations

APEC	Asia-Pacific Economic Cooperation Forum
ASEAN	Association of South-East Asian Nations
ECE	Economic Commission for Europe
EEZ	exclusive economic zone
EIA	environmental impact assessment
EIS	environmental impact statement
ESCAP	Economic and Social Commission for Asia and the Pacific
ESD	ecologically sustainable development
EU	European Union
FAO	Food and Agriculture Organization
g	gram
GATT	General Agreement on Tariffs and Trade
GDP	Gross domestic product
GEF	Global Environment Facility
GIS	geographic information system
glc	ground level concentrations
h	hour
ha	hectare
HDI	Human Development Index
IAEA	International Atomic Energy Agency
IFAD	International Fund for Agricultural Development
ILO	International Labour Organization
IMF	International Monetary Fund
IMO	International Maritime Organization
IPCC	Intergovernmental Panel on Climate Change
ISA	International Seabed Authority
IUCN	International Union for the Conservation of Nature and Natural Resources (now the World Conservation Union)
IWC	International Whaling Commission
km	kilometre
m	metre
MW	Megawatt
NAFTA	North American Free Trade Association
NGO	non-government organization

OECD	Organization for Economic Cooperation and Development
ppbv	parts per billion by volume
ppmv	parts per million by volume
s	second
SD	sustainable development
t	tonne
UN	United Nations
UNCED	United Nations Conference on Environment and Development
UNDP	United Nations Development Program
UNEP	United Nations Environment Program
UNESCO	United Nations Educational, Scientific and Cultural Organization
UNFPA	United Nations Population Fund
UNICEF	United Nations Children's Fund
UNIDO	United Nations Industrial Development Organization
US/USA	United States of America
US EPA	United States Environmental Protection Agency
WB	World Bank
WCU	World Conservation Union (previously the International Union for the Conservation of Nature and Natural Resources)
WFP	World Food Program
WHO	World Health Organization
WMO	World Meteorological Organization
WRI	World Resources Institute
WTO	World Trade Organization
WWF	World Wide Fund for Nature

'Although there are hundreds of millions of poor and unfortunate people in all countries, the overall global trend is of immense human progress.'

Gro Harlem Brundtland (1988) *Shifting Attitudes in a Changing World*, Britannica Book of the Year, New York and London

1
Economics and the Environment

1.1 Introduction

The chapter opens by briefly traversing the subject of economics, the superpower of the social sciences; the functions of an economic system; the nature of the market and the market economy; the profit motive; market efficiency and market failure; the laws of supply and demand; the concept of Pareto optimality and the compensation principle; and the relationship and importance of the gross domestic product (GDP) in relation to economic growth, economic development, and the quality of life. Much of this will be familiar to some readers.

The chapter then turns to the meaning of the word 'environment'; the nature of macro- and microenvironmental problems; and the evolution and role of environmental economics.

Such topics as the U-shaped hypothesis, the choice of environmental policy instruments, the optimization of pollution abatement, pollution control and international trade, and the role of substitution are then examined. The chapter concludes with an examination of some USA costs and benefits in relation to pollution control.

The subject of environmental economics encompasses all the costs of environmental damage and environmental control and all the benefits of environmental and resource protection considered within a global cost–benefit framework, with a balancing of costs and benefits within each segment of human endeavour, sustaining the resource base in one form or another on which the present and future generations will rely. The concept of 'Spaceship Earth' has much to commend it, while its need for careful management is an obligation falling collectively upon humanity.

1.2 Economics

Economics is a social science concerned with how people, either individually or in groups, attempt to accommodate scarce resources to their wants through the processes of production, distribution, substitution, consumption and exchange.

There have been many definitions of economics in the past. The earliest definitions regarded economics as the study of wealth. Indeed, Adam Smith (1723–90) the father of modern economics, entitled his great work *An Inquiry into the Nature and Causes of the Wealth of Nations* (1776). Some seventy years later, John Stuart Mill (1806–73) in his *Principles of Political Economy* (1848) defined economics as 'the practical science of the production and distribution of wealth'.

The Victorian economist, Alfred Marshall (1842–1944) in his *Principles of Economics* (1890) defined economics as 'the study of mankind in the ordinary business of life; it examines that part of individual and social action which is most closely connected with the attainment and with the use of the material requisites of wellbeing. Thus it is on the one side a study of wealth; and on the other, and more important side, a part of the study of man .... Economics is a study of men as they live and move and think in the ordinary business of life.' Lionel Robbins (1898–1984) in his *An Essay on the Nature and Significance of Economic Science* (1932) defined economics as 'a science which studies human behaviour as a relationship between ends and scarce means which have alternative uses'.

These early definitions stressed only the alternative and competitive uses of resources. Today, the science of economics is regarded as being concerned with the allocation of resources, priced or unpriced, between alternative individual and social uses; the distribution of output among individuals and groups; the ways in which production and distribution change over time; the efficiencies and inefficiencies of economic systems; and the implications of sustainable development. Box 1.1 seeks to define the nature of an economic law or tendency, while Box 1.2 further illustrates this with a statement of the law of diminishing marginal utility. Box 1.3 seeks to define the functions of an economic system, which goes far beyond the buying and selling of goods and services in a local market. Box 1.4 completes this section with the comments of *The Economist* on the uses and abuses of economics.

Box 1.1 Economic Law

Economic law is a statement of what will occur in the economic world under specified conditions. George J. Stigler (1911–91), in his *Theory of Price* (1966), offers the following example of an economic law:

If: (a) An entrepreneur seeks maximum profits;
(b) His marginal cost curve does not fall so fast as (or more rapidly than) his marginal revenue curve;
(c) The curves are continuous.

Then: He operates at the output where marginal revenue equals marginal cost.

Alfred Marshall (1842–1924), in his *Principles of Economics* (1920), took the view that

Continued on page 3

Continued from page 2

there are no definite and universal propositions in economics which can compare with the Laws of Gravitation and of Conservation of Energy in physics, but there are many which rank with the secondary laws of those natural sciences which resemble economics in dealing with the complex actions of many heterogeneous and uncertain causes ... An economic law is nothing but a general proposition, or statement of uniformity, more or less certain, more or less definite.

A specified condition, often assumed, is that of *ceteris paribus* or the holding of other factors constant.

Box 1.2 Law of Diminishing Marginal Utility

The law of diminishing marginal utility is the proposition that the marginal utility of a commodity or service (that is, the total utility of consumption of a good or service which results from increasing or decreasing the quantity consumed by one unit) to anyone, other things being equal, diminishes after some point with every increase in the supply of it. Water, which has a very high total utility, life depending on it, usually has a very low marginal utility bordering on zero; in the desert, under trying conditions, the marginal utility of water may be very high indeed. Alfred Marshall (1842–1924) stressed an implicit condition in this law: 'that we do not suppose time to be allowed for any alteration in the character or tastes of the individual concerned' (Marshall 1920). This law explains why most demand curves slope downward to the right, indicating that the higher the price, the fewer goods will be sold.

It is generally held that the marginal utility of money diminishes as the quantity of money possessed by an individual increases. Also, the larger the amount of a thing which a person has, the less that person will pay for a little more of it (other things being equal), i.e. the marginal demand price diminishes. The concept of marginal utility also explains why the selling of an additional unit of output may require a lowering of price. It also explains why each additional investment in pollution-control equipment, which may be increasingly expensive, yields less and less marginal benefit to the public, as lower concentrations of pollution are achieved.

The factors of production are defined as natural resources (traditionally land), labour including managerial and entrepreneurial skills, and capital. Today, four kinds of capital are recognized, as described below:

- *Human-made capital*: traditionally, one of the factors of production, defined as wealth used for the production of further wealth, or simply a commodity used in the production of other goods and services. The term is also commonly used for the the money subscribed by shareholders, or lent by financial institutions for use in a business, or with reference to savings. Capital is described as real when it consists of machines, factories, aircraft, stocks of raw materials, and so on. It may be subdivided into fixed, floating or circulating, specific

Box 1.3 The Functions of an Economic System

The functions of an economic system are as follows:

- to match supply to effective demand for goods and services, at home and overseas, in an efficient manner and to an acceptable quality;
- to allocate scarce resources among industries and activities producing and distributing goods and services;
- to organize wholesaling, retailing, trading, exporting and importing facilities for all kinds of goods and services;
- to provide channels for fruitful capital investment and transactions in financial instruments;
- to fully utilize the human resources of society, endeavouring to provide adequate employment opportunities, both full-time and part-time, for men, women and school-leavers, with adequate training facilities;
- to utilize the natural resources and environmental assets of society within a framework of sustainable development;
- to convey to consumers information on which goods and services are available for best value, within a socially responsible framework;
- to promote research and development aiming at improved, safer, environmentally friendly products and services, better designed to meet society's objectives;
- to increase productivity in all sectors, easing distress resulting from downsizing, restructuring, devolution, off shoring, and competitive failure;
- to conserve raw materials, achieve cleaner production, reduce waste, encourage reuse and recycling, minimize waste disposal problems, and generally protect the environment at all stages of activities;
- to control all emissions from activities to meet statutory requirements and social aspirations;
- to relate to the affected community through regular meetings and open-days;
- to promote satisfactory industrial relations, equal pay for equal work, and equality of opportunity;
- to promote fair trading in all markets;
- to co-operate with government in improving the performance of the economic and financial system and promoting the interests of the nation.

Box 1.4 The Use and Abuse of Economics

For all its imperfections, economics has turned itself into the superpower of the social sciences, one that exercises a powerful influence on the lives of ordinary people. Arguably, the discipline has never been more influential. And yet economists themselves have seldom been more derided. They are accused of being irrelevant, wrong, and in perpetual disagreement ... the chief explanation for the contradictory standing of economics is that although it is economists who produce economic ideas, it is other people who use them. And abuse them.

Source: The Economist, 25 November 1995, p. 15.

and non-specific capital. It represents current consumption foregone in order to obtain future enhanced production and consumption.

- *Natural capital*: our natural endowment or the stock of environmentally provided assets such as soil, atmosphere, forests, water, oceans, wetlands, fossil fuels, minerals, woodland, biodiversity, fauna and flora, and ecosystems, which provide a flow of goods and services of various kinds, renewable or non-renewable, marketable or non-marketable, including commercial, existence, ecological and spiritual values. The concept has been extended to cultivated natural capital such as agriculture and aquaculture, and the sink functions of ecosystems. Revised national accounts tend to reflect the well-being of natural assets.
- *Human capital*: the sum of all the skills and energy embodied in the community at large or, more narrowly, in the workforce. Investment in people is now more readily recognized as being a high-return investment, especially in developing countries. The concept of development has been extended to include investment in human resources as an important ingredient of development strategies. This includes investments in health, nutrition, education and literacy, socially valuable employment, equal opportunities for women in education and employment, intragenerational equity, and good working conditions.
- *Social capital*: factors that make a society more than the sum of a collection of individuals, reflecting cohesion, basic unity, common ideals, democracy, and common pride. Failures are to be seen in Rwanda, Yugoslavia, Somalia, and a dozen other countries with horrific internal conflicts. Social capital ensures a functioning, rewarding, social order with an improving quality of life. It is an essential component of sustainability, political stability and human welfare. It is the basis of international respect and recognition. Investment in social capital depends very much on the contributions of individuals at the local, regional, national and international levels.

These four forms of capital hang together to make and break nations and societies. However, the term 'social capital' troubles some distinguished authorities as it does not lend itself readily to the processes of accumulation, depletion, depreciation and financial investment, unlike the three other forms of capital.

The term 'neoclassical economics' is descriptive of economic science as it evolved from the work of the Classical School during the last quarter of the 19th century. Those belonging to the Classical School were the first to formulate a systematic body of economic principles, seeking to explain uniformities in economic activity which are not the result of deliberate design by a central planning body, but the product of interplay of separate decisions of individuals and groups. The principal members of the group included Adam Smith (1723–90), Thomas Robert Malthus (1766–1834), David Ricardo (1772–1823), Nassau William Senior (1790–1864), James Mill (1773–1836) and John Stuart Mill (1806–73). Other members of the School included John Elliot Cairns, John Ramsay McCullock, Jean Baptiste Say, David Hume and Robert Torrens. The name of Karl Marx is sometimes included. Numerous schools of economics have emerged over the years including the Austrian School, the Chicago School, the Keynesian School, the Lausanne School and the Neoclassical School.

The Neoclassical School, which developed after the Second World War, follows in the tradition of the Classical School. Essentially, the Neoclassical School has been concerned with the problems of equilibrium and growth at full employment, in contrast with the Keynesian School which was preoccupied primarily with the underemployment of resources. Based on the writings of Marshall, neoclassical economics has involved a shift in emphasis from issues such as the distribution of income towards the study of the principles that govern the optimal distribution of scarce resources to a range of wants. This theme is central to environmental economics, a branch of welfare economics.

Welfare economics is the study of the social desirability of alternative arrangements of economic activities involving allocations of resources; in effect the analysis of the optimal behaviour of consumers at the level of society as a whole. It involves judgements about how the economy should perform and how society should look, to maximize economic welfare.

1.3 What is a Market?

A market is an area, however large or small, where buyers and sellers are in sufficiently close contact with each other to ensure that the price of a commodity tends to be the same in all parts of the market, allowance being made for transportation costs, trade barriers, lack of information, and other obstacles. There are markets for all kinds of goods and services: foodstuffs, dwellings, currency, furniture, information technology, literature, property rights, labour, entertainment, pharmaceuticals, emission credits, contraband, armaments, securities, travel, imports and exports, sex, and so on. A market may be confined to a village, town or suburb, or be regional, national or international in scope. A market may promote or demean the suppliers and the consumers. A former prominent part of national and international trade was slavery and opium. Until 1865, the southern states of the US regarded the buying and selling of men, women and children as a property exchange, without moral implications. Slavery in a strictly legal sense elsewhere in the world ended probably in 1970, but it exists today in similar or modified forms in all economies and in the great prison encampments of dictatorships.

1.4 The Market Economy

The market economy is one in which the crucial economic decisions and choices are made in a decentralized manner by numerous private individuals and firms, operating through a free market mechanism. Equilibrium prices and quantities are determined in a market economy through the laws of supply and demand, consumer preferences expressed through the price mechanism influencing the kinds of goods and services produced. The market economy is an admixture of monopoly (a sole seller), oligopoly (a few sellers), oligopsony (a few buyers), and open competition (many suppliers, many consumers); however, the market is far removed from the concept of *perfect competition* as originally conceived for theoretical purposes by economists, despite its merit in that role. John Kenneth Galbraith in his *Economics and the Public Purpose* (1974) has described the model as 'not implausible as a decription of a society that once existed. Nor is it entirely unsatisfactory as a picture of that part of the economy called the market system ... it is a rough description of half the economy; it has lost touch with the other and in many ways decisive half.'

Since the collapse of the Soviet Empire and the centralized planning system, the world has moved towards a comprehensive global market system; even the last major communist country, the People's Republic of China, has moved uncertainly towards a freer system. In this process, the merits of the planned system have been lost, with materialism alone dominating the scene throughout.

1.5 The Profit Motive

A traditional assumption in the theory of the firm is that the firm or business corporation attempts to maximize returns of profits. In other words, subject to the constraints of knowledge, energy and ambition, the firm in the market system, competitive or monopolistic, does in fact maximize profits in the short or long run. Special criteria have been set for public sector industries which have both an economic and social function. However, it has been argued in recent years that the assumption of profit maximization in the private sector may be something of a simplification; security of earnings, growth, extension of markets, and increase in market share are also important. A growing firm is better able to control its technology, its costs and its prices, avoiding more successfully the threat of bankruptcy and take-over. Shareholders may be willing to accept a reasonable level of dividends, while the company pursues some of these alternatives with strengthened policies. Control passes effectively to management, which sees satisfaction and rewards as much linked to size as to profits.

More recent theory suggests that firms do not seek to maximize anything, but only aim to reach satisfactory levels of sales, profits, growth and security.

The participants or stakeholders today include shareholders, creditors, directors, managers, customers, suppliers, the workforce, marketing and advertising specialists, outlets, consultants and strategists, members of the local community, and possibly government at all levels. Some will have a defined contractual relationship, others a financial interest with certain claims in the event of liquidation; many will have direct personal concerns by way of employment and careers. The local, state and provincial governments may have encouraged the enterprise by granting planning and tax concessions, installing infrastructure at

Box 1.5 The Invisible Hand

> . . . every individual necessarily labours to render the annual revenue of the society as great as he can. He generally, indeed, neither intends to promote the public interest, nor knows how much he is promoting it . . . he intends only his own gain, and he is in this, as in many other cases, led by an invisible hand to promote an end which was no part of his intention. Nor is it always the worse for the society that it was no part of it. By pursuing his own interest he frequently promotes that of the society more effectually than when he really intends to promote it. I have never known much good done by those who affected to trade for the public good.

On the other hand:

> People of the same trade seldom meet together even for merriment and diversion, but the conversation ends in a conspiracy against the public, or on some contrivance to raise prices.

Source: Adam Smith, An Inquiry into the Nature and Causes of the Wealth of Nations (1776), reprinted by Ward Lock, London, pp. 116 and 354.

public expense, providing subsidies and other means of financial support such as grants for research and development. Adverse trading conditions, high-interest rates, errors of business judgement, changes in taxation, variations in levels of protection, mergers and take-overs endanger the interests of all stakeholders. All non-enforceable contracts may suddenly count for nothing. Survival, rather than profit maximization, becomes the primary concern. The whole process is essentially amoral, being concerned solely with financial considerations. Ethics may be taken into account only when it can be demonstrated that the principle has a 'bottom line', affecting the future prospects of the firm. Environmental considerations fall into this category where legal proceedings by government may lead to the fining or imprisonment of directors, increasing resistance to the firm's future expansion, or damage to a marketing image. In this way, a company may acquire a socially responsible image. The theme of profit maximization becomes modified in many ways, yet profit is essential not only to survival, but to growth, rewards to shareholders, the satisfaction of creditors, and the ability to render continuing and improving service to the biggest stakeholder of all, the purchasing public (see Box 1.5).

1.6 Market Efficiency and Market Failure

Market efficiency prevails when no one can be made better off without making someone worse off. Market efficiency often implies three other types of efficiency: *productive efficiency*, in which the output of goods and services is being provided at the lowest cost; *allocative efficiency*, in which resources are being allocated for the production of the goods and services the community requires;

and *distributional efficiency*, in which output is distributed in such a way that consumers would not wish (given their disposable income and market prices) to spend their incomes in any other way, and in which the degree to which a market economy meets the needs of society, contributes to the quality of life, encourages efficient production, and responds over time to changing requirements and tastes. Another measure is the extent to which markets clear, the quantities of goods and services balancing those demanded, matching supply and demand, particularly in respect of labour; a situation broadly in which marginal costs (of production) equal marginal benefits (consumer satisfaction). The theoretical conditions for market efficiency are very demanding, and are rarely to be met completely in the real world.

The theory requires *perfect competition*: many buyers and sellers, homogeneous competing goods, buyers and sellers who are fully informed, all firms aiming to maximize profits, firms readily entering and leaving the industry; a perfectly elastic supply of the factors of production; and markets which offer prices for the future as well as for immediate delivery. The essential features of perfect competition are rarely encountered, save to some degree in world commodity markets.

The market mechanism may sometimes prove defective in that it does not necessarily succeed in regulating supply as required, it does not necessarily operate without inflation, and it does not automatically adjust social and private marginal net products. There are also areas in which the price mechanism could not operate, the state finding it necessary to provide low-income and welfare housing, institutions for the disabled, art galleries and museums, courts and most correctional centres, many lighthouses, post offices and telephone systems, schools and many universities, bridges and most roads, sewerage and sewage-treatment systems, garbage collection and recycling centres, many railways and bus services, and some airlines. In addition, the world commodity markets have proved at times singularly ineffective weapons for achieving reductions in output. Small diminutions in price or demand have produced gluts of astonishing proportions, due to the small peasant increasing output when prices fall in order to maintain income. Falls in prices may be accompanied by falls in demand, if potential buyers believe that prices will fall further. Markets also fail when natural resources are over-exploited leading to salination and soil degradation, over-fishing, the extinction of rare and endangered species, the exhaustion of the resource base, and when important social costs such as air and water pollution are not reflected in prices. Markets take no account of externalities.

Monopolies and oligopolies (markets with a small number of suppliers) also present situations in which effective competition is absent and the community may be exploited. In cases where it is believed that a serious maldistribution of resources has occurred, it is often argued that government intervention is justified. A politically important issue is the question of unemployment where the market system has shed much labour through technological and structural improvement

Box 1.6 Pareto Optimality

Pareto optimality is a situation from which it is impossible to deviate so as to make one person economically better off, without making some other person worse off. This Paretian criterion, as usually conceived, does not concern itself with the distribution of income; hence a result may be Pareto optimal even if the distribution it represents is totally unacceptable on other grounds. The rule does not identify the best or optimal social state. It is named after Vilfredo F.D. Pareto (1848–1923), who succeeded to the Chair of Economics at Lausanne University in 1892.

In an attempt to deal with cases that cannot comply with the Pareto criterion, cases in which there are losers as well as gainers, Nicholas Kaldor (1908–86) developed a compensation principle; this states that if those who gain from a policy could fully compensate those who lose, and still remain better off, then the policy should be implemented. However, the compensation is hypothetical and not actually paid. A public policy resulting in say the well-to-do becoming better off, while the poor become worse off, even if overall the gains exceed the losses, could have unfortunate repercussions of both a social and a political kind. Such concepts of economic welfare may contravene the principle of intragenerational equity.

yet has not provided sufficient development in new areas to absorb most of the unemployed. The entire system becomes essentially unbalanced without, for a lengthy period, any signs of a self-correcting mechanism. The relationship of markets and sustainability remains one for constant observation and analysis, sustainability involving such difficult concepts as intragenerational equity and intergenerational equity. Pareto optimality (see Box 1.6) has an elusive quality.

1.7 The Laws of Supply and Demand

The laws of supply and demand are the basic laws in respect of the marketing of goods and services in a market economy. Five laws may be identified:

1. *Ceteris paribus* (other things remaining the same), as demand exceeds supply, the market price tends to rise;
2. conversely, when supply exceeds demand, the price tends to fall;
3. an increase in price tends to lead to a contraction of demand;
4. conversely, a fall in price tends to lead to an expansion of demand and a contraction of supply;
5. prices tend to move towards a level at which demand is equal to supply, that is, to an equilibrium price, the market clearing.

The term 'demand' refers to effective demand (or the ability to pay), and not to need, however urgent and acute. Short-term exceptions may occur in the operation of these laws. For example, during a period of growing or feared shortages, an increase in price may be accompanied by an increase in demand; also a fall in

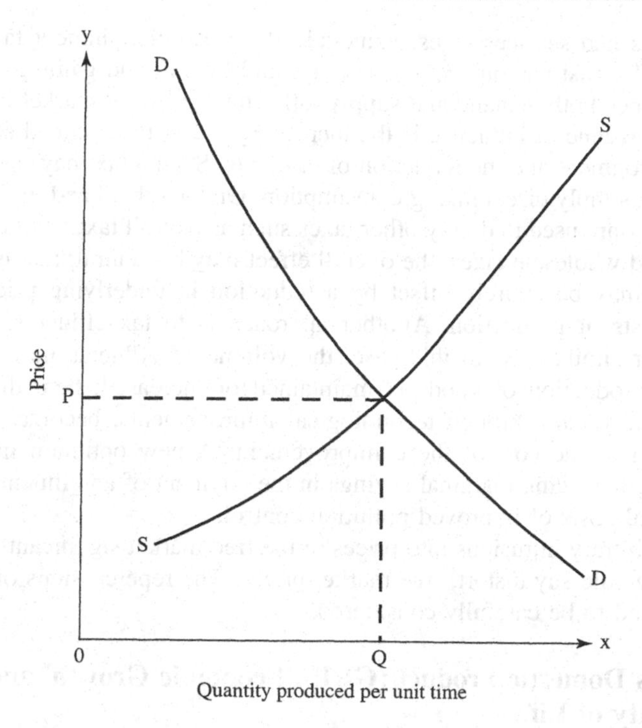

Figure 1.1 *Supply and demand curves, with the equilibrium price shown at P in respect of quantity Q*

share prices might not be accompanied by an immediate increase in demand, if potential buyers are confident that prices will fall still lower.

See Figure 1.1 for an illustration of supply and demand curves. Elasticity plays a key role in the formation of such curves and the equilibrium price. Many factors influence the elasticity of demand for a commodity, that is the response of demand to a change in the price of a commodity. A commodity having no close substitutes is likely to have an inelastic demand; conversely a commodity with close substitutes will have an elastic demand. A commodity that commands only a small share of consumers' total expenditure is likely to have an inelastic demand. Demand will also respond to shifts in income; with shifts in income the demand for some commodities will fall, while other demands will increase. The supply of a commodity is likely to be elastic if a given increase in the output of firms results in only a small increase in the marginal costs of production, and if it is relatively easy for new firms to enter the industry. A supply will be inelastic when rapid increases cannot be technically achieved, there being shortages in the factors of production. Many factors influence elasticity (Gilpin 1986).

Other complications to the simplistic model of demand and supply curves include the imposition of taxes on selling prices such as wholesale and retail sales taxes, goods and services taxes, value-added taxes, entertainment taxes, import duties, tariffs, customs duties, excise duty, and betting and gaming taxes. Such taxes influence both demand and supply, offering significant market interference.

Another avenue of influence is the increasing use of taxes aimed at protection of the environment and the reduction of effluents. Such a tax may raise the price of products, simply discouraging consumption. On the other hand, if the proceeds of such a tax are used to defray other taxes such as payroll taxes and other labour charges, and wholesale taxes, the overall effect may be neutral; that is, the effect of the tax may be entirely offset by a reduction in underlying prices through reduced costs of production. Another approach is to tax effluents, by way of a carbon or similar tax. In this case, the volume of effluents may be reduced while the production of goods is maintained or increased; here the reduction in tax actually paid through technological improvements, becomes an indirect contribution to the cost of those improvement. A new optimum may then be established, balancing marginal savings in the payment of an effluent tax against the marginal costs of improved pollution control.

These arbitrary intrusions into prices in the free market significantly influence, and some would say distort, free market prices. The repercussions of such price controls need to be carefully considered.

1.8 Gross Domestic Product (GDP), Economic Growth, and the Quality of Life

The most commonly used indicator of the national output of goods and services has been the gross domestic product (GDP), which is the value of final goods and services produced within a nation during a specified period of time, usually one year. GDP may be measured at market prices (the expenditure method) or at factor cost (the income method).

GDP at market prices is an estimate of the total market value of goods and services produced, excluding goods and services used up in the process of production, and before allowing for the depreciation of fixed capital. GDP at factor cost is an estimate of the total cost of producing those goods and services; it is the sum of the gross payments made to the factors of production. GDP per capita is calculated by dividing GDP by the population of the country, while GDP at constant prices may be measured over a series of years by discounting changes in the value of money.

These modes of calculation have been criticized on the grounds that the value of domestic and voluntary work, which is considerable, is excluded; also GDP does not reflect changes in the quality of goods and services. The adverse effects on the underlying resource base are also ignored, though attempts by the UN to introduce satellite accounts relating to resources may go some way to rectifying this. However, other adverse effects on the environment generally and the prospects for sustainable development generally are not to be measured through the national accounts system. Other criticisms of GDP relate to its technical and statistical accuracy when the pattern of consumption changes and amendments to GDP calculations come later, and to the fact that as a concept it holds centre stage with governments and the media. Everything tends to be measured by the likely effects on GDP and economic growth. Certainly, GDP has tended to be regarded as a measure of welfare, though at best it is a rough measure of material welfare and a guide to the distribution of wealth throughout the community. It remains central to economic planning. In respect of the quality of life, it can never be more than a partial though important guide to better living.

The concept of the quality of life includes those characteristics of human existence that may promote or retard the broader well-being of the individual, group, community, city or nation. Viewed as a parcel, it embraces such matters as sustainable income and employment; housing and working conditions; health and educational services; domestic harmony; recreational opportunities; physical and mental well-being; travelling times; and access to electricity, water, sewerage, communications, entertainment, literature and modern domestic facilities. Other highly relevant matters include the form of government and its integrity; community, race and cultural relationships; civil liberties and pride; compassion and justice; freedom and fair play; safety and security; law and order; general tolerance; and environmental circumstances. Many individuals and groups lack some, many or all of these elements. On the other hand, over time, there is evidence of a great movement forward in respect of the material standard of living and hence the quality of life, even in the poorest countries.

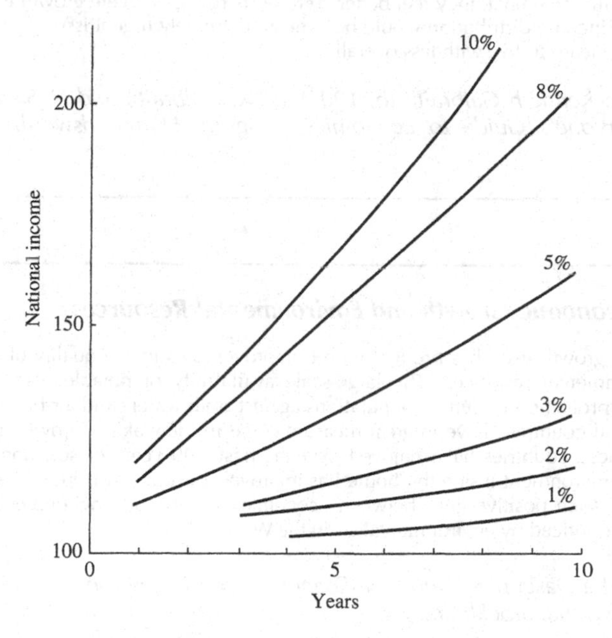

Figure 1.2 *Rates of growth (per cent per annum)*

Economic growth then is the growth per head of population in the production of goods and services of all kinds available to meet final demands, such as goods and services for domestic consumption, capital goods for accumulation, exports to pay for imports and so on. An acceleration of economic growth requires as much emphasis on such elements as better management, better education and training, and better industrial relations, as on higher capital investment.

Furthermore, the quality of investment may count for as much as the quantity of it. Capital investment, though crucial, is but one significant factor in the growth rate of an economy. The *growth rate* is the percentage annual growth compared with the previous year of productive capacity in the community. Figure 1.2 illustrates the effects on GDP of a range of growth rates from 1 to 10% per annum; it illustrates beyond doubt the power of compound interest on long-term prospects. Growth in GDP is a rough measure of material welfare, but it has been assailed in recent years by those who have identified a divergence between material welfare and the quality of life. Some have urged less emphasis on economic growth and

Box 1.7 Galbraith on Growth

Sooner or later people were certain to see what unconsidered, unplanned economic growth was doing to their air, water, scenery or ears, would see that economic life had a dimension of quality as well as of quantity . . . though I have never quite agreed with the advocates of zero growth. Growth and increasing income ease a good many social problems. People compare their income this year with their income last year, and if they find some improvement, they feel better. Tension is reduced. Were growth to come to an end . . . income distribution would become an extremely urgent issue . . . but we do not have to learn to live with less overall.

Source: John Kenneth Galbraith (b. 1908) in J.K. Galbraith and N. Salinger (1978) Almost Everyone's Guide to Economics, Penguin, Harmondsworth, Middlesex, p. 155.

Box 1.8 Economic Growth and Environmental Resources

Economic growth itself has brought with it improvements in the quality of a number of environmental resources. The large-scale availability of potable water, and the increased protection of human populations against both water- and air-borne diseases in industrial countries, have in great measure come in the wake of growth in national income these countries have enjoyed over the past 200 years or so. Moreover, the physical environment inside the home has improved beyond measure with economic growth . . . such positive links between economic growth and environmental quality often go unnoticed by environmentalists in the West.

Source: Partha Dasgupta (1996) The Economics of the Environment, Beijer Discussion Paper No. 80, Stockholm.

more emphasis on the quality of life; a few have commended the concept of *zero economic growth*. A policy of zero economic growth world stabilize material production per head, without the prospects of improvement.

Box 1.7 conveys the views of John Kenneth Galbraith on the subject of zero economic growth, while Box 1.8 conveys the views of Partha Dasgupta on the relationship between economic growth and a range of environmental resources.

1.9 The Meaning of the Word 'Environment'

Essentially and in its broadest sense, the word 'environment' embraces the conditions or influences under which any individual or thing exists, lives or develops. These surrounding influences may be placed into three categories:

1. the combination of physical conditions that affect and influence the growth and development of an individual or community;
2. the social and cultural conditions that affect the nature of an individual or community; and
3. the surroundings of an inanimate object of intrinsic social value.

The concept here is essentially anthropogenic or human-related, though many would stress the importance of nature for its own sake, with or without benefits to humanity.

The environment of humanity includes the abiotic factors of land, water, atmosphere, climate, sound, odours and tastes; the biotic factors of other human beings, fauna, flora, ecology, bacteria and viruses; and all those social factors that make up the *quality of life*. The concept has emerged of the environment as an assembly of people and things which render a stream of services and disservices to the individual or other segments of the environment, and which take their place alongside the stream of services rendered by real income, commodities, homes, infrastructure, and markets generally.

The European Commission has defined the environment as 'the combination of elements whose complex interrelationships make up the settings, the surroundings and the conditions of life of the individual and of society, as they are or as they are felt'. This approach also places humanity at the centre of things, with all environmental values traced back to the welfare of human beings. Clearly, the evolution of the word 'environment' has some way to go.

However, we should not overlook how far the word has already come; for it is only in comparatively recent years that it has become identified as an entity to be addressed by world conferences, to command segments of national legislation and international conventions, and to have received allocations of financial resources in the public and private sectors of the economy.

On the one hand, early origins may be traced to the 19th century in the areas of public health, insanitary dwellings and streets, contaminated public water supplies, drains and sanitation, public nuisances, unhygienic food processing and handling, overcrowding, noxious effluvia, offensive trades, refuse dumps, epidemics, contagious diseases, and the breeding of vermin. The responses were largely in the area of public water supply and sewerage schemes, slum clearance programmes, food hygiene drives, regular sanitary inspections, elementary pollution control, and vermin control.

On the other hand, increased interest was being taken in nature, the conservation of natural areas and the creation of national and marine parks and reserves. These appeared at the time to be two completely separate developments with little communication between the two.

It was not until the 1960s that the two paths began to coalesce, new strands being added, new agencies being created for environment protection as a whole. The first world conference on the human environment was held in Stockholm in 1972, the second in Nairobi in 1982, and the third in Rio de Janeiro in 1992.

While pollution control and nature conservation have merged into the general subject of environment protection, culminating in the all-embracing process of environmental impact assessment, some embryonic subjects have matured, such as environmental economics, risk assessment, and relevant cultural, social and heritage considerations. International environmental concerns and concepts such as ecologically sustainable development have also entered the arena.

Box 1.9 categorizes the microenvironmental problems immediately affecting the lives of citizens, while Box 1.10 outlines the macroenvironmental problems

Box 1.9 Microenvironmental Problems

- Unsafe, inadequate and remote water supplies
- Inadequate or non-existent sewerage systems; nightsoil collection systems
- Stagnant water and vector breeding
- Noise pollution
- Fumes, odours, waste gases and vibration from industrial and commercial processes
- Dereliction, slums, shanties and blight in certain locations
- Hazards from traffic in the street; congestion and delays
- Unsealed streets and lanes; lack of stormwater drains
- Littering, noxious accumulations and abandoned vehicles
- Poorly located industrial plant, schools and shops
- Severance of neighbourhoods by highways, railways, large-scale developments, or traffic management schemes
- Abandoned dwellings, quarries, business premises and factories
- Lack of space for play or recreation; lack of parks and walkways
- Loss of heritage buildings and the special character of areas
- Loss of privacy, views and vistas; overcrowding
- Loss or deterioration of natural assets and ecosystems in the immediate neighbourhood
- Visual squalor, including overhead wirescape
- Loss of fauna and flora; destruction of trees and habitats

Box 1.10 Macroenvironmental Problems

- Unsafe, unclean, inadequate, remote and uncertain water supplies
- Inadequate or non-existent sewerage systems
- Inadequate or non-existent means of solid-waste disposal
- The pollution of streams, lakes, rivers, harbours, waterways, estuaries and oceans
- Urban air pollution from industrial, commercial and domestic sources
- Transboundary and global air pollution; global warming; threats to the ozone layer
- Floods, fires, eruptions, drought and famine
- Soil erosion and desertification
- Dereliction, slums, shanties and urban and rural blight
- Mortality and morbidity arising from environmental sources
- Population growth in relation to natural resources; national settlement patterns
- Threats to biodiversity; fauna and flora
- Failures of environmental or town planning
- Vector breeding
- Threats to natural resources, including forests, marshes, wetlands, woodland, bush, mangroves, ecosystems, coasts, sand dunes, marine resources, waterways, reserves and heritage areas
- Noise pollution from road traffic and aircraft
- The disposal of toxic and nuclear wastes
- Landfills and incineration
- The location of hazardous industries
- The implementation of international conventions and agreements; the international movement of hazardous wastes
- The identification and preservation of heritage areas
- The promotion of alternative energy sources
- The promotion of sustainable development, intragenerational and intergenerational equity
- Implementation of the polluter-pays principle and the precautionary principle
- Visual pollution
- Assistance to less-developed countries of an environmental kind
- The promotion of international dialogue on environmental issues, including climate change and biodiversity

affecting humanity. There is of course some overlap of the two, but the fundamental division is important. Box 1.11 summarizes some of the more striking individual events and disasters occurring this century.

1.10 The Role of Environmental Economics

Environmental economics, a specialized branch of economics, has evolved in relatively recent years. It embraces the issues of pollution control, climate change, protection of the natural environment, conservation of scarce resources, biodiversity and economic instruments; issues in the resolution of which markets play

Box 1.11 Some Major Environmental Disasters

1930 *Meuse Valley incident*: an air pollution incident in the Meuse valley, Belgium, in which several hundred people became ill and 60 died.

1948 *Donora Valley incident*: an air pollution incident occurring in Donora, Pennsylvania, affecting some 42% of the population of 14 000, with 18 deaths.

1950 *Poza Rica incident*: an air pollution incident at Poza Rica, Mexico, arising from the malfunction of an oil refinery sulphur recovery unit, resulting in 22 deaths, and much illness.

1952 *London smog disaster*: an incident over several days in which 4000 people died as a direct consequence. In 1962, a similar smog occurred with a much reduced death roll of 700, due to smoke control measures.

1953 *Cadmium and mercury pollution in Japan*: cadmium pollution caused a painful malady to occur in the town of Fuchu, north-west of Tokyo, with many deaths. A mining company was responsible. Acute illness in the Minimata Bay area was attributed to mercury poisoning emanating also from heavy industry around the Bay. The consequences of these events have been felt to the present day.

1957 *Windscale nuclear reactor incident*: a major accident which occurred at the nuclear plant at Windscale, Cumbria, UK, in which radioactive fission products escaped to the atmosphere. Iodine-131 and tellurium-132 escaped over a wide area.

1966 *Aberfan disaster*: a disaster at a mining village in Mid-Glamorgan, Wales, in which the liquefaction of coal waste overwhelmed a school and houses. Of the 144 dead, 116 were children.

1967 *The loss of Lake Pedder, Tasmania*: the flooding of a lake of rare beauty due to hydroelectricity development. However, a later proposal to flood the Franklin River for similar purposes was soundly defeated, the area being incorporated in a world heritage park.

1967 *Torrey Canyon Disaster*: an incident in which the oil tanker *Torrey Canyon* went aground on the Seven Stones Reef near the Scilly Isles, off Lands End, UK, liberating her cargo of crude oil, much of the oil landing on the Cornish beaches, with some drifting across the English Channel towards France.

1969 *Santa Barbara blow-out*: an incident off the coast of California, in which large quantities of oil spurted from the ocean floor following drilling operations. The blow-out continued for 10 days, the oil ultimately reaching the beaches, promenades and jetties of Santa Barbara. Santa Barbara was declared a Federal Disaster Area. The oil well was ultimately sealed with oil and cement.

1973 *Michigan episode*: a tragic incident in which livestock all over Michigan State were fed with contaminated feed; from the milk, people suffered a wide range of disabilities. The effects of this error were felt for many years.

1976 *Seveso incident*: a major incident at a herbicide plant at Seveso, near Milan, Italy. A vapour cloud of a highly toxic dioxin was accidentally released to the atmosphere, with 500 people showing evidence of poisoning. There were no human deaths, but many domestic animals had to be destroyed.

Continued on page 19

Continued from page 18

1977 *Love Canal*: an area of Niagara Falls township, whose residents were found to be suffering a high rate of cancers, birth defects, liver and kidney damage, and respiratory ailments. The area had been a dumping ground for toxic chemical wastes. Love Canal was declared a disaster area.

1979 *Three-Mile Island nuclear incident*: the release of radioactive material from a nuclear power plant located 16 km from Harrisburg, the capital of Pennsylvania. There were no casualties from this incident, but much damage was done to the reputation of the nuclear industry. The year also witnessed a huge blow-out from an oil well off the coast of Yucatan, Mexico, with an oil slick stretching 500 km across the Gulf of Mexico.

1983 *Times Beach*: a town in Missouri found to be too contaminated with dioxins to be safe for habitation. The US Government offered to buy out all 2400 residences.

1984 *Bhopal disaster, India*: A catastrophic gas leak at a pesticide plant, as a result of which over 2000 people, mainly children and older people, died, and some 50 000 suffered from blindness, temporary and permanent. The year also saw a fire and explosion in Mexico City at a natural gas plant, resulting in several hundred deaths and many more injured. A fire also occurred at Cubatao in Brazil when a leaking gasoline pipeline ignited; about 500 people died.

1985 *Stava tailings dam disaster*: the collapse of a tailings dam in northern Italy resulted in a devastating mud slide, killing more than 250 people.

1986 *Chernobyl nuclear disaster*: Located 104 km north of Kiev in the Ukraine, Chernobyl was a nuclear power station comprising four reactors. During a test, a chain reaction went out of control blowing the steel and concrete lid off the No. 4 reactor. Large amounts of radioactive material were released into the atmosphere and were carried over Europe. Local inhabitants were evacuated. The immediate death rate was 30; this grew over time to 12 000. The event strengthened worldwide opposition to nuclear power generation.

1989 *Exxon Valdez disaster*: a major incident in which the tanker *Exxon Valdez* ran aground in Prince William Sound, Alaska, on the Bligh Reef. Some 250 000 barrels of oil poured into the Sound, and 2400 km of beach were fouled by the spill. Claims for damages followed. In the same year the London Conference on Climatic Change discussed the protection of the ozone layer and the need for a major reduction in the emissions of carbon dioxide.

1994 *Komi pipeline oil disaster*: major leakages from a pipeline in Russia's far north, in the republic of Komi.

1996 *Sea Empress disaster*: the *Sea Empress* ran aground near the entrance to Milford Haven harbour in Wales, spilling about 70 000 tonnes of oil. It was Britain's worst oil spill since the *Torrey Canyon* disaster in 1967. In the same year, WHO reported that air pollution claimed 350 lives a year in Paris. Another report from the University of Michigan said that most children in African cities had blood-lead levels high enough to cause neurological damage, with many suffering lead poisoning.

1997 *South-east Asia air pollution crisis*: the worst episode of air pollution in half-a-century, with smoke and photochemical smog from forest fires settling over parts of Malaysia, Singapore, Brunei, Indonesia, Thailand, Hong Kong and the Philippines.

little or no part, but in which vast natural assets need to be allocated sensibly to the common good.

Environmental problems are frequently characterized by the existence of externalities, the presence of free-access natural assets such as the atmosphere and large stretches of ocean, ill-managed public goods and assets, under-valuation of common property resources, and inadequate environmental management either through direct regulation or economic instruments.

The subject is very much concerned with the ways and means of achieving a sensible allocation of resources through such channels as emission and effluent charges, user charges for the treatment and disposal of wastes, environmental taxes, product charges, deposit refunds, tradable pollution rights, performance bonds, natural resource accounting, road pricing, and the economic implications of intragenerational equity and intergenerational equity, together with the economic implications of sustainable development. The subject is concerned therefore with costs and benefits on a social and a global scale, transcending the narrower approaches of the past. The subject drives a path to a larger concept of optimization.

Several specialized journals serve the subject, such as the Beijer journal *Environment and Development Economics* and the *Journal of Environmental Economics and Management*. The subject has prospered in university courses, being found not only in economics courses, but widely in the social sciences generally, in geography, and in civil, chemical and environmental engineering. Box 1.12 attempts to outline the scope of the subject, but not exhaustively.

Box 1.12 The Scope of Environmental Economics

Acoustics
Aesthetics
Agenda 21
Air pollution
Allocative efficiency
Alternative sources of energy
Amenity
Assimilative capacity
Atmosphere
Attitudinal changes
Balance of nature
Beneficial uses
Best practicable means
Biodiversity
Biosphere
Buffer zones
Capital
Carbon tax
Carrying capacity
Cleaner production
Climate change
Common property resources
Conservation
Contingent valuation
Cost–benefit analysis
Cost–effective analysis
Cumulative effects
Debt-for-nature swaps
Demand management
Demographic transition
Depletion theory
Deposit–refund systems
Development
Direct controls
Discounting
Drought and famine
Ecology
Economic–environmental–ecological interactions

Continued on page 21

Continued from page 20

Economic impact assessment
Economic instruments
Economic systems
Economic welfare
Ecosystem
Ecotaxes
Emission trading programmes
Endangered species
Energy conservation
Environment
Environmental assurance bonds
Environmental audits
Environmental health impact assessment
Environmental heritage
Environmental impact assessment
Environmental impact statement
Environmental law
Environmental levies
Environmental planning
Externalities
Fauna impact statement
Fisheries
Forestry
Food production and distribution
Geographical information system
Global environmental facility
Greenhouse gases and effects
Gross domestic product (GDP)
Habitat
Hedonic price technique
Heritage conservation
Human development index
Hydrological cycle
Ice ages
Infant mortality
Intergenerational equity
International conventions
Intractable wastes
Intragenerational equity
Land-use planning
Market failure
Maximum sustainable yield
Mitigating measures
Motor vehicle congestion
Multicriteria analysis
Multiplier
National and marine parks
Natural resource accounting
Noise fees
Non-renewable resources
Official aid
Oil pollution compensation
Open access regimes
Opportunity cost
Polluter-pays principle
Pollution control strategies
Population growth
Precautionary principle
Present worth
Property rights and duties
Public goods and services
Quality of life
Recycling
Regulatory impact statement
Renewable resources
Road pricing schemes
Sanitation and public health
Shadow prices
Social impact assessment
Substitution
Sulphur tax
Surrogate market techniques
Sustainable development
Sustainable yield
Taxes
Town planning
Transferable quotas and development rights
Travel cost approach
UN Commission for Sustainable Development
UN Development Program
UN Economic Commission for Europe
UN Educational, Scientific and Cultural Organization
UN Environment Program
UN Food and Agriculture Organization
UN World conferences
Urbanization and cities
User pays
U-shaped hypothesis
Valuation of environmental assets
Visual pollution
Water pollution
World Bank
World conservation strategy
World Health Organization
World heritage
World population
World Trade Organization

Source: Derived from Gilpin (1996).

Box 1.13 Munasinghe on the Role of Environmental Economics

Environmental economics plays a key role in identifying efficient natural resource management options for sustainable development. It is an essential bridge between the traditional techniques of decision-making and the more environmentally sensitive approach, now emerging. Environmental economics helps us to incorporate ecological concerns into the conventional framework of human society ... thereby improving decision-making at the economy-wide, sectoral and micro-levels.

Source: M. Munasinghe (1994) Protected Area Economics and Policy, The World Bank, pp. 18 and 29.

Box 1.14 Cropper and Oates on Environmental Economics

... over the past two decades, environmental economists have reworked existing theory, making it more rigorous and clearing up a number of ambiguities; they have devised new methods for the valuation of benefits from improved environmental quality; and they have undertaken numerous empirical studies to measure the costs and benefits of actual or proposed environmental programs and to assess the relative efficiency of incentive-based and command-and-control policies ... economists now know more about environmental policy and are in a position to offer better counsel on the design of measures for environmental management ... and there are encouraging signs in the policy arena of a growing receptiveness to incentive-based approaches.

*Source: M.L. Cropper and W.E. Oates (1992) Environmental economics: a survey, Journal of Economic Literature, **XXX** (June): 675–740.*

Box 1.13 indicates the views of Mohan Munasinghe on the role of environmental economics while Box 1.14 follows with the views of Maureen Cropper and Wallace Oates,

Box 1.15 outlines the approaches of different disciplines, economics and ecology, to the subject of environment. The economist's approach is essentially anthropogenic, or human-related, intent on serving the needs and wishes of the community. The ecologist's approach stands somewhat apart, giving additional emphasis to other forms of life and their relationship with the environment. Animal rights forms part of this. The views of Lester Brown in Box 1.16 sharpen up the contrast. Box 1.17 illustrates the history of the term 'Spaceship Earth'. Occasionally, other claims to the title are made.

1.11 The U-Shaped Hypothesis

The U-shaped hypothesis reflects a widely held view that with economic growth and development all environmental indices initially deteriorate with increasing pollution of air and water, congestion, squalor and desolation. However, it

Box 1.15 The Approaches of Different Disciplines

Economists are principally concerned about resource allocation and efficiency. They seek to identify the optimal arrangement of economic activities and resource-uses in society, and emphasize the role of markets in achieving efficiency in resource-use. While some point to market failures as indicative of an important role for government in achieving better outcomes others believe that governments are unable to achieve such outcomes efficiently and prefer well-specified property rights and the use of incentives to achieve desired outcomes.

Ecologists are primarily concerned about the relationship between living organisms (including human beings) and their environment. They often focus on issues relating to change and survival. Some ecologists believe that equilibrium models can be applied to ecological processes; they tend to analyze resource development options in terms of carrying capacity and are comfortable with prescriptive planning. Others emphasize the dynamic nature of the environment and the importance of chance events; they tend to encourage an adaptive management approach, whereby development occurs in an exploratory fashion, and effects are closely monitored and taken into account in ongoing management decisions.

Source: Resource Assessment Commission (1992) Methods of Analysing Development and Conservation Issues; The RAC's Experience, Research Paper No. 7, December, RAC, Canberra, p. 20.

Box 1.16 Lester R. Brown: Two Views of the World

Anyone who regularly reads the financial papers or business weeklies would conclude that the world is in reasonably good shape and that long-term economic trends are promising ... Yet on the environmental front, the situation could hardly be worse ... every major indicator shows a deterioration in natural systems: forests are shrinking, deserts are expanding, croplands are losing topsoil, the stratospheric ozone layer continues to thin, greenhouse gases are accumulating, the number of plant and animal species is diminishing, air pollution has reached health-threatening levels in hundreds of cities, and damage from acid rain can be seen on every continent.

These contrasting views of the state of the world have their roots in economics and ecology: two disciplines with intellectual frameworks so different that their practitioners often have difficulty talking to each other.

Source: L.R. Brown in State of the World 1991: A Worldwatch Institute Report on Progress Toward a Sustainable Society, W.W. Norton, New York and London.

is argued, a social and political reaction occurs that ensures a diversion of resources to pollution control and environmental objectives. Hence, with increasing material prosperity (rising GDP per capita) environmental excesses are curbed and environmental objectives pursued as a matter of public policy. Future developments are closely scrutinized. Hence private affluence abates public squalor. Evidence of this may be found in the OECD countries, and increasingly in the developing nations.

Box 1.17 Spaceship Earth

A few days before his death, Adlai E. Stevenson (1900–65), chief US representative at the UN, addressing the UN Economic and Social Council in Geneva, said, 'We travel together, passengers on a little spaceship, dependent on its vulnerable supplies of air and soil; all committed for our safety to its security and peace, preserved from annihilation only by the care, the work, and I will say, the love we give our fragile craft.'

In 1966, Barbara Ward (Baroness Jackson, 1914–81), in her work *Spaceship Earth,* said, 'The most rational way of considering the whole human race today is to see it as the ship's crew of a single spaceship on which all of us, with a remarkable combination of security and vulnerability, are making our pilgrimage through infinity ... this space voyage is totally precarious ... rational behaviour is the condition of survival.'

In the same year, Kenneth E. Boulding (b. 1910) wrote: 'The closed earth of the future requires economic principles which are somewhat different from those of the open earth of the past ... I am tempted to call the open economy the "cowboy economy", the cowboy being symbolic of the illimitable plains and also associated with reckless, exploitative, romantic, and violent behaviour ... The closed economy of the future might similarly be called the "spaceman" economy, in which the earth has become a single spaceship, without unlimited reservoirs of anything, either for extraction or for pollution, and in which ... the essential measures of success are not production and consumption at all, but the nature, extent, quality and complexity of the total capital stock. In the spaceman economy we are primarily concerned with stock maintenance ...' (Boulding 1966).

This U-shaped hypothesis is probably sound in respect of many elements in the environment such as gross industrial and domestic air and water pollution, the provision of safe and clean drinking water, drainage and sanitation, garbage removal, building standards, cleaner streets, litter removal, traffic management, the management of many toxic wastes and zoning of activities.

The U-shaped hypothesis is less true in terms of the volumes of wastes to be collected, handled, processed and disposed of; general traffic and aircraft noise; traffic congestion; intractable wastes; marine pollution; and overuse of the commons.

It is an incorrect hypothesis in terms of greenhouse gases, particularly carbon dioxide, in relation to global warming without Draconian measures; the problem of radioactive wastes; the depletion of the resource base (e.g. forests, fisheries, water, fuelwood and fossil fuels) and the stresses on the carrying capacity of the Earth, in respect of population, fauna and flora, biodiversity and endangered species (see Figure 1.3).

1.12 The Optimization of Pollution Abatement

As pollution intensifies and ground-level concentrations increase, the degree of damage or the risk of damage progressively endangers comfort, welfare and

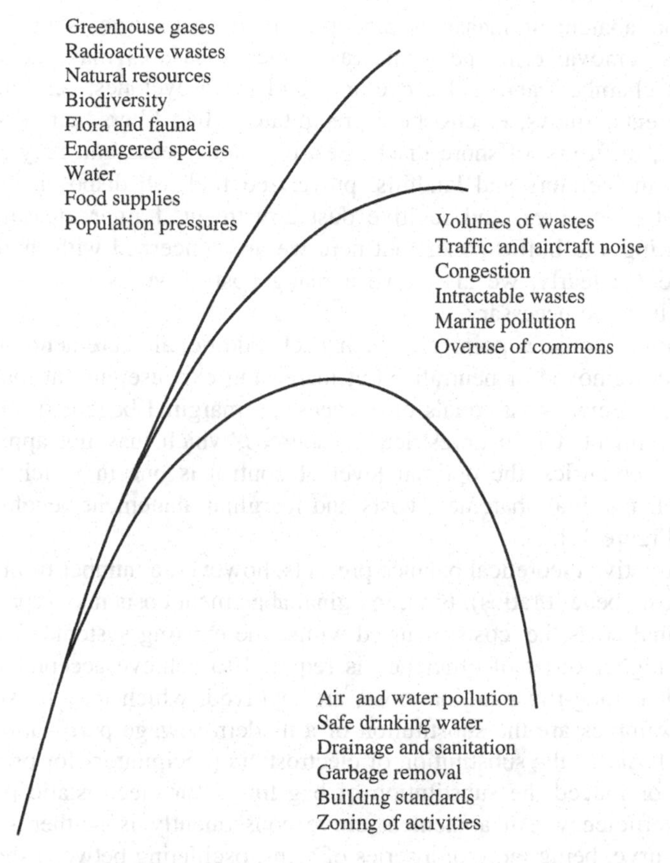

Figure 1.3 *The U-shaped hypothesis*

health. The characteristics of emissions are of considerable importance, for some may solely give rise to nuisance (such as grit and dust, or odours), while others lead to health effects such as respiratory problems and high lead levels in the blood retarding the development of children. Others again may kill fish and damage crops, or pollute potable water and impair recreational value. Visibility may be affected and noise may irritate. Insecticides may enter the food chain. Many effects may be small but cumulative. Some effects may be regional and transboundary. Some effects, such as greenhouse gases (most notably carbon dioxide), may have no local effects at all, but contribute to consequences in the upper atmosphere such as ozone depletion or global warming. The Chernobyl disaster, which resulted in over 12 000 deaths, exemplifies what may occur at local, regional and international levels, with immediate and prolonged consequences for all. Thus the damage curve tends to increase from left to right with increases in pollution levels, depending on their character.

Pollution abatement measures also present a cost curve, sometimes rising sharply as removal efficiencies increase. Such measures may include simple settlement chambers and tanks, cyclones and multi-cyclones, scrubbers, chemical treatments, sprays, electrostatic precipitators, bag filters, aerators, purification ponds, diffusers, offshore discharges, digesters, reuse and recycling, waste reduction, incinerators and landfills, pulverized fuel ash disposal, exhaust and entrapment procedures, and fugitive dust arrestment. Larger measures include suitable siting and buffer zones, but here we are concerned with industrial plan management. Clearly, we also have a rising cost curve as more sophisticated measures become necessary.

There must come a point at which each additional increment of pollution can only be removed or neutralized at increasing expense, and at that point the marginal abatement cost equals or exceeds the marginal benefit from that additional investment. Given unrestricted resources, which may not apply in many developing countries, the optimal level of control is one in which there is an equality of marginal abatement costs and marginal abatement benefits, as illustrated in Figure 1.4.

This attractive theoretical balance presents, however, a number of problems (as with all cost–benefit ratios). First, marginal abatement costs may represent short-run marginal costs, i.e. costs incurred within the existing system. If, however, a step to a higher order of efficiency is required to achieve acceptable levels of control, then long-run marginal costs are involved, which may be significantly higher. Examples are the substitution of a modern sewage purification plant for a sewage farm, or the substitution of electrostatic precipitators for primitive dust cyclones, or indeed the substitution of bag filters for electrostatic precipitators of lower efficiency. An abatement curve consequently is neither smooth, nor indeed a curve, being more of a series of steps, oscillating between short-run and long-run marginal costs (see Box 1.18).

Marginal damage cost curves present much greater problems. The nature of damage presents the first problem. Respiratory problems may affect a

Box 1.18 Marginal Cost: Short Run and Long Run

The marginal cost is the increase in the total cost of producing each successive increment of an output, i.e. the cost of producing $M+1$ units, minus the cost of producing M units. As a concept it may be regarded as including social and environmental costs, although it is often construed more narrowly in accountancy terms.

The short-run marginal cost is the cost incurred in making marginal (small) changes, say in the energy output of a system, within existing capacity. Long-run marginal costs include the marginal costs of changes in the capacity of the system. References to marginal costs need to be quite clear on this distinction, particularly when referring to marginal cost pricing.

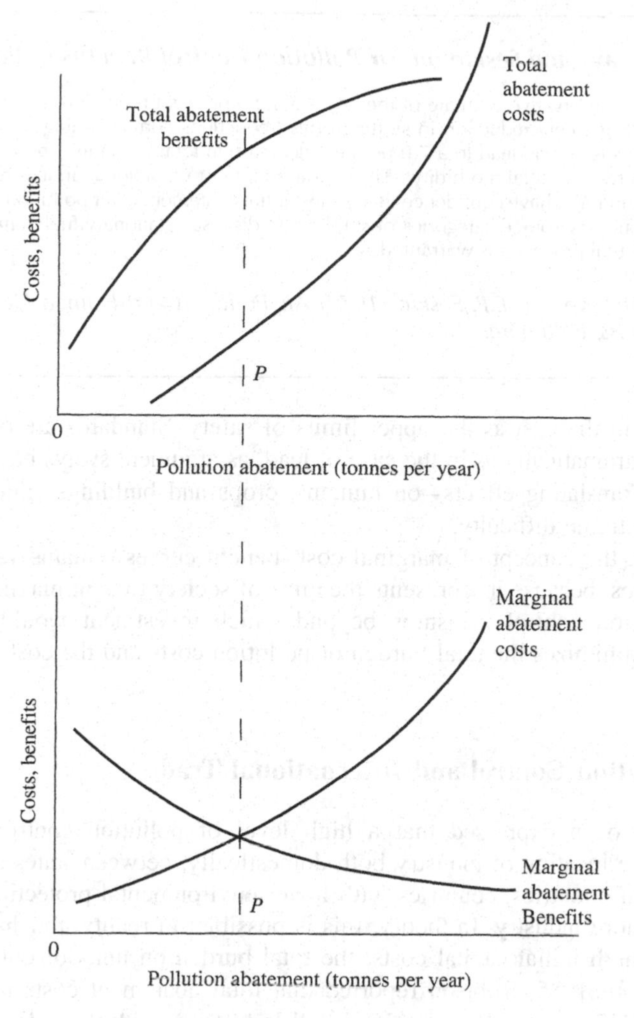

Figure 1.4 *Optimal level of pollution abatement, with funding restraints*

population but may not be due solely or even mainly to emissions from factories and automobiles; many may be smokers and working conditions may not be beneficial. Effects may be short, sharp and devastating, as with the Bhopal disaster; while other effects may be devastating but also extended over many years, as in the case of Chernobyl. Other effects may be insidious, such as acid rain or the effects of lead on children, impáiring intellect. Research by Lave and Seskin (1977) reveals the difficulties of a national study, with valid findings, yet with little guidance for the design of an individual plant (see Box 1.19). There are also continuing debates about the existence or non-existence of thresholds, often

Box 1.19 Lave and Seskin on Air Pollution Control Benefits in the USA

Our most conservative estimate of the effect of a 58 per cent reduction in particulates and an 88 per cent reduction in sulfur oxides (reductions corresponding to proposed control levels) would lead to a 7.0 per cent decrease in total mortality (and at least an equal decrease in total morbidity). This would amount to a national annual benefit of $16.1 billion. We have confidence that a substantial abatement of air pollution from the major stationary source categories of solid waste disposal, stationary fuel combustion, and industrial processes is warranted.

Source: L.B. Lave and E.P. Seskin (1977) Air Pollution and Human Health, Johns Hopkins Press, Baltimore.

proclaimed in the past as the upper limits of safety. Standards are often varied, sometimes dramatically as in the case of lead, as argument sways backwards and forwards. Translating effects – on humans, crops and buildings – into costs is a matter of extreme difficulty.

However, the concept of marginal cost–benefit curves remains valid, despite all difficulties, because it represents the aims of society to gain maximum benefit from pollution control measures beyond which investment would be simply wasted. It minimizes the total burden of pollution costs and the cost of pollution control.

1.13 Pollution Control and International Trade

Fears have been expressed that a high level of pollution control costs will influence the location of industry both domestically, between states and regions, and between countries, countries with lower environmental protection standards attracting more industry. In theory, this is possible; in reality, much less so.

Despite high initial capital costs, the total burden on units of output remains reasonable. In 1985, Robison reported that total abatement costs per dollar of output in 1977 were well under 3% in all industries, with the sole exception of electric utilities where the costs were 5.4% per kilowatt-hour.

In 1988, Leonard, in a case study of foreign trade and investment flows for several key industries and countries, found little evidence that pollution control measures had exerted a significant effect on international trade and investment. Leonard concluded that 'the differentials in the costs of complying with environmental regulations and in the levels of environmental concerns in industrialized and industrializing countries have not been strong enough to offset larger political and economic forces in shaping aggregate international comparative advantage.' Tobey (1989, 1990) also looked at the same issue in a large econometric study of international trade patterns in pollution-intensive goods and could not find any effects resulting from stringent domestic environmental policies.

Another important factor has been that nearly all industrialized countries have introduced environmental measures (and within a similar time period) so that such measures have not been a source of significant cost differentials among major competitors. In developing countries, a fear of major disasters has developed, while major economic and political uncertainties have influenced location decisions much more decisively than any marginal savings in environment protection. There is, therefore, little force to the view that a high standard of environment protection will weaken competitiveness in overseas markets.

1.14 Substitution: A Key to Continued Progress

Substitution is simply the exchange of the services of one article, substance, service or role for another, as a consequence of a change in relative prices or as a consequence of legislation and standards. Substitution may involve the use of more common, cheaper resources in place of scarcer, more expensive resources. Examples include the use of aluminium in place of copper in a variety of products, the use of plastics in place of metals for a variety of purposes, and the substitution of electronic devices for mechanical devices; all such changes may involve increasing capital intensity and the displacement of people. Legislation has discouraged the use of asbestos and lead for a variety of purposes. Energy substitutions occur when natural gas becomes cheaper for power generation compared with other fossil fuels; when public opinion turns against nuclear power; and when solar energy, hydroelectricity and hot rock displace fossil fuels in whole or part, in response to the greenhouse effect. Substitutions occur continuously in the market-place, such as the substitution of capital for labour, the development of new industries and services at the expense of outdated activities, or shifts in production in response to changes in the character of consumer demand, export markets or financial services. Shifts occur as a consequence of falling tariffs and free trade agreements.

Potentially large-scale substitutions are likely to flow from the application of the principles of sustainable development, such as large-scale shifts in the means of transportation in moving to a car-free society, with rationing in place in the developed countries.

Substitutions are likely to arise from slow changes in cultural attitudes such as declining markets for the products of endangered species and wildlife generally, products associated with exploited labour or foreign despotism, and the implementation of international conventions relating to timber. The trade in human beings has been substantially reduced.

Generally, substitutions are engaged in by everyone as producer, purchaser, employee, entrepreneur, craftsperson, housekeeper, child carer, public servant, factory worker, farmer or construction worker, when marginal costs get too much out of line with marginal benefits.

Box 1.20 Market Responses to Scarcity

- Other things being equal, as particular priced resources become scarce, their prices increase.
- Other resources then become more viable substitutes.
- Price increases stimulate the search for new deposits.
- Price increases also stimulate reuse, recycling and greater economy in use.
- Where lower-grade resources are still available, these become profitable with prevailing technology.
- Price increases stimulate progress in technological and processing techniques, reducing costs.
- Rising prices may influence the use of these resources in the final products through substitution, partially or completely.
- Variations in technology may also influence the final demand for goods and services.

Box 1.21 Partha Dasgupta on substitution

> I began by suggesting that it is not specific resources we care for, it is certain characteristics we seek, certain services we need and desire. A wide variety of resources may offer the same set of services. This is precisely why *substitution* is the key concept in resource economics. Timber, coal, oil and natural gas are all sources of energy. So is uranium 238 ... moreover, if at some future date nuclear fusion comes to be controlled at an acceptable risk, or if solar energy comes to be tapped in large amounts at reasonable cost, we will have at our disposal vast new deposits of energy, so vast that to all intents and purposes we will have a renewable resource ...

Source: P. Dasgupta (1989) Exhaustible resources, in The Fragile Environment: the Darwin College Lectures, University of Cambridge Press, Cambridge, Chapter 6.

The likely responses of the market to the emergence of genuine scarcities of natural resources, particularly non-renewable resources, are set out in Box 1.20. The views of Dasgupta (1989) on substitution are reflected in Box 1.21.

1.15 The USA: Some Costs and Benefits

The economic costs of pollution control in the USA increased approximately fourfold between 1972 and 1990, reaching a level of $115 billion ($115 thousand million) in 1990, i.e. about 2.1% of GDP. This vast increase in spending mirrors the growth in the extent of federal environmental regulations. In 1975, the US Code of Federal Environmental Regulations consisted of 2763 pages; by 1990, the same section of the US Code contained 11 087 pages.

This information was incorporated in the USA National Report to the 1992 UN Conference on Environment and Development held in Rio de Janeiro (see

Table 1.1 *Total annualized environmental compliance costs by medium, 1990 (millions of 1986 US dollars), assuming full implementation of the law*

Medium	Costs	Major statutes
Air and radiation, total	28 029	
Air	27 588	Clean Air Act (CAA)
Radiation	441	Radon Pollution Control Act
Water, total	42 410	
Water quality	38 823	Clean Water Act (CWA)
Drinking water	3 587	Safe Drinking Water Act
Land, total	26 547	
RCRA	24 842	Resource Conservation and Recovery Act (RCRA)
Superfund	1 704	Comprehensive Environmental Response, Compensation and Liability Act (CERCLA)
Chemicals, total	1 579	
toxic substances	600	Toxic Substances Control Act (TSCA)
pesticides	979	Federal Insecticide, Fungicide and Rodenticide Act (FIFRA)
Total costs	100 167	

Source: US EPA (1990).

Table 1.2 *Pollution control costs and expenditures (thousand-millions of 1990 dollars)*

	1972	1987	1990	2000 'Present'	2000 'Full'
Total annualized costs	30	98	115	171	185
% of GNP	0.9	1.9	2.1	2.6	2.8
Federally mandated annualized costs	21	77	93	NA	158
% of GNP	0.6	1.5	1.7	NA	2.4
Capital expenditures	23	35	47	35	45
% of total capital investments	2.5	2.3	2.8	1.7	1.9

Source: US EPA (1990) Environmental Investments: The Cost of a Clean Environment US EPA, Washington, DC.

Table 1.1). Costs are expressed as *total annualized costs* for all pollution control activities in the USA, including the operating costs and the interest and depreciation associated with cumulative pollution control investments (at a 7% rate of interest). Although total annualized costs are increasing, the rate of increase is slowing. Pollution control costs that are required to meet federal regulations represent about 80% of total expenditure. Pollution control investment data for selected years are summarized in Table 1.2.

It will be noted that the total capital investment reached a high in the mid-1970s of about 3.4% but has been trending downwards since then. The Table presents two scenarios for the year 2000. Estimated costs for *present* implementation in the year 2000 assumes that existing programmes are implemented to the same degree as in 1987. Estimated costs for *full* implementation in 2000 assumes that investments are undertaken to bring about a nationwide attainment of the national ambient air quality standard for ozone (as provided for in the Clean Air Act Amendments of 1990) and the fishable/swimmable goals of the Clean Water Act. Pollution control costs would thus sum to nearly 3% of GDP by the year 2000.

The USA report to the UN Conference offered seven general statements about the cost of a clean environment:

- National environmental pollution control expenditures were increasing and had approached one-quarter of those for national defence and medical care, and one-fifth of those for housing.
- The total annualized pollution control costs paid by the federal government would increase, primarily as a result of the cost of hazardous and nuclear waste clean-ups at sites managed by the Departments of Defence and Energy. All other shares, particularly in the private sector, were expected to decrease.
- Costs to be borne by local government in real terms were expected to increase significantly.
- A major reallocation was expected over the next decade to the year 2000 in the percentage of pollution control expenditures in each environmental medium, with a shift from water and air pollution to land pollution control. Expected pollution control expenditure by media are indicated in Table 1.3.
- National expenditure on environmental pollution control measured as a percentage of GDP has been higher in the USA than in many West European nations, due to the large size of the US economy and the relative share of GDP devoted to the problem (see Figure 1.5).
- In a partial shift away from direct regulation (command and control), US applications of market-oriented environmental policies have included the use of tradable credits for lead during the phase-out of leaded gasoline, the use of taxes to provide an incentive for the phase-out of chlorofluorocarbons, the use

Table 1.3 *Pollution control expenditures by media (share of total US expenditures)*

	1987	1997
Air and radiation costs	28.9%	27.1%
Water costs	42.9%	35.7%
Land costs	26.0%	33.9%
Chemical control costs	1.2%	1.9%
Multimedia costs	1.1%	1.5%

Source: US EPA (1990) Environmental Investments: The Cost of a Clean Environment, US EPA, Washington, DC.

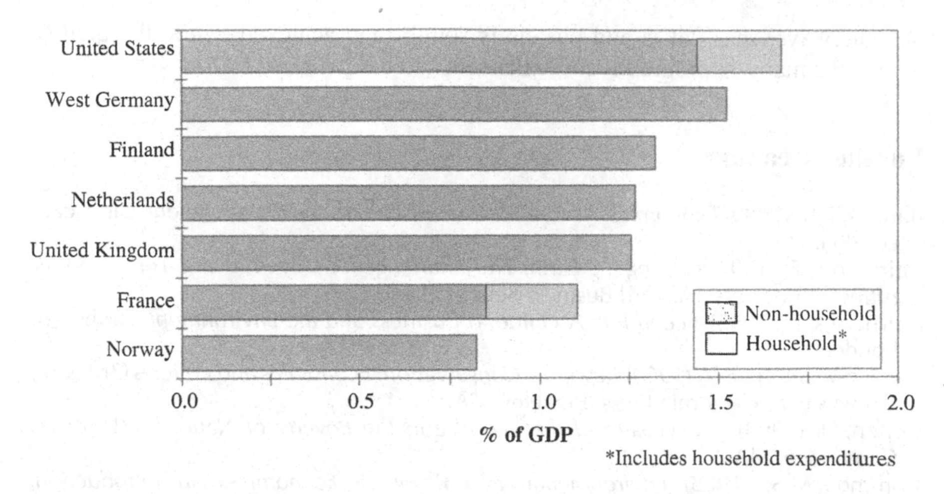

Figure 1.5 *International pollution control expenditure as a percentage of GDP in 1985. Reproduced by permission from US EPA (1990) Environmental Investments: The Cost of a Clean Environment*

of increased financial liability for damage to natural resources, and the use of tradable allowances for industrial emissions of sulphur dioxide. The emissions trading programme under the 1990 Clean Air Act amendments, to achieve the same environmental quality targets, was expected to save $1 billion ($1000 million) per year.

- Environmental expenditure has resulted in decreases in emissions compared with estimates of what would have been emitted using 1970s technology only.

1.16 Summary

Environmental economics sees all activity within a grand framework of costs and benefits, some costs and many benefits being intangible. The allocation of resources overall is a central task of government at all levels, whatever the political system, for good or ill. Within this framework, social and environmental concerns receive their share of attention, opinion varying on the quality of the outcomes. Community opinion in all countries is polarized and variable in strength and direction. But over time there has been a fundamental shift in attitudes. Whales matter, national and marine parks are appreciated, potable water is increasingly expected, food hygiene is fundamental, codes of best practice are increasingly recognized, catchment management is promoted, conventions transcend national frontiers, international conferences bring nations together, health progressively improves, life spans extend, diets become more balanced, air and water pollution diminishes at least in the richer countries, increasing protection is extended to the great biological reserves of the world, collective improvement of waterways, seas, lakes and oceans is sought and achieved, while the quality of life for many is progressively enhanced.

Further Reading

Barbier, E.B. (1989) *Economics, Natural Resource Scarcity and Development*, Earthscan, London.
Cairncross, F. (1992) *Costing the Earth: The Challenge for Governments: The Opportunities for Business*, Harvard Business School, Boston.
Cairncross, F. (1995) *Green Inc: A Guide to Business and the Environment*, Earthscan, London.
Ciriacy-Wantrup, S.V. (1968) *Resource Conservation: Economics and Policies* (3rd edn), University of California Press, Berkeley, CA.
Cohen, D. (1998) *The Wealth of the World and the Poverty of Nations*, MIT Press, Cambridge, MA.
Common, M.S. (1988) *Environmental and Resource Economics: An Introduction*, Longman, Harlow.
Cropper, M.L. and Oates, W.E. (1992) Environmental economics: a survey, *Journal of Economic Literature*, **XXX**: 675–740.
Daly, H.E. and Cobb, J.B. (1989) *For the Common Good: Redirecting the Economy towards Community, the Environment, and a Sustainable Future*, Beacon Press, Boston, MA.
Dasgupta, P. (1989) Exhaustible resources, in *The Fragile Environment* (eds L. Friday and R. Laskey), The Darwin College Lectures, Cambridge University Press, Cambridge, pp. 107–23.
Dasgupta, P. (1996) *The Economics of the Environment*, Beijer International Institute of Ecological Economics, Beijer Discussion Paper Series No. 80, Stockholm.
Field, B.C. (1997) *Environmental Economics: An Introduction* (2nd edn), McGraw-Hill, New York.
Folmer, H., Gabel, H.L. and Opschoor, H. (eds) (1995) *Principles of Environmental and Resource Economics: A Guide for Students and Decision-makers*, Edward Elgar, Aldershot and Brookfield.
Hardin, G. (1968) The tragedy of the commons, *Science*, **162**: 1243–8.
Landes, D. (1998) *The Wealth and Poverty of Nations*, Norton, New York.
Owen, L. and Unwin, T. (eds) (1997) *Environmental Management: Readings and Case Studies*, Blackwell, Oxford.
Rich, B. (1994) *Mortgaging the Earth: The World Bank, Environmental Impoverishment and the Crisis of Development*, World Bank, Washington, DC.
Siebert, H. (1998) *Economics of the Environment: Theory and Policy* (5th edn), Springer Verlag, Germany.
Thampapillei, D.J. (1991) *Environmental Economics*, Oxford University Press, Oxford.
Turner, R.K., Pearce, D.W. and Bateman, I. (1993) *An Introduction to Environmental Economics*, Harvester Wheatsheaf, Brighton, UK.

2
Luminaries

2.1 Introduction

Throughout the history of environmental interests and concerns, a few individuals have marked out the boundaries of those issues that constitute the controversial subject of the environment, and consequently fall within the boundaries of environmental economics. For example, Rachel Carson is forever associated with the issue of insecticides; likewise Garrett Hardin is associated with 'The Tragedy of the Commons', and the Ehrlichs are associated with the population explosion. More such names appear in Box 2.1, spelling out the evolution of environmental awareness and policies. Significant names rush forward, yet in the end the list is arbitrary and incomplete. Fortunately, many other significant contributors and contributions are identified in the further reading sections at the end of each chapter, in the Bibliography, and in the many memorable quotes that appear in boxes throughout the text. Even then there remain many national and international figures of note, some of whom have lost their lives in the cause of the environment, who are not included here. Omissions from this role of honour are regretted.

Brief accounts of the contributions of over thirty individuals should enable the reader to gain insight into the nature and character of this complex subject, including the subjects of air and water pollution, global warming, the visual environment, population pressures, national parks, the value of nature for its own sake, the economics of 'spaceship Earth', the limits of resources, the costs of economic growth, the divergence of private and social interests, the allocation of resources, and the emergence of national and international policies.

2.2 John Evelyn (1620–1706)

John Evelyn was an English country gentleman, famous for his diary, which is considered to be an invaluable source of insight and information on the political, social, cultural and religious life of 17th century England. He also wrote some thirty books on the fine arts, forestry and religious topics. At the Restoration of the monarchy in 1660, Evelyn was well received by Charles II, serving on a variety of commissions such as those concerned with London street improvement (1662), the Royal Mint (1663) and the repair of St Paul's (1666). More importantly, he

Box 2.1 Evolution of Environmental Awareness and Policies

1661	John Evelyn's thesis on London smog published.
1782	William Gilpin published the first of a series of books on the visual environment.
1791	William Bartram's *Travels* (the son of John Bartram, 'the father of American botany').
1798	Thomas R. Malthus publishes his *Essay on the Principle of Population.*
1805	Ferdinand Bauer's botanical paintings.
1817	David Ricardo on the theme of subsistence in his *The Principles of Political Economy and Taxation.*
1827–38	John J. Audubon, US ornithologist and artist, published *Birds of America.*
1832	George Catlin, US artist and author, proposed the idea of national parks in which Indian culture and wild country could both be preserved.
1836	Ralph W. Emerson, US essayist and poet, published *Nature.*
1854	Henry D. Thoreau, US poet and philosopher, an advocate of wilderness conservation and basic living, published *Walden.*
1864	George P. Marsh published *Man and Nature,* a significant advance in ecology and resource management.
1876	John Muir, an emigrant from Scotland, urged the US federal government to adopt a forest conservation policy. The Sequoia and Yosemite national parks were created in 1890. Muir also influenced the large-scale conservation programme of President Theodore Roosevelt. He founded the Sierra Club in 1892.
1908	President Theodore Roosevelt proclaimed, in a State of the Union message, on the need for the conservation of natural resources; later transferred large areas to the public domain. Gifford Pinchot, who first used the term 'conservation', became chairperson of the National Conservation Commission. Pinchot developed the US Forest Service and founded the Yale School of Forestry.
1913	Victor E. Shelford, ecologist, published *Animal Communities in Temperate America.*
1922	Sir Alexander Carr-Saunders, biologist and sociologist, published *The Population Problem,* a significant historical study in demography. In 1936, his *World Population* contained demographic data on numerous countries that had never before been the subject of such study. As Vice-Chancellor of the University of London, Carr-Saunders established many colleges and universities overseas.
1927	Charles S. Elton, an English biologist, published his landmark *Animal Ecology,* establishing the basic principles of the food chain and the nutrition cycle, the characteristics of foods, ecological niches, and the pyramid of numbers. In 1930 his provocative book *Animal Ecology and Evolution* appeared. In 1932 he established his Bureau of Animal Population at Oxford University.
1933	Aldo S. Leopold's textbook on *Game Management.*
1946	Start of the Pittsburgh clean-up; International Whaling Commission founded.
1947	The Los Angeles anti-smog campaign begins.

Continued on page 37

Continued from page 36

1948	WHO established; World Conservation Union.
1952	London smog disaster; RFF established.
1954	The Report of the Committee on Air Pollution (The Beaver Report) into the London smog disaster.
1956	The British Clean Air Act.
1960	River Thames clean-up begins.
1962	Rachel L. Carson publishes *Silent Spring*, an attack on the misuse of pesticides and insecticides and other chemicals.
1964	Revelle founds the Harvard Center for Population Studies.
1965	White House Conference on Restoring the Quality of our Environment; Ralph Nader publishes *Unsafe At Any Speed*.
1966	Kenneth E. Boulding publishes *The Economics of the Coming Spaceship Earth*.
1967	Edward J. Mishan publishes *The Costs of Economic Growth*.
1968	Garrett Hardin publishes 'The tragedy of the commons' in *Science*; Paul Ehrlich and Anne Ehrlich publish *The Population Bomb*.
1969	The US National Environmental Policy Act (NEPA), introducing EIA at federal level and creating the Council on Environmental Quality.
1970	Paul Ehrlich and Anne Ehrlich publish *Population, Resources and Environment: Issues in Human Ecology*; British Royal Commission on Environmental Pollution created.
1971	The International Institute for Environment and Development established; Greenpeace founded.
1972	The UN Conference on the Human Environment, Stockholm; Barbara Ward and René Dubos publish for the UN *Only One Earth: The Care and Maintenance of a Small Planet; The Ecologist* publishes *Blueprint for Survival*; the Club of Rome publishes *The Limits to Growth*; and Barry Commoner publishes *The Closing Circle: Nature, Man and Technology*.
1973	Green bans in Sydney; Canada introduces EIA.
1974	Australia introduces EIA; the UN World Population Conference.
1975	The UNESCO World Heritage List.
1976	The UN Conference on Human Settlements (Habitat I).
1977	The UN Conference on Desertification; the UN Water Conference; Clark publishes *Population Growth and Land Use*.
1979	Baumol and Oates publish *Economics, Environmental Policy and the Quality of Life*.
1980	The World Conservation Strategy; the Global 2000 Report.
1981	Julian Simon publishes *The Ultimate Resource*.
1985	The European Union introduces EIA for all members.
1986	Climate Institute founded, Washington, DC.
1987	The World Commission on Environment and Development (the Brundtland Commission) report: *Our Common Future*.
1988	Malaysia introduces EIA.
1989	The Langkawi Declaration on the Environment, issued by the Commonwealth Heads of Government.
1990	The ASEAN Kuala Lumpur Accord on Environment and Development; Coombs publishes *The Return of Scarcity*; the Beijer International Institute established.

Continued on page 38

Continued from page 37

1992	The UN Conference on Environment and Development (the Rio Earth Conference); the UN Conference on Nutrition; creation of the UN Commission for Sustainable Development; Framework Convention on Climate Change; Convention on Biological Diversity.
1994	UN Conference on Population and Development; British Environmental Agency created.
1995	UN World Conference on Women; UN World Summit for Social Development.
1996	UN Conference on Human Settlements (Habitat II).
1997	Kyoto summit on global warming.
1998	Buenos Aires summit on global warming.

was concerned with a commission for the sick and wounded mariners and prisoners of war involved in Charles II's Dutch wars. In this investigation, Evelyn exposed himself to plague while incurring considerable personal expenses. He was helped by Samuel Pepys, with whom he formed a lasting friendship. Evelyn continued to hold high office for many years, including commissioner for the privy seal. A formidable indictment of air pollution was published by Evelyn in 1661, entitled *Fumifugium: or the Smoake of London Dissipated.* In an introductory paragraph, Evelyn commented upon the adverse effects on health, of pollution in London at that time:

> Her inhabitants breathe nothing but an impure and thick Mist, accompanied by a fuliginous and filthy vapour, which renders them obnoxious to a thousand inconveniences, corrupting the lungs, and disordering the entire habit of their Bodies, so that Catarrs, Phthisicks, Coughs and Consumption rage more in this one City, than in the whole Earth besides.

Evelyn's treatise enlarged on the subject of the control of smoke and vapours, suggesting that obnoxious trades should be moved downwind. The pamphlet was addressed to Charles II.

2.3 William Gilpin (1724–1804)

Known as the Apostle of the Picturesque, William Gilpin was born in Westmorland, England, in 1724. He gained a degree at Oxford and was ordained in 1745. In 1777, he took over a school at Cheam, Surrey, where he remained for thirty years. Later in life he took up the vicarage of Boldre in the New Forest, his home for the remainder of his life, building and endowing a parish school.

During the long summer vacations, Gilpin undertook sketching tours for which he became famous. He was one of the earliest to recognize the importance of the picturesque qualities of landscape, not only of nature itself but of human artifacts.

He recognized the importance of what we now call the visual environment, in a world bent on material, industrial progress.

In 1782, Gilpin published his *Observations on the River Wye and Several Parts of South Wales*. This was the first of a series of five books with similar titles, illustrated with his own aquatint drawings. The work on the Wye went into five editions, appearing later in French. Later works focused on the mountains and lakes of Cumberland and Westmorland, and the highlands of Scotland, which were equally successful.

In 1790, one of his best known works, *Remarks on Forest Scenery and Other Woodland Views*, was published, including scenes of the New Forest. For this work, his brother Sawrey etched a set of drawings. In 1792, Gilpin published three essays: on picturesque beauty, on picturesque travel, and on sketching landscapes, with a poem on landscape painting.

Carl Paul Barbier described Gilpin as a 'master of the picturesque and the man who more than anyone alive helped to shape the picturesque vision'. He added that 'our heritage, whether in the guise of ruins or noble gothic churches, was seen with an understanding which had hitherto been absent' (Barbier 1963).

Gilpin died in 1804 at Boldre, leaving behind a history of the Gilpin Family, published by the Cumberland and Westmorland Antiquarian Society. An exhibition of the Reverend William Gilpin's works was held by the London County Council in 1959, over 150 years after his death. Several biographies have been published.

2.4 Ferdinand Lucas Bauer (1760–1826)

Ferdinand Bauer was born in Feldsberg, Austria, his father being court painter to the Prince of Liechtenstein. The three sons, Franz, Ferdinand and Joseph, all became eminent artists, Ferdinand turning to botanical illustration. He became the natural choice to be artist to the expedition of *HMS Investigator* to Australia in 1801. In 1805, on return to England, Sir Joseph Banks (1743–1820) was overwhelmed at the sight of the 11 packing cases of drawings that came back with Bauer.

The botanical paintings of Ferdinand Bauer have been highly praised both at the time and since. The British Natural History Museum has stated that Bauer 'created a standard of artistry combined with critical botanical observation that has never been equalled'. After the circumnavigation of Australia, Bauer returned to Austria, undertaking many European excursions to paint and draw. He has been considered perhaps the world's greatest botanical artist.

A selection of Bauer's work was published in 1976 by Basilisk Press under the title *The Australian Flower Paintings of Ferdinand Bauer*. In 1998, Nokomis Press published another selection of Bauer's work, known as the *Ferdinand Bauer Collection*.

2.5 Thomas Robert Malthus (1766–1834)

Thomas Robert Malthus was an English clergyman who subsequently became Professor of History and Political Economy at the East India Company's Haileybury College. His famous *Essay on the Principle of Population as it affects the Future Improvement of Society* was published in 1798, reaching a sixth edition in 1826. His theme was pessimistic, asserting that (1) population would soon outstrip the means of feeding it, if it were not kept down by vice, misery or self-restraint; (2) in a state of society where self-restraint does not act at all, or only acts so little that we need not think of it, population will augment until the poorest class of the community have only just enough to support life; and (3) in a community where self-restraint may act eventually, each class of the community will augment until it reaches the point at which it begins to exercise that restraint. Population, Malthus declared, tended to increase in a geometrical progression (1, 2, 4, 8 and so on); whereas the means of subsistence increased in only an arithmetical progression (1, 2, 3, 4 and so on). The increase in population, therefore, always tends to outrun the food supply. Most people would therefore be condemned to live in misery and poverty, with wars, epidemics and famines serving to slow the growth of population. Malthus thus presented economics as 'a dismal science'.

Malthus did not foresee the dramatic improvements in agricultural techniques that were to occur, nor the tendency for population growth to modify itself markedly in developed countries. Many countries have passed through a demographic transition in which the dire predictions of Malthus have been dramatically modified by family planning, social services, increased food production per capita, education, and generally improved living standards. The Malthusian theory of population acted as a brake on economic optimism, helping to justify subsistence wages and discouraging traditional forms of charity. He saw workhouses for the poor and distressed as places where 'fare should be hard'. Yet he saw public works and private luxury expenditure as palliatives for economic distress, increasing effective demand (a term he invented) and prosperity. He anticipated some of the thoughts of John Maynard Keynes.

See Box 2.2 for the views of Schumpeter.

2.6 John Muir (1838–1914)

John Muir was a naturalist and advocate of US forest conservation. He was largely responsible for the creation of the Sequoia and Yosemite national parks in California (1890). Early in 1897, President Grover Cleveland designated 13 national forests to be protected from commercial exploitation. However, business interests induced Congress to defer this measure. Subsequently, two persuasive magazine articles by Muir swung Congressional and public opinion behind the reservations. He also influenced the large-scale conservation programme that was

Box 2.2 Schumpeter on Malthus

> The falling birth-rate seems to me to be one of the most significant features of our time ... it is of cardinal importance both as a symptom and as a cause of changing motivation ... forecasts of future populations, from those of the seventeenth century on, were practically always wrong. For this, however, there is some excuse. There may be even for Malthus' doctrine. But I cannot see any excuse for its survival. In the second half of the nineteenth century, it should have been clear to anyone that the only valuable things about Malthus' law of population are its qualifications. The first decade of this century definitely showed that it was a bogey.

Source: Joseph A. Schumpeter (1883–1950) in J.A. Schumpeter (1943) Capitalism, Socialism and Democracy, George Allen and Unwin, London, p. 115.

initiated by President Theodore Roosevelt. In 1903, Roosevelt accompanied Muir on a camping trip to the Yosemite region. In 1908, the Congress established the Muir Woods National Monument in Marin County, California. The monument, which lies 24 km north-west of San Francisco, is a wooded area of 204 ha, in which some of the redwood trees stand more than 90 m high.

John Muir became the first president of the Sierra Club, a US voluntary organization founded in 1892 for the conservation of natural resources. Muir involved the Club in political action to further its objectives. The Sierra Club has its headquarters in San Francisco, but it has branches in all American states. It continues to lobby for environmental legislation and to educate the public on environmental matters.

2.7 Theodore Roosevelt (1858–1919)

Theodore Roosevelt was the 26th President of the United States (1901–08), although it was not until 1904 that Roosevelt was elected as President overwhelmingly in his own right, having previously been President only through the assassination of William McKinley. Following this success, Roosevelt began expanding the role of the White House, bringing pressure to bear on Congress for new powers. He gained substantial powers to regulate interstate railroad rates. The next step was the passage of the Pure Food and Drugs and the Meat Inspection Acts (1906), which introduced the modern concept of consumer protection. Then, responding to the rapid disappearance of federal lands, the President used existing powers to convert remaining land to national forest. Under Roosevelt's predecessors only about 16 million ha had been so transferred. Roosevelt accelerated the pace of transfer and, with broadened powers, reserved for future generations parks and mineral, oil and coal resources, as well as hydroelectric power sites. During seven years, an additional 80 million ha of the federal domain were closed to commercial development. Box 2.3 conveys the President's words on the subject of resource conservation during his last year in office.

Box 2.3 President Theodore Roosevelt on Resource Conservation

The time has come to enquire seriously what will happen when our forests are gone; when the coal, the iron, the oil, and the gas are exhausted; when the soils shall have been still further impoverished and washed into the streams, polluting the rivers, denuding the fields, and obstructing navigation. These questions do not relate only to the next century, or to the next generation. It is time for us now as a nation to exercise the same reasonable foresight in dealing with our great natural resources that would be shown by any prudent person in conserving and wisely using the property which contains the assurance of well-being for the family and its children. The nation behaves well if it treats its natural resources as assets which it must turn over to the next generation increased and not impaired in value.

Source: Theodore Roosevelt, May 1908.

2.8 Arthur Cecil Pigou (1877–1959)

Arthur Cecil Pigou was an English economist who developed the theory of welfare economics in his book *The Economics of Welfare* (1919). He argued that the obvious divergence between private returns and social returns could often be brought about only through taxes and subsidies, to achieve an optimal allocation of resources. He presented the case for government intervention to improve the efficiency of resource allocation. Pigou warned that there was no precise line between economic and non-economic satisfaction; hence economic welfare alone does not serve as a barometer or index of total welfare. An economic benefit could affect non-economic welfare in ways that cancel its beneficial effect. Thus a society may boost real income in the short term by cutting down all its trees and selling the timber, without any attempt at reafforestation. Through such activities a society could, in the end, destroy both its economic and non-economic welfare.

Taxes on the polluter, based on damage caused, are often known as Pigovian taxes. By placing a monetary cost on each unit of pollution, a Pigovian tax attempts to send the proper price signals to the polluter. The polluter may avoid the tax, in whole or in part, by installing pollution control equipment or varying the mode of production. If the polluter fails to respond, the tax may be raised. The concept presents difficulties in respect of the setting of a tax that will match public injury to health and welfare; the unpredictability of the polluter; and the fact that the tax may not be used to benefit those affected but may flow into general revenue. Assuming compensation is undertaken, consideration must be given to the incredibly difficult task of distributing benefit to those suffering from the effects of air and water pollution, noise, and the adverse effects of say lead or radiation poisoning. Many take the view that inducement to behave is second best as a solution to pollution, direct regulation and enforcement being more effective.

See Box 2.4 for the views of Pigou on the duties of government.

Box 2.4 Pigou on the Duties of Government

It is the clear duty of Government, which is the trustee for unborn generations as well as for its present citizens, to watch over, and if need be, by legislative enactment, to defend, the exhaustible natural resources of the country from rash and reckless spoliation. How far it should itself, either out of taxes, or out of State loans, or by the device of guaranteed interest, press resources into undertakings from which the business community, if left to itself, would hold aloof, is a more difficult problem. Plainly, if we assume adequate competence on the part of governments, there is a valid case for some artificial encouragement to investment, particularly to investment the return from which will only begin to appear after many years.

Source: Arthur Cecil Pigou (1946) The Economics of Welfare (4th edn) Macmillan, London.

2.9 Karl Gunnar Myrdal (1898–1987)

Karl Gunnar Myrdal was a Swedish economist, social scientist, statesperson, reformer, dissenter, pacifist, opposer of inequality, and an architect of the Swedish welfare state. He was awarded the Nobel Prize in Economics (with Frederick von Hayek) in 1974. In 1944, Myrdal published his *An American Dilemma*, a classic study of the black minority in the USA. In 1968, he published another famous study, *Asian Drama: An Inquiry into the Poverty of Nations*. Here he argued that land reform and a narrowing of social inequalities were prerequisites for eradicating poverty. In 1970, he published *The Challenge of World Poverty*. In more recent times, Myrdal argued that economists should accept the need to make explicit value judgements, without which their theoretical structures are unrealistic. He regarded economics as a soft science, inevitably loaded with value judgements. Gunnar Myrdal was one of the distinguished speakers to address delegates to the UN Conference on the Human Environment 1972, held in Stockholm. The title of his address was 'Economics of an Improved Environment'.

He argued that, contrary to many gloomy forecasts, the era from Malthus up to the present had been one of historically unprecedented economic growth, with rising standards of living even in the lower income brackets. This had happened in spite of periodic depressions and devastating wars. In respect of fears about impending shortages of raw materials, there have been never-ending discoveries of new supplies of raw materials, particularly of oil and various metal ores. This had been reinforced by technological inventions and substitution. The growth of population alone remains a continuing anxiety. He then referred to the Club of Rome report, *Limits to Growth: A Global Challenge*. He criticized the report on the grounds that it uncritically accepted the concept of GDP; aggregated in a most careless way extremely uncertain data particularly relating to threatening pollution and depletion of resources; and excluded inequalities in distribution, on the grounds that these are social problems. He also said that the report placed attitudes and institutions outside the world model, and used mathematics and a huge computer for calculations with little, if any, scientific validity. In respect of any threats of pollution or depletion, he argued that there are two instruments of policy: direct regulation and the use of charges and subsidies. Myrdal considered population growth as the key factor in the environmental and resource problem, and that what was needed here was the perfection of an already vastly improved technology of birth control.

Myrdal had also come to the conclusion that the trend to greater inequality should be broken; egalitarian reforms would not only be beneficial but a precondition for sustained and rapid growth. Up until now, most development in the undeveloped countries has mostly enriched a tiny upper stratum, and left the masses of the people where they have always been.

While governments have generally been committed to steps to improve the environment, Myrdal argued that the ideology and psychology of unrestrained economic growth retained its hold over people's minds as powerfully as ever. Everyone is intent on raising their incomes and standard of living. Further, the owners and would-be owners of cars are everywhere by far the biggest political party. Myrdal saw the basic problem in the economics of the environment as being that everyone wants a progressively improving environment, but many find difficulty in paying for it.

2.10 Constantinos Apostolos Doxiados (1913–75)

Constantinos Apostolos Doxiados was a Greek architect and city planner, and a major contributor to the science of human settlements, which he called *'ekistics'*. Doxiados first used the term 'ekistics' in his lectures of 1942 at the Athens Technical University. Another term attributable to Doxiados was *'ecumenopolis'*, or the universal city, the logical consequence of unlimited population growth. He traced development through the metropolis, conurbation and megalopolis to the ecumenopolis, the all-encompassing urban city. This remarkable concept of the universal city would incorporate areas reserved for recreation and agriculture, as well as desert and wilderness conservation areas; however, in essence it would be a web of interconnected cities throughout the world, closely linked by rapid transport and electronic communication. Ecumenopolis would cover the entire land area of the Earth.

The goal of ekistics is to develop a system and a methodology (a) to study all kinds of settlements in order to draw general conclusions about them; (b) to study each as a whole to solve the specific problems of that settlement; and (c) to solve the economic, social, political, technological and cultural problems in a balanced way.

In the early 1960s, Doxiados called for the establishment of a UN special agency for human settlements; the Centre for Human Settlements was established in Nairobi by the UN a year after his death. Doxiados' famous book, *Ecology and Ekistics*, appeared in Britain in 1977 and in Australia in 1978. It was a major attempt to describe the relationship between ecology, the science concerned with the relationships between living organisms and their surroundings, and ekistics, the science of human settlements. The book does not attempt to solve the problems of ecological and ekistic conflict, but rather to start the process in a logical and scientific way. He envisaged in the end a global ecological balance. Doxiados saw the need for the wise use of our natural resources, the prevention of waste and despoilment, an enhancement of the quality of life, and the reconciliation of the aspirations of humanity with the limitations of a finite world.

2.11 Lewis Mumford (1895–1990)

Lewis Mumford was an authority on American architecture, art and urban planning. As a student he was influenced by the writings of Patrick Geddes (1854–1932), one of the modern pioneers of the concept of town and regional planning. In his wide-ranging books, Mumford criticized the dehumanizing tendencies of modern technological society, and urged that it be brought into harmony with humanistic goals and aspirations. One of Mumford's most important works was *The City in History* (1961), a sweeping study of the city's role in human civilization. For this work he received the National Book Award in 1962. In later years, Mumford angrily attacked everything from highways and automobiles to aircraft and electronics. Yet in his *Renewing of Life*, he explored ways of improving cities.

2.12 Resources for the Future (RFF)

Resources for the Future, established in 1952, is a non-profit corporation for research and education in the development, conservation and use of natural resources and the improvement of the quality of the environment. It has made substantial contributions in these areas. The RFF is based in Washington, DC. In 1964, the RFF published *World Prospects for Natural Resources: Some Projections of Demand and Indicators of Supply to the year 2000* (by Joseph L. Fisher and Neal Potter). This was in response to growing concerns about the capacity of the natural environment and its resources to sustain desired rates of economic growth, particularly in a context of continuing increases in population, which in some less developed areas was then running at 3% per year. Box 2.5 summarizes the projections to the year 2000. The central finding was that the data available did not support a generalization that the world is about to run out of raw materials. This may be contrasted with the later (1972) finding of the Club of Rome.

2.13 The Beaver Committee on Air Pollution (1954)

In 1954, the Beaver Committee on Air Pollution reported to the UK Minister of Housing and Local Government on its findings into the London smog disaster

Box 2.5 RFF Projections of Demands for Natural Resources, 1964–2000

- Population projections from Malthus to the recent past have been notoriously wide of the mark. Projections of natural resources have, if anything, been worse.
- Historical data do warn against easy generalizations that the world is about to run out of raw materials.
- The picture is mixed: quite favourable for energy commodities, much less so for food.
- Conventional energy resources equal to about 900 years of use at the 1960 rate of consumption are available.
- By the year 2000, some 10–20% of the world's energy consumption may come from nuclear sources.
- Studies indicate that for iron, aluminium and manganese, the known and inferred reserves are large enough, world-wide, to supply projected demands for at least the next 40 years without significant increases in costs.
- In recent years, major new sources of iron ore have been discovered in Venezuela, Canada, Liberia, Brazil and Australia.
- For copper, lead and zinc, projected demands are likely to exhaust these materials by the year 2000. It should be kept in mind that there are satisfactory substitutes for these metals in most uses.
- As far as forest products are concerned, the world scene offers encouraging signs.
- The standard of living in most countries will increase over the coming years, with diets improving slowly. Shortages of natural resources are unlikely to retard this process.
- It is nevertheless likely that a slower rate of population increase would be associated with a faster increase in per capita standard of living; there would be fewer mouths to feed compared with hands to work.
- Changing attitudes may check the present population growth rates long before there is 'standing room only'.
- There are certain escape hatches from any tendency towards increasing shortages, such as the substitution of more plentiful, convenient and cheaper materials for the ones becoming scarcer; and programmes of research, conservation and better management offsetting incipient tendencies towards scarcity.
- As time goes on, we should be able to answer with less and less uncertainty the central question as to whether resources are becoming scarcer world-wide.

Source: summarized from J.L. Fisher and N. Potter (1964) World Prospects for Natural Resources: Some Projections of Demand and Indications of Supply to the Year 2000, Resources for the Future, Washington, DC.

of 1952, during which there were some 4000 deaths more than the normal death rate for such a period. The report declared that air pollution on the scale familiar in Britain was a social and economic evil that should no longer be tolerated. The committee was confident that the carrying out of its proposals would secure happier and more healthy living conditions for millions of people. Also, on all counts, the cost of the cure would be far less than the national loss in allowing the evil to continue.

The committee was satisfied that the most serious immediate problem to be tackled was smoke, grit and dust. An objective of the recommendations was to reduce smoke in the most heavily polluted areas of the UK by 80% within 10–15 years. This would mean relief from air pollution not known in many parts of the country for more than a century. It recommended a Clean Air Act, with appropriate measures against industrial and domestic pollution, largely through the introduction of smoke control areas.

The committee examined the economic cost of air pollution, referring to many earlier studies, and assessing both direct and indirect costs. The findings were published in considerable detail in Appendix II of its report (Committee on Air Pollution Report 1954; Gilpin 1963). Sir Hugh Beaver subsequently reviewed these estimated costs and considered them too low.

Smoke control areas were introduced under the Clean Air Act 1956, the concept being a geographical area that was proclaimed smoke-free. West Bromwich in the industrial Midlands was the first area to proclaim such an order, and experience the first public inquiry into public objections. The coal industry fought these orders tooth-and-nail. By 1970, some 75% of all premises in the Greater London area were covered by smoke control orders. By 1985, more than 5000 smoke control orders were in operation, covering more than eight million premises. Smoke was reduced by some 80% and the amount of sulphur dioxide in the air from coal had been halved. December sunshine in London increased by over 70% and visibility had improved considerably. London smog incidents such as those of 1952 and 1962 became a thing of the past.

See Box 2.6 relating to the economic losses caused by air pollution, and Box 2.7 relating to pollution over the industrial areas of Britain.

2.14 Ronald Coase

Winner of the Nobel Prize in Economics in 1991, Ronald Coase (b. 1910) made an initial mark in economics with his article 'The Nature of the Firm' (*Economica* 1937), which explained why individuals group together in firms, rather than working through contracts with each other. In the US he made a second major contribution in his 'The Problem of Social Cost' (*Journal of Law and Economics*, October 1960). The Coase theorem, as it is called, relates to the efficiency of property rights as a means of allocating resources. Coase argued that economic efficiency would be achieved as long as property rights were fully allocated, with free trade in such rights. Trade will place resources in their highest-value role, eventually. The theorem has been used to show that a solution to the problem of externalities is the allocation of property rights. Pollution can then be dealt with through negotiations between owners, and the problem resolved in some suitable settlement. Market forces find the best solution to a dispute. He argued for dispute resolution by negotiation, outside of the courts, avoiding litigation.

Box 2.6 Beaver on the Economics of Air Pollution

The Beaver Report (1954) placed economic losses due to air pollution in Britain in the following categories:

(a) direct medical losses
(b) lost income resulting from absenteeism from work
(c) decreased productivity
(d) increase of travel costs and time of travel due to reduced visibility
(e) increase of costs of artificial illumination
(f) repair of damage to buildings and other structures
(g) increased costs of cleaning
(h) losses due to damage to crops and ornamental vegetation
(i) losses due to injury to animals of economic importance
(j) decrease of property values
(k) extra costs of manufacture because of pollution from outside sources
(l) losses due to the inefficient combustion of solid, liquid and, possibly gaseous fuels

Various monetary estimates of these losses have been made, and all of them have been of a high order of magnitude. The Beaver Committee, in assessing the economic cost of air pollution in Britain, arrived at the following main conclusions:

(1) The direct costs of pollution:

	£ m. per annum
Laundry	25
Painting and decorating	30
Cleaning and depreciation of buildings other than houses	20
Corrosion of metals	25
Damage to textiles and other goods	52.5
	152.5

(2) The indirect costs of pollution. The following figures were given as 'a rough guess at possible orders of magnitude': £10 million a year loss to agriculture; £55 million a year loss of working efficiency in manufacturing industry; £60 million a year in other sections, including transport – altogether 'the loss of efficiency resulting from pollution might be as great as the direct costs'. To avoid the dangers of over-estimation, a total figure of £100 million a year was suggested.

(3) The value of the fuel lost under conditions which produce smoke was estimated to be £25–50 million a year, although it was considered that this should not be included as one of the costs of air pollution.

The total estimated cost was given as £250 million a year, equivalent to £10 per head in the black areas and £5 per head elsewhere throughout Britain, plus the value of the fuel lost. This estimate did not include an evaluation of the cost of medical services.

Subsequently, during the closure of the Conference on the Mechanical Engineer's Contribution to Clean Air held in London in February 1957, Sir Hugh Beaver stated that he had spent some months trying to check some of his Committee's calculations and had come to the conclusion that £250 million was a very considerable understatement: 'in various directions it could be said that the figures that the Committee had put against particular sections of the cost of pollution were considerably too low, by perhaps 200 per cent'.

Continued on page 49

Continued from page 48

Scorer has attempted to apportion the costs of air pollution between classes of fuel consumer. His studies were based on the assumption that the effluents from different sources are not dispersed in the same way, consequently the damage done by pollution does not depend only, or even mainly, on the amount emitted. Five independent factors which determine how damaging the pollution will be were estimated for the seven main classes of fuel consumer. The factors were multiplied together and the total estimated damage of £250 million a year for the whole country was apportioned among the classes. The factors considered were:

(a) height, temperature and bulk nature of effluent
(b) radiative effects of suspended pollution
(c) concentration of output into times of adverse weather
(d) closeness to site of damage
(e) damaging chemical nature of effluent

The estimates of the damage done per ton of fuel consumed by various classes of consumer were indicated. The most striking estimate was that the amount of damage to be attributed to the average domestic ton of coal was about £6. Scorer estimated that the saving to the nation resulting from the universal use of smokeless fuel in domestic grates would be between £100 and £140 million a year. These estimates were made for the country as a whole and should not be applied without modification to particular areas.

A later estimate by Scorer states that a ton of bituminous coal burnt at home does about £10 worth of damage, and that the total damage experienced in the whole country amounts to about £500 million a year.

Source: Report of the Committee on Air Pollution (1954) Cmd 9322, HMSO London.

Box 2.7 Extract from The Smoke Inspector's Handbook, 1923

For it must be obvious to any observer that, in spite of many welcome improvements, the boiler chimney still remains a factor of enormous magnitude in so far as our great manufacturing towns are concerned. The railway passenger passing through Lancashire, the West Riding of Yorkshire, the Black Country or other similar areas, can see for himself the extent of the evil; he can watch the black masses emitted from individual chimneys collecting to form a dark canopy spreading over miles of country. The experts say that such clouds are unnecessary and can be avoided by intelligent care.

Source: Cyril Banks' Foreword in H.G. Clinch (1923) The Smoke Inspector's Handbook or Economic Smoke Abatement, H.K. Lewis, London.

Coase argued that, as a rule, no form of government action was required to deal with externalities or public goods. There was no need for taxes, subsidies and public provision; property rights, not government action would resolve problems. The argument was against government intervention. The Coase theorem sought to demonstrate (as in the case of bees and lighthouses, two of Coase's analytical studies) that externalities (both positive and negative) and public goods did not automatically call for government intervention. The theorem inspired further critical examination of these issues. It put regulation somewhat out of fashion. Economists now think harder on how to achieve green ends through market forces, rather than statutory regulation. There is clearly some scope for this, even though such resolutions need to take place, as with all transactions within a regulatory framework. There remains scope also for the resolution of disputes through out-of-court dispute resolution; a growing practice in all areas. It seems that property rights may improve situations over and above free access regimes, but fail significantly when it comes to global issues and transboundary disputes; nor does the theorem deal with Los Angeles smog, Pittsburgh's 'darkness at noon', the London smog disaster, the pollution of the tidal Thames, or the destruction of tropical forests. The scope for the Coase theorem in the real world is limited.

2.15 The Pippard Committee on the Tidal Thames (1961)

In 1961, a committee chaired by Sutton Pippard reported to the Minister of Housing and Local Government (Mr Henry Brooke) on the effects of heated and other effluents on the condition of the tidal Thames, London, both at that time and in the context of future developments. The committee concluded as follows:

- There had been a fairly steady deterioration in the condition of the tidal river water since 1930–1935, this deterioration becoming much more marked since the Second World War.
- The belt of worst pollution had been gradually extending both upstream and downstream, and now stretched from Bermondsey to Thames Haven.
- The temperature of the water in the middle reaches had risen since 1920, heated effluents from power stations clearly playing an important part.
- The stretch between London Bridge and Gravesend was often offensive, the predominant source of pollution being the combined discharge of inadequately treated sewage effluent from the northern and southern outfall works; a drastic reduction in this pollution load would improve the river substantially.
- The committee noted the contribution of sulphite to the river from the Battersea and Bankside power stations where the flue gases were washed to remove sulphur.

Following the committee's report, measures were put in hand to clean up the Thames. By 1969, the sand goby, flounder and smelt were being found on the screens at Fulham power station, which was situated 16 km above London Bridge and 64 km from the sea. In 1985, a seal reached central London in search of fish. The Thames had not been cleaner for more than a century.

2.16 Restoring the Quality of our Environment (1965)

In November 1965, in the USA, the report of the Environmental Pollution Panel of the President's Science Advisory Committee was presented to President Lyndon Baines Johnson (1908–73). The report embraced the many aspects of pollution including the health effects, the effects on other living organisms, the impairment of water and soil resources, the polluting effects of detergents, the deterioration of materials, and the climatic effects of pollution, notably global warming.

The report examined the sources of pollution in the US, including municipal and industrial sewage, animal wastes, urban solid wastes, mining wastes and consumer wastes. The report ended with a wide range of recommendations. One of the recommendations was a careful study of tax systems in which all polluters would be subject to effluent charges in proportion to their contribution to pollution. Of the many measures recommended to minimize pollution from all sources, one was to intensify the measurement of carbon dioxide in the atmosphere, and to monitor temperature changes in the upper atmosphere. It was noted that fossil fuels that have lain in the Earth for over 500 million years could all be burnt within a few generations. About half the carbon dioxide released remains in the atmosphere and the effects on our climate are likely to be significant. It was estimated in the report that atmospheric carbon dioxide would increase by 25% by the year 2000; this may be sufficient to produce measurable and perhaps marked changes in climate deleterious to human welfare. The report vigorously attacked water pollution problems and hazardous wastes.

2.17 Edward Mishan

Mishan of the London School of Economics, in his 1967 work *The Costs of Economic Growth*, focused on what he called the 'neighbourhood effects' generated by a wide range of economic activities. These are otherwise known in the literature as external economies or diseconomies, externalities, side-effects, spillover effects or spillovers. A power station may be something of an external diseconomy in itself, but electricity at the point of use has an important external economy, for unlike other fuels and sources of energy, it is pollution-free. Mishan stressed that the operation of firms, or the doings of ordinary people, frequently have significant effects, good or bad, on others of which no account need be taken by the firm or individual responsible for them, unless constrained by law or custom. Consequently, the market price of a good or service is not necessarily an index of its marginal value to society. Hence the social value of a good or service (i.e. the value remaining after subtracting from its market price the estimated value of the damage inflicted on others by producing or using the good or service) may be well below its market price. It may be argued that prices should be equal to social marginal cost, although the estimation of damages (or benefits) can pose considerable practical difficulties. For example, the more general social afflictions, such as industrial noise, dirt, stench, ugliness, urban sprawl and other features that jar the nerves and impair the health of many, are difficult both to measure and to impute to any single source.

Mishan argued that the growing incidence of the external diseconomies generated by certain sectors of the economy and suffered by the public at large may be regarded as the most salient factor responsible for the misallocation of the nation's resources. However, government intervention is not necessarily desirable for, unless the injuries are large, the cost of intervening and administering a satisfactory scheme may exceed the apparent social gain. Examples of major spillovers include adverse effects on flora and fauna; soil degradation and salinity; deforestation without replanting; traffic congestion and road accidents; noise from industry, aircraft and road traffic; air and water pollution; hazardous wastes; unregulated landfills; dereliction and slums; unsafe and unreliable water supplies; inadequate drainage and sewage treatment; and visual pollution.

It should not be forgotten, however, that many external effects are positive, as described in Alfred Marshall's *Principles of Economics* (1890). These include the mutual benefits arising from the concentration of businesses in a particular locality, improved transport systems, specialists serving a number of companies, easier communications and information systems, and common services such as electricity, water, gas, shopping and recreational centres, cheaper imports, lower interest rates and inflation.

2.18 Charles Sutherland Elton (1900–91)

Charles Sutherland Elton was an English biologist, credited with framing the basic principles of modern animal ecology. Graduating in zoology in 1922, Elton's first book, *Animal Ecology*, appeared in 1927. It became a landmark in the establishment of the basic principles of ecology, notably the concepts of food chains and the nutrition cycle, the characteristics of foods, ecological niches, and the pyramid of numbers. In 1930, his provocative work *Animal Ecology and Evolution* appeared, in which he denied the concept of the 'balance of nature', stressing the role of the migration of animals in which the environment is chosen by the animal as opposed to the adaptation of the animal to the environment, via natural selection. In 1932, Elton established his Bureau of Animal Population at Oxford, which became a world centre for data and research. He became editor of the new *Journal of Animal Ecology*. He was elected a fellow of the Royal Society in 1953 and awarded the gold medal of the Linnean Society in 1967, the year of his retirement. At all times, Elton stressed the need for nature conservation (Hardy 1968).

2.19 Rachel Carson (1907–64)

Rachel Carson was a US marine biologist and conservationist, whose work helped to launch the environmental movement world-wide. She was born in Springdale, Pennsylvania, the youngest of three children. In 1929, she took her degree in zoology, winning a graduate scholarship to Johns Hopkins University. She then joined the US Bureau of Fisheries (later the US Fish and Wildlife Service), researching new developments in the science of oceanography. In 1951, she published *The Sea Around Us*. It spent 86 weeks on the best-seller list, winning the John Burroughs Medal and the National Book Award. Carson was now able to leave her government job, and began an ecological study of the seashores of the Atlantic coast. In 1955, she published *The Edge of the Sea* and in 1962 *Silent Spring*, the opening salvo of the environmental movement. Drawing on hundreds of scientific studies, she described the serious challenge to health and life arising from the indiscriminate use of DDT and other toxic chemicals that accumulate in the environment, entering the food chain and poisoning animals and humans. Serialized in the *New Yorker, Silent Spring* instilled a sense of outrage in readers, alerting Americans to the collusion of big business, research institutions and government regulatory agencies in the pollution and depletion of natural resources. Carson's findings were confirmed by President Kennedy's Science Advisory Committee, notwithstanding the hysteria of the chemical industry. She was awarded the Audubon Medal and was elected to the American Academy of Arts and Letters. With growing afflictions, she died at the age of 56 at Silver Spring, Maryland. Much Congressional legislation followed her revelations. Two years after the publication of *Silent Spring*, the US Public Health Service had positive proof that the pesticide endrin was responsible for the deaths of ten million fish in the lower Mississippi River and the Gulf of Mexico, as a result of runoff from farms. It confirmed Rachel Carson's assertion that those who make extensive use of pesticides are upsetting the balance of nature, and in doing this we are fighting a losing battle. The UN Food and Agriculture Organization held a world conference in Rome to study how pesticides could be used effectively, without harming people. In the emergence of the middle ground, it is now recognized that chemistry, biology, wildlife and humanity can coexist.

2.20 Roger Randall Dougan Revelle (1909–91)

Roger Revelle was a US oceanographer who spearheaded many scientific endeavours that resulted in a greater understanding and appreciation of the Earth. He has been credited with pioneering research into global warming and plate tectonics. During the Second World War, he commanded the oceanographic section of the US navy's Bureau of Ships, and directed the post-war oceanographic investigations for the 1946 atomic bomb tests at Bikini Atoll. In 1949, Revelle returned to the Scripps Institution of Oceanography (where he had earlier received a doctorate) as professor of oceanography and became increasingly concerned with the levels of carbon dioxide in the atmosphere.

He maintained that the higher levels of carbon dioxide in the atmosphere produced by the burning of fossil fuels and the burning of forests to make way for crops, could not be absorbed by the oceans, and were accumulating in the atmosphere and trapping the heat of the Sun. Revelle was thus one of the first to warn of the effects of global warming.

After serving as science adviser to the US Secretary of the Interior (1961–63), Revelle became interested in helping the poorer countries of the world. In 1964, he founded the Harvard Center for Population Studies, and was its director until he moved to the University of California at San Diego in 1976. Revelle wrote more than 200 scholarly papers, receiving the National Medal of Science in 1990.

Roger Revelle had become known as 'Dr Greenhouse', yet wrote just before his death in 1991, 'The scientific basis for greenhouse warming is too uncertain to justify drastic action at this time' (*The Economist*, 20 December 1997, p. 21).

2.21 The Ehrlichs

Paul R. Ehrlich and Anne R. Ehrlich, both of Stanford University, are most noted for two compelling publications: *The Population Bomb* (1968) which became a best-seller, and *Population, Resources, Environment: Issues in Human Ecology* (1970). These books were the first comprehensive, detailed analyses of the world-wide crisis of overpopulation and the resulting demands on food, resources and the environment. The Ehrlichs demonstrated that problems such as environmental deterioration, hunger, resource depletion and war are closely interconnected and that together they constitute a challenge without precedent in human history.

The Ehrlichs reached a range of conclusions:

- Our planet is grossly overpopulated now.
- The large absolute number of people and the rate of population growth are major hindrances to solving human problems.
- The limits of human capability to produce food by conventional means have very nearly been reached.
- Some 10–20 million people are starving to death annually.
- Attempts to increase food production further will tend to accelerate the deterioration of our environment, which in turn reduces the capacity of the Earth to produce food.
- There is reason to believe that population growth increases the probability of a lethal world-wide plague and of a thermonuclear war.
- Technology properly applied in such areas as pollution abatement, communications, and fertility control could provide massive alleviation.
- The fundamental solutions involve dramatic and rapid changes in human attitudes, especially those relating to reproductive behaviour, economic growth, technology, the environment, and conflict resolution.

- They wished to emphasize their belief that the problems can be solved.
- Population control is absolutely essential if the problems now facing humanity are to be solved; but it is not a panacea: *whatever your cause, it is a lost cause without population control.* Zero population growth should be the objective; a maximum of two children per family.
- Perhaps the major necessary ingredient that has been missing from a solution to the problems of both the USA and the rest of the world is a goal, a vision of a Spaceship Earth and the kind of crew it should have aboard.

2.22 The Club of Rome

The Club of Rome is a voluntary association formed in 1968 by a group of 30 individuals from 10 countries, including scientists, educators, economists, humanists, industrialists and civil servants, following an initial meeting in the Accademia dei Lincei in Rome. The meeting was at the instigation of Aurelio Peccei (1908–84), an Italian industrialist. He was one of the distinguished speakers at the special series of lectures associated with the UN Conference on the Human Environment 1972.

The purpose of the Club of Rome was to foster understanding of the varied but interdependent components that make up the global system and to promote new policy initiatives. The Club initiated a Project on the Predicament of Mankind, a research project, which resulted in the publication in 1972 of *The Limits to Growth* (Meadows *et al.* 1972). The report concluded that, even under the most optimistic assumptions, the world cannot support present rates of economic and population growth for more than a few decades; however, with suitable policies a state of equilibrium might be achieved. Table 2.1 indicates the Club of Rome predictions on the life of non-renewable natural resources, with the expected years of exhaustion. It was thought in 1972 that by the year 2000, petroleum and natural gas, lead, mercury, silver, tin, tungsten and zinc would no longer be available.

The Club's second report, *Mankind at the Turning Point*, was published in 1975. Further reports followed, *Goals for Mankind* appearing in 1977 (Laszlo 1977).

A sequel to *The Limits to Growth* appeared 20 years later in 1992 (Meadows *et al.* 1992) entitled *Beyond the Limits: Global Collapse or a Sustainable Future?* The position taken in this work closely resembles that taken in the earlier work, endorsing the same main conclusions:

- The human use of many essential resources and the generation of many kinds of pollutants have already surpassed rates that are physically sustainable. Without significant reductions in material and energy flows, there will be in the coming decades an uncontrolled decline in per capita food output, energy use and industrial production.

Table 2.1 *Club of Rome predictions on the life of non-renewable natural resources 1970*

Resource	Expected year of exhaustion
Bauxite (aluminium)	2001
Chromium	2065
Coal	2081
Cobalt	2030
Copper	2091
Gold	2079
Iron	2063
Lead	1991
Manganese	2016
Mercury	1983
Molybdenum	2004
Natural gas	1992
Nickel	2023
Petroleum	1990
Platinum	2017
Silver	1983
Tin	1985
Tungsten	1998
Zinc	1988

Source: Adapted from Meadows et al. (1972) The Limits to Growth, Pan Books, London.

- This decline is not inevitable. Two changes are necessary: a comprehensive revision of policies and practices that perpetuate growth in material consumption and population; and a drastic improvement in the way in which materials and energy are used. A sustainable society is still technically and economically possible.

The conclusions are described as a 'conditional warning, not a dire prediction'. It should be noted, however, that for many resources estimated reserves are now much larger, and prices lower, than they were in 1972. Indeed the global production of most metals has tended, in recent years, to exceed consumption. In 1997, lead was the only metal in deficit. The supplies of copper, zinc, nickel, aluminium and tin have continued to grow. Supplies of petroleum and natural gas remain buoyant.

See Table 2.2 for current predictions on petroleum.

2.23 Barbara Ward (Baroness Jackson of Lodsworth) (1914–81)

Barbara Ward was educated at Oxford University, later working as a university lecturer and as an Assistant Editor with *The Economist*. She became a Governor

Table 2.2 *Petroleum: life of known reserves at 1996–97 rate of output (ranked according to size of reserves)*

Country	Life of reserves	
	Years	Expiry
Saudi Arabia	83	2079
Iraq	over 100	
United Arab Emirates	over 100	
Kuwait	over 100	
Iran	69	2065
Venezuela	58	2054
Mexico	43	2039
Russia	22	2018
USA	10	2006
Libya	56	2052
China	21	2017
Nigeria	20	2016
Norway	9	2005
Algeria	19	2015
Kazakhstan	48	2044
Canada	9	2005
Angola	21	2017
Oman	16	2012
Indonesia	9	2005

NB New discoveries and improved extraction technologies mean that figures for proven reserves underestimate the true quantity of the world's reserves of petroleum. The picture is also complicated by the continual progressive rates of increase of consumption. Also petroleum is often accompanied by useful reserves of natural gas. Known reserves by region are as follows:

Middle East	65%
Latin America	12%
Europe	9%
Africa	6%
Asia-Pacific	4%
North America	4%

Source: BP and The Economist, 2 August 1997.

of Sadler's Wells and the Old Vic (1944–53). She ultimately became Schweitzer Professor of International Economic Development at Columbia University, and President of the International Institute for Environment and Development. Her published works included *Spaceship Earth* (1966) and (with René Dubos) *Only One Earth: The Care and Maintenance of a Small Planet* (1972), a preparatory document for the UN Conference on the Human Environment in that year.

The Stockholm Conference, commencing on 5 June 1972 (henceforth to be known as World Environment Day), was accompanied by a series of lectures by distinguished guests. The first of these was entitled 'Speech for Stockholm' by Barbara Ward, presented in a plenary session. She described the difficulties experienced by herself and René Dubos in writing *Only One Earth*, for they had been deluged by a cataract of ideas and concepts about the planet and its prospects. She thought, therefore, that Stockholm had come at one of the moments in human affairs when people were beginning to reconsider how they should look at their life on Earth. Such times of rethinking do recur in history, when central ideas are overturned, such as in the time of Confucius and Copernicus. Barbara Ward drew attention to today's concerns: the risk of irreversible planetary damage, the vulnerability of the oceans, doubts about the economics of growth, the growth of the world's population, and that we cannot run a functioning planetary society on the basis of the irresponsibility of 120 different governments. The difficulties presented were graphically described, but Barbara Ward concluded:

> I would therefore put to you what I could call a modest hope. It is possible that in the latest age of turbulence and disaster, our science and our wisdom are coming together and our faith and our reality are beginning to coincide. And if Stockholm is a place where this process begins, let us all thank God that we were here when it started.

2.24 René Jules Dubos (1901–82)

René Dubos was a French-born American microbiologist and environmentalist whose research into antibacterial substances produced by certain soil micro-organisms led to the discovery of major antibiotics, followed by their commercial production. He also studied organisms that caused dysentery, pneumonia and tuberculosis. In his later years, his interest shifted to humanity's relationship with the natural environment. Among many works, he published in 1968 *Man, Medicine and the Environment*. He sought to demonstrate that early environmental influences can have lasting effects on the anatomical, physiological and behavioural characteristics of adult animals and humanity. In 1972, in conjunction with Barbara Ward, he published *Only One Earth: The Care and Maintenance of a Small Planet*, an unofficial report commissioned by the Secretary-General of the UN Conference on the Human Environment, supported by consultants from 58 countries. This report was considered an integral part of the preparations for the UN Conference held in Stockholm in June 1972.

2.25 Roger Tory Peterson (1908–96)

Roger Tory Peterson was a US ornithologist, conservationist and wildlife artist, born in Jamestown, NY. He wrote the pocket-sized field guides that stimulated US and European interest in the study of birds. Following some teaching experience, he worked for the National Audubon Society (1934–43). In 1934, his first *Field*

Guide to the Birds was published, after being rejected by at least four publishers. The initial run sold out within two weeks. Many other books followed. In 1954, his *Field Guide to the Birds of Britain and Europe* appeared. Peterson's books sold by the million, and were translated into many languages. Among other awards, he was awarded the World Wildlife Fund Gold Medal and was twice nominated for the Nobel Peace Prize. In 1986, the Roger Tory Peterson Institute of Natural History was founded in Jamestown.

2.26 Blueprint for Survival

Blueprint for Survival, a publication by the editors of *The Ecologist* (1972), reviewed the environmental problems of humanity, setting out a radical programme for immediate action. It predicted that the reserves of all but a few metals would be exhausted within 50 years, at current exponential growth rates. New discoveries and advances in mining technology would be likely to provide only a limited stay of execution. The mineral resources threatened in the short term included bauxite (the basis for aluminium), chromium, cobalt, copper, gold, iron, lead, manganese, mercury, molybdenum, nickel, platinum, silver, tin, tungsten and zinc. With exponential growth in consumption, petroleum and natural gas would be exhausted before the year 2000. Coal would last much longer, about 300 years.

The authors of *Blueprint for Survival* also argued that with continued population growth, continually increasing per capita consumption, and threats to food supplies and to all ecosystems, clearly 'indefinite growth of whatever type cannot be sustained by finite resources'. Without radical changes in curbing consumption, stabilizing population, redistributing wealth and conserving ecosystems, the inevitable outcome would be a succession of famines, epidemics, social crises and wars. *Blueprint for Survival* was endorsed by 34 distinguished biologists, ecologists, medical practitioners and economists. It endorsed completely the work and recommendations of the Club of Rome. Its final conclusion was, 'Our task is to create a society which is sustainable and which will give the fullest possible satisfaction to its members.'

2.27 Barry Commoner

Barry Commoner's *Science and Survival* appeared in 1967, when he was Professor of Plant Physiology and Director of Washington University's Centre for the Biology of Natural Systems. The book was an urgent warning. Science had unleashed vast forces without knowledge of what the long-range effects would be on the environment. He asserted it was only through moral judgement and political choice that we could take the proper steps to avoid self-extinction. He believed that continued pollution of the Earth, if unchecked, would eventually destroy the fitness of this planet as a place for human life.

If we are to survive, Commoner argued, we need to become aware of the damaging effects of technological innovations, determine their economic and social costs, and balance these against the expected benefits. But he warned that never before in the history of the planet had its thin life-supporting surface been subjected to such diverse, novel and potent agents. He believed that the cumulative effects of these pollutants, their interactions and amplifications, could be fatal to the complex fabric of the biosphere. However, anyone who proposes to cure the environmental crisis undertakes to change the course of history.

In 1972, Commoner followed up with *The Closing Circle: Nature, Man and Technology*, during the preparations for the UN Conference on the Human Environment. Commoner remained of the opinion that the course of environmental degradation at that time, at least in industrialized countries, represented a challenge to essential ecological systems that was so serious that, if it continued, it would destroy the capability of the environment to support a reasonably civilized human society.

To resolve the environmental crisis, Commoner saw a need to forego the luxury of tolerating poverty, racial discrimination and war; and a need for a rational, social organization of the use and distribution of the Earth's resources. The alternative was a new barbarism.

He concluded that without seriously reducing the present level of *useful* goods available to the individual, a programme to control pollution could improve significantly the quality of life. So far there had been apparently hopeless inertia in the economic and political system, allowing a huge fraud to be perpetrated on the people of the world. In the end, the social, global nature of the ecosphere must determine a corresponding reorganization of the productive enterprises that depend on it.

2.28 Colin Clark (1905–89)

Colin Clark, in the second edition of his seminal work *Population Growth and Land Use* (1977), critically examined the evidence relating to the growth of the world's population and the growth of population in a range of countries. He referred to the famous theory propounded in 1798 by Robert Malthus, as a theory conflicting with a great deal of evidence, both geographical and historical. Indeed, Clark demonstrated that in a great many times and places population had been undesirably low, increasing at a very low rate. He also noted that Malthus, throughout quite a long lifetime, was apparently unaware of the agricultural, commercial and industrial revolution going on all around him, with a great increase in the production of food.

Clark observed that a world population growth of nearly 2% per year (and in some countries 3–4%) were rates that never prevailed in the past. At the beginning of the Christian era the order of magnitude of the world's population was some 250 million, and in 1650 had reached some 500 million. The average rate of increase over this period could only have been something like 0.04% per year, in stark contrast to present rates of growth. By 1800, the world population had become 890 million; by 1900, 1668 million; and by 1962, 3036 million (over three billion). An increase of several billion more was expected by 2000. However, Clark rejected the idea of a population explosion, noting the extraordinary fall in reproductivity in so many countries towards the close of the century. We are presented with a picture of a decline of unprecedented rapidity and severity, whose effects will be principally felt in the early decades of the coming century, when drastically reduced numbers of men and women of working age will have to support very large numbers of old people. Clark envisaged in the coming century a demographic disaster comparable with that caused by the Black Death. However, on a more cheerful note, Clark remarked:

> The world has immense physical resources for agricultural and mineral production still unused . . . the principal problems created by population growth are not those of poverty, but of exceptionally rapid increases of wealth in certain favoured regions of growing population, their attraction of further population by migration, and the unmanageable spread of their cities.

Clark proposed measures to cure these evils.

Until his death, Colin Grant Clark was Research Fellow at Monash University, Melbourne, Australia, having been for many years Director of the Institute for Research in Agricultural Economics, Oxford.

2.29 William J. Baumol

The writings of William J. Baumol stretch back at the very least to his significant paper, 'On the Social Rate of Discount', published in the *American Economic Review* in 1968. This was followed in 1971 by his Swedish Wicksell lecture entitled 'Environmental Protection, International Spillovers and Trade'. In 1971 also, co-authored by Wallace Oates, 'The Use of Standards and Prices for Protection of the Environment' appeared in the *Swedish Journal of Economics*. In 1972, Baumol's 'Economic Theory and Operations Analysis' and 'On Taxation and the Control of Externalities' were published.

In 1979, together with Wallace Oates, Baumol produced the seminal work *Economics, Environmental Policy, and the Quality of Life*, with a second edition being published in 1988. The preparations for this work compelled the writers to revise an earlier simplistic view that environmental deterioration had been a universal, accelerating process whose source was industrialization and population growth. It was found that the trends in environmental quality ran the gamut from steady deterioration to spectacular improvement. Some environmental damage was found to be the result of natural forces, as in the loss of oxygen in the central waters of the Baltic Sea. Also, the trends in environmental deterioration were found to be varied and uneven; though various types of environmental damage associated with human activity have grown rapidly and without interruption for a very long period of time. Some of them have produced serious consequences. Effective countermeasures are justified.

A central theme of the book is that the economy, left to its present course, will produce more and more consumer goods but will offer them to a society in which filth, noise and other forms of pollution grow and in which public services continue to deteriorate. It is another theme of the book that a clean environment and a range of services essential to a good society can be achieved with a relatively modest sacrifice of private consumption. They do not require halting all economic growth or dismantling our industrial system. Indeed what is needed is the design and enactment of a set of policies that will provide direct incentives to consumers, government agencies and business, to protect rather than abuse the environment. Economics can help to achieve both a continuation of abundance and improvement in the quality of life.

The message of the book was basically optimistic. By making pollution costly to both individuals and business firms, we can make it directly profitable to avoid activities that damage our air, rivers and natural preserves. We should correct the malfunctioning of the economic system by making the use of environmental resources appropriately expensive. We must enlist the profit motive in the cause of environmental protection. However, there remain important roles for direct controls by government, and even moral suasion in a comprehensive and effective programme for the enhancement of the environment. An improved environment can be attained, however, only at a substantial cost, and people must bear this cost in one form or another.

2.30 Julian Lincoln Simon (1933–98)

Julian Simon, an optimistic economist, was famous for his book *The Ultimate Resource*, published in 1981. He regarded people as the ultimate resource: 'Skilled, spirited and hopeful; people who will exert their will and imagination for their own benefit, and so, inevitably for the benefit of us all.' This had been, Simon argued, the experience of humanity since Eden and there was no reason for it not to continue indefinitely.

Simon's response to the argued 'limits to growth', with its threats of scarcity and rising prices for seemingly essential materials, was that the search for new supplies and alternatives would simply be encouraged. Simon was then simply regarded as a heretic. To make matters worse, Simon appeared unconcerned about the growth of the world's population and even the threat of global warming.

Simon stressed that India, once seen by Malthusians as a hopeless case, had now become able to feed itself. The world, Simon argued, could accommodate a lot more people. The genius of the people would solve all problems. In respect of birth control, he asked how many Mozarts, Michelangelos and Einsteins had been lost to the world.

He wrote some 30 books and was in demand as a speaker. He called his critics 'doomsayers'. In *Fortune* magazine he was listed as one of the '150 great minds of the 1990s'. He predicted that humanity's condition will improve in just about every material way, whether it was life expectancy, the price of natural resources, or the number of telephones in China. He had a bet with Paul Ehrlich that the prices of several key resources (copper, chrome, nickel, tin and tungsten) would fall over the coming ten years. By 1990, as predicted by Simon, all five metals had fallen in price. Dr Ehrlich paid up in settlement. Julian Simon's obituary appeared in *The Economist* on 21 February 1998.

2.31 Gro Harlem Brundtland

The World Commission on Environment and Development was created by the UN General Assembly in 1983 for the purpose of examining conflicts between environment protection and economic growth. The commission was chaired by Gro Harlem Brundtland, Prime Minister of Norway. Earlier she had served on the Independent Commission on Disarmament and Security which published its report, *Common Security*, in 1982. She actively promoted equal rights and the role of women in politics, becoming Norway's first woman prime minister. The 21 members of the World Commission were drawn from a range of nations with widely different economic and political backgrounds. The commission held a large number of meetings and travelled widely.

Its report *Our Common Future* was presented in April 1987 to the Governing Council of UNEP. It envisaged a new era of economic growth, based on the policies of sustainable development coupled with an expansion of the environmental resource base. Such growth was seen by the Commission as essential to the relief of poverty in much of the developing world, and to sustain coming generations. In respect of the greenhouse effect, the Commission recognized greater efficiency in the use of energy as one weapon with which to combat the problem.

Sustainable development was defined by the Commission as 'development that meets the needs of the present without compromising the ability of future generations to meet their own needs'. It requires an integration of environmental, social and economic considerations in decision-making. It introduces the concepts of intragenerational and intergenerational equity.

The Commission drew attention to much current success: infant mortality is falling, human life expectancy is increasing significantly, the proportion of the world's adults who can read and write is increasing, and global food production has outstripped population growth. On the other side, in terms of absolute numbers, there are more hungry people than ever before, more people who cannot read and write, and more people without clean and safe water and acceptable dwellings. The gap between rich and poor nations is widening, and the gaps within societies are also widening. Each year, large areas of productive dryland turn into worthless desert. Extensive areas of forests are destroyed yearly. Acid precipitation kills forests and impairs lakes. The burning of fossil fuels emits carbon dioxide, resulting in gradual global warming. In the end, sustainable development means a process of change in which the exploitation of resources, the direction of investments, the orientation of technical development, and institutional change are made consistent with future as well as present needs.

In 1993, Prime Minister Gro Harlem Brundtland delivered the Rafael M. Salas Memorial Lecture to the UN in New York. In it she defined three priority tasks:

- The need to change production and consumption patterns in industrialized countries to use fewer natural resources and create less pollution.
- The need to harness development in poor countries so as to eliminate poverty, meet basic human needs and protect the environment.
- The need to slow population growth so that sustainable development may be achieved.

She emphasized that countries that have succeeded relatively well in limiting their population growth have several features in common. Many have deliberately tried to combine economic growth with an equitable distribution of income. They have given priority to the development of human resources and focused on health and education and especially on improving the status of women, the employment of women, and mother and child care. Above all, they have given strong support to family planning and family planning services.

In 1998, Dr Gro Harlem Brundtland became Director-General of the WHO. She is now known as the Thatcher of the United Nations. Time servers are heading for the door.

2.32 Herbert Cole 'Nuggett' Coombs (1906–97)

An adviser to Australian Governments for some 36 years, Coombs' interests ranged over economics, the arts, social philosophy, the environment and Aboriginal affairs. He spent much time with Aboriginal communities throughout Australia, with a genuine commitment to reconciliation. In the economic field he became the first Governor of the Australian Reserve Bank. He became also a Chancellor of the Australian National University, Canberra, and Chairman of the Australian Council of Aboriginal Affairs, the Australia Council and the Elizabethan Theatre Trust. He also held office as President of the Australian Conservation Foundation. His last book was entitled *The Return of Scarcity: Strategies for an Economic Future* (1990), which addressed the key questions of a sustainable society.

Coombs believed that it was necessary to abandon the conviction that 'bigger is better' and to qualify seriously the belief that maximum growth is the *sine qua non* of economic wisdom. He considered that sustainability should be the primary objective for enterprises exploiting or using natural resources. Also community action should be taken to reassert, and where necessary re-establish, the public ownership of the natural resources of the Australian continent, its minerals, forests, seas, soils, vegetation and wildlife; Coombs believed title to these resources should be vested in an authority independent of corporate and political control. The granting of licences, with conditions and restrictions on the use of these assets, would ensure proper management and a competitive rent for the community. The income from licensing could then be used for social purposes contributing to the quality of life, supporting research and development, and providing a national dividend for all citizens. He envisaged a gradual reduction in foreign ownership and external indebtedness. He saw access to a healthy, stimulating and dignified lifestyle for all citizens as the prime objective of economic policy. Coombs saw nothing inherently impracticable in these proposals, preserving an economic system based upon private ownership of economic enterprises motivated by profit and the accumulation of private and corporate wealth. A monastic lifestyle among the community would not be necessary, but a modification of values was necessary to a better life. Simplicity and frugality should once again be seen as virtues.

2.33 Jack Mundey

Social development in the inner cities of Australia, as a result of huge property developments in the central business districts and neighbouring areas, has caused great stress to communities compelled to move. In certain circumstances, as in the British slum-clearance programmes, it may be possible to move a great many quite readily, difficulties lying mainly with the aged and those with very strong neighbourhood attachments. In general, the ground is prepared through a general social acceptance that slum dwellings, classified as unfit for human habitation, should be cleared as soon as possible and the residents rehoused by the public housing authority. To get a new house in compensation, even at a higher rent and further out of town, is the aspiration of most in such circumstances, while the holding of independent public inquiries offers a channel for remaining objections by landlords and tenants. Schemes are often modified as a result, or the basis for compensation varied.

However, where settled communities are to be transferred from areas with strong community ties, and the exodus does not arise from a change in the social attitudes of residents, who are reasonably content, the effect can be traumatic. A great upsurge in office and hotel construction in the central business district can be seen as an impersonal force of considerable magnitude brought to bear on people who then feel merely pawns in a game played by others for profit, power and prestige.

The Australian city in which the building boom of the early 1970s had its most devastating effect was Sydney, the capital of New South Wales. The resources of the building industry appeared to be devoted in substantial part to the construction of commercial buildings, with little activity in the way of housing for low-income people. The central city was encircled by low-income communities; as the city expanded and encroached upon these communities, plans were prepared to begin demolition either in whole or in part of all these areas. To try and stem the tide of this development, people began to form Residents' Action Groups. They sought allies in the construction trade unions, requesting work bans on the projects. These work bans were known initially as 'black bans' and later as 'green bans'.

The most prominent of the unions involved in the imposition of green bans were the Builders Labourers Federation and the Federated Engine Drivers Association. Among these, Jack Mundey was prominent. As a result of the support of these unions, The Rocks Residents' Action Group endeavoured to secure better terms for existing residents (The Rocks being that area of Sydney first settled in 1788). From November 1971, a green ban was placed on demolition work.

October 1973 became a time of crisis, for in that month, a developer and a team of non-union demolishers moved into The Rocks. For two weeks, on and off, the residents stood their ground against police and bulldozers. Some 80 residents barricaded themselves on the site where demolition was taking place, and around 200 police were ordered into the area to arrest them, which they did, pensioners and housewives alike. People who had never before had a blemish on their characters became, through conviction, criminals.

On 18 October 1973, some 500 construction workers picketed the demolition site. Arrests led to further close-downs which resulted in the complete shut-down of the building industry in Sydney for two and a half weeks. Workers in the ports of Newcastle and Wollongong, and other industries, also struck, supporting the people of The Rocks, the birthplace of European Australia.

Jack Mundey claimed that this proved conclusively that not only workers in the construction industry, but a wide section of people, supported the philosophy that workers have a right to a say in what is the end result of their labours; a right to determine the social value of their labours.

The New South Wales Government paused and recast the policies of the Sydney Cove Redevelopment Authority, with an emphasis on heritage conservation and low-income housing, redevelopment of selected areas, and tourist attractions meeting many needs and purposes. Today, The Rocks is a major attraction with its roots in a struggle with which the name of Jack Mundey will always be associated.

In 1998, Jack Mundey, then Chairman of the NSW Historic Houses Trust, received the honorary degree of Doctor of Science from the University of New South Wales.

2.34 The Green Parties and the NGOs

The Green parties and the NGOs encompass all of the various environmentally or ecologically oriented groups and political parties which formed in European and other countries during the 1980s. An umbrella organization known as the European Greens was formed in Brussels, Belgium, in 1984 to co-ordinate the activities of the various European Green parties, while Green representatives in the European Parliament formed the Rainbow Group. The first Green Party was formed in (West) Germany in 1979, arising out of a merger of some 250 ecological groups. The Party was headed by Herbert Gruhl and Petra Kelly, and was to prove the most successful. The Party sought public support against nuclear energy and air and water pollution, and called for the dismantling of both NATO and the Warsaw Pact, the demilitarization of Europe and the breaking up of industrial conglomerates. From 1979 onwards, the Greens won some seats in the various state (*Land*) elections, and in 1983 the Party won a 5.6% share of the national vote for the election of the Bundestag (the Federal Diet), returning members for the first time. Environmental issues came to the fore.

Within a decade, almost every European country had a Green party, e.g. the Green Alliance in Ireland and Finland, the Green Alternatives in Austria, the Green List in Italy, the Green Ecology Party in Sweden, and the Ecological Party in Belgium.

The Australian Greens were established in 1992 as a national entity, combining several existing state parties. Bob Brown, one of the founders, led the Tasmanian Wilderness Society and the campaign that saved the Franklin River. Now a Senator, in 1987 he won the UNEP Global 500 Award, and later the Goldman Environmental Prize USA. Of course, many in Australia have contributed to state and national environmental and conservation policies and legislation, another notable being the Honourable Jack Beale, for some time an environmental minister, organizing the Australian delegation to the UN Conference on the Human Environment 1972, and later becoming a senior adviser to UNEP and UNDP. In 1989, the Australian National University established the Jack Beale Chair of Water Resources.

Green parties have emerged in some 60 countries including New Zealand, Canada, Chile, Argentina, Brazil, and throughout the countries of Eastern Europe. While such parties may not gain mass support, they often gain seats which can swing or influence votes in national or state parliaments and sway opinion through the media.

However, while the Green parties have entered the political arena, many voluntary groups (non-government organizations, or NGOs) have been active for a great many years throughout the world. The Sierra Club was formed in 1892; the National Audubon Society in 1900; the World Conservation Union in 1948; the World Wide Fund for Nature in 1961; the Australian Conservation Foundation in 1970; Greenpeace and the Friends of the Earth in 1971; the World Resources Institute in 1982; and the Worldwatch Institute since 1984. Numerous environmental NGOs thrive at local levels throughout the world, growing annually. They have held separate but parallel conferences at all the international UN conferences on the human environment.

2.35 Marion Clawson (1905–98)

Marion Clawson was among the first generation of RFF (Resources for the Future) research fellows, a pioneering institution which he joined in 1955, studying agriculture, park and forest use, outdoor recreation and land development. Methods he developed to measure the demand for and value of outdoor recreation have formed the basis for several hundred studies throughout the world. Clawson was a prolific writer authoring some 40 books in all, producing at one stage 20 books in 20 years. His works included the *Economics of Outdoor Recreation* (1966), and *Forests for Whom and for What?* (1975). Earlier works included *Federal Lands: Their Use and Management* (1957), a classic on the history of US land administration, and later, *Federal Lands Revisited* (1983). He covered the environmental problems associated with forest harvesting, insecticide use and water pollution, and the development of public concern regarding these matters.

In respect of the nation's renewable resources generally, Clawson considered that over his lifetime the situation had improved, not deteriorated, with far higher levels of material consumption. Land was producing far more per unit area than ever before. The volume of wood produced had increased greatly, with little loss in timber acreage. Further, the scars on the environment were now no worse than at the turn of the century, though there was scope for much improvement. He argued that the true revolutionaries of the last half-century had been the agricultural scientists.

Clawson's professional career spanned 70 years; he still drove to work at the age of 92.

2.36 Summary

This chapter offers an insight into a range of contributions over the last three hundred years or more, through the time that the word 'smog' was first used by Dr H.A. Des Voeux, Founder-President of the British National Smoke Abatement Society in 1905, to the present, in which pollution, global warming, world population and resource issues are of international concern.

In 1972, in Stockholm, representatives from around the world met for the first time to discuss environmental problems. Since then a series of world conferences have addressed various aspects of the human condition. All nations have environmental agencies of various strengths and abilities, and all nations have legislation relating to national parks and heritage, pollution control, city planning and, increasingly, environmental impact assessment. Public inquiries, once rare, are now widespread events throughout many democratic countries for the examination of controversial social, planning and environmental issues.

It should be noted that every step in the development of environmental consciousness, legislation and policy has been controversial. The British Clean Air Act of 1956 was vigorously opposed by the coal merchants' associations.

Box 2.8 Portney and Oates on The Economist article on environmental scares of 20 December 1997

As a matter of fact, the prophets of environmental doom do have a very bad record. Their forecasts have, as *The Economist* says, been "invariably wrong" ... *The Economist* has erred, however, in not acknowledging that doomsayers have mobilized political forces for needed environmental protection. The principal reason that concentrations of air pollutants have fallen in virtually every US metropolitan area over the last twenty-five years was the enactment of federal air pollution controls in 1970. These controls were prompted in large part by the dire warnings of environmentalists (and some economists) who foresaw the likely effects of unchecked industrial growth. Similar warnings and subsequent measures have reversed the deterioration of many streams, rivers, lakes and estuaries, and have awakened us to the folly of the careless disposal of hazardous wastes.

Source: P.R. Portney and W.E. Oates (1998), Resources, Spring, published by Resources for the Future, Washington, DC.

Measures against pollution from the car were vigorously opposed by the motor vehicle manufacturers. Opposition against the removal of lead from gasoline was organized at an international level. The introduction of town planning clashed with property interests. Growing evidence against insecticides led to orchestrated opposition. Rachel Carson was maligned. The consumer advocate Ralph Nader was conspired against. John Sinclair, who had organised the opposition to sand mining on Fraser Island, off the Queensland coast with complete success, was driven out of Queensland, Australia.

Other activists have faced arrest, harrassment, dismissal, persecution, discrimination, ridicule, defamation and treachery. Three activists in Australia disappeared without trace. In 1993, two Brazilian conservationists were murdered, one being opposed to the extraction of sand from beaches, dunes and salt marshes; the other opposed to the logging of mahogany in tribal land and ecological reserves.

In 1985, the *Rainbow Warrier* was sunk in Auckland Harbour, New Zealand, a Greenpeace photographer being killed. The French Government publicly admitted responsibility for this outrage. Dai Qing was imprisoned in China for opposition to the Three-Gorges hydroelectric project.

The world has gradually adapted to the fact that it is shrinking in terms of available space, natural resources and time. But the process has been traumatic.

See Box 2.8 for the views of Portney and Oates on the broad arena of environmental concerns.

Further Reading

Carson, R. (1962) *Silent Spring*, Houghton Mifflin, Boston.

Commoner, B. (1971) *The Closing Circle: Nature, Man and Technology*, Knopf, New York.

Doxiados, R.B. (1977) *Ecology and Ekistics*, Elek Books, London.
Dubos, R.B. (1982) *Man, Medicine and the Environment*, Penguin, Harmondsworth.
Ehrlich, P.R. and Ehrlich, A.H. (1970) *Population, Resources, Environment*, Freeman, San Francisco.
Lear, L. (1998) *Rachel Carson: Witness for Nature*, Allen Lane, UK; Henry Holt, USA.
Malthus, T.R. (1798) *Essay on the Principle of Population*, Penguin, Harmondsworth.
Meadows, D.H., Meadows, D.L., Randers, J. and Behrens, W.W. (1992) *Beyond the Limits: Global Collapse or a Sustainable Future?* The Club of Rome, Earthscan, London.
Ward, B., Dubos, R., Heyerdahl, T. *et al.* (1973) *Who Speaks for Earth?* Seven distinguished lectures at the UN Conference on the Human Environment, 1972, Norton, New York.
World Commission on Environment and Development (the Brundtland Report) *Our Common Future*, Oxford University Press, Oxford.

3

Institutions

3.1 Introduction

The evolution of national legislation and policy formulation on environmental matters has been accompanied by and, in varying degrees, influenced by a range of far-reaching world UN conferences. These are summarized in Box 3.1, starting with the Lake Success Conference in 1949 on the subject of resource scarcity, and later conferences on population, settlements, water resources, desertification, the law of the sea, new and renewable sources of energy, environment and development, social development and world food supplies. The outcome of these conferences, and others outside the UN system, have resulted in the creation of a range of institutions by the UN, and others such as the World Trade Organization and the International Whaling Commission. Other relevant organizations have sprung into being from private initiatives such as the Beijer International Institute of Ecological Economics and the International Institute for Environment and Development.

Box 3.1 United Nations Conferences on the Human Condition

UN Conference on the Conservation and Utilization of Resources 1949
The UN Conference on the Conservation and Utilization of Resources was held at Lake Success, New York, in 1949. It reviewed the world's critical shortages of minerals, fuel and energy, forest resources and food. Fears were expressed that many materials upon which the industrial world depended might become exhausted before the year 2000. Particular concern was expressed about supplies of copper, tin, zinc and lead.

UN Conference on the Human Environment 1972
The UN Conference on the Human Environment was held in Stockholm in 1972, with representatives from some 113 governments and agencies. The purpose of the conference was to review environmental problems and provide guidelines to protect and improve the human environment. The conference achieved a Declaration on the Human Environment, agreement upon an extensive programme of international activity, and the creation of a permanent secretariat in Nairobi, Kenya, to be known as the United Nations Environment Program (UNEP). A second environment

Continued on page 72

Continued from page 71

conference was held in Nairobi in 1982, to review progress, and a third conference was held in Rio de Janeiro in 1992 on environment and development.

UN World Population Conference 1974

The UN World Population Conference held in 1974 involved representatives from 130 countries, and was held in Bucharest, Rumania. The purpose of this first world conference on population was to consider the population policies and programmes that were needed to promote human welfare and development. The principal achievement of the conference was a World Population Plan of Action and a number of related resolutions on matters such as the status of women, food production and the environment. Delegates from 132 countries attended a second World Population Conference held in Mexico City in 1984; a third conference was held in Cairo, Egypt, in 1994. The 1994 conference sought means for modifying the growth rate of the world's population which had reached 5.7 million people, and might well reach between 7.9 billion and 12 billion by the year 2050. The conference promoted the provision of contraceptive and family planning services, sex education and safe abortion, and sought a definite improvement in the status of women throughout the world.

UN Conference on Human Settlements 1976

A UN Conference on Human Settlements held in Vancouver in 1976 was attended by representatives from 131 governments. It was concerned with the urgent problems of housing shortages, crises of urban and rural communities, the proper use of land, access to essential services such as clean and safe water, and public involvement in efforts to improve the living conditions of people throughout the world. The conference agreed on a Vancouver Declaration on Human Settlements and a Vancouver Plan of Action. The outcome was the establishment in Nairobi of a Centre for Human Settlements. A second UN Conference on Human Settlements (Habitat II) was held in 1996, to review progress and set priorities for the next decade.

Habitat II, held in Istanbul, was warned that most cities in the developing world will face extreme water shortages by 2010 simply because they are not adequately prepared for the huge influx of people from the rural districts. The conference sought to address the urban ills of poverty, homelessness, and social and environmental decay. The European Union backed demands that housing be given the status of a human right, needing international financial help.

The conference listed Cairo, Lagos, Dhaka, Beijing, Calcutta and São Paulo among the developing cities facing the greatest water problems. However, cities such as Houston, Los Angeles, Warsaw, Cardiff and Tel Aviv also faced acute shortages. The water crisis had emerged not only because of a lack of rain in some regions, but also from the inability of governments to make the necessary investments in a timely manner to ensure that water is available to all cities. More than one billion people could not get clean and safe drinking water. About 100 million people world-wide, mostly women and children, were homeless, and up to 600 million people poorly and unhealthily housed.

UN Conference on Desertification 1977

The UN Conference on Desertification, held in Nairobi in 1977, was the first world-wide effort to consider the problem of the advancing deserts. The outcome of the conference was a World Plan to Combat Desertification, which made little progress. Governments with desertification problems tend to have

Continued on page 73

Continued from page 72

limited financial and human resources, and insufficient financing has plagued the programme. About one-fifth of the world's farmlands suffer from the effects of desertification. Later, in 1994, a Convention to Combat Desertification was signed in Paris, aimed at tackling the spread of deserts generally, giving fresh impetus to the programme.

UN Water Conference 1977

The UN Water Conference held in Mar del Plata, Argentina, in 1977, attempted to focus the attention of policy-makers on the water needs of society up to the year 2000, the steps that could be taken to meet them, and the difficulties likely to be experienced by those who failed to make adequate provision. The conference noted that less than one-fifth of the world's population can get water simply by turning a tap; for the remaining four-fifths, the getting of water is part of the daily struggle for existence. The conference urged better water management at local, regional and national levels. In 1992, a UN Conference on Water and the Environment was held in Dublin, Ireland, to review progress.

UN Conference on the Law of the Sea, 1974–82

The UN Conference on the Law of the Sea comprised a series of international discussions aimed at establishing a revised legal regime for the oceans and their resources, while maintaining the right of ships to free passage. One outcome was the creation of 200-mile (320 km) exclusive economic zones for each country, where practicable; others were the creation of an International Seabed Authority to administer the rules regarding the management of minerals and food supplies as yet unexploited, and the prevention of pollution and over-exploitation. The conference ended with a Convention on the Law of the Sea (the Montego Bay Convention).

UN Conference on New and Renewable Sources of Energy 1981

A UN Conference on New and Renewable Sources of Energy was held in 1981. It adopted a programme of action for increased use of new and renewable sources of energy in a socially equitable, economical and technically viable, and environmentally sustainable manner. The plan particularly emphasized the need to consider the environmental aspects of programmes for the exploration, development and utilization of new and renewable sources of energy.

UN Conference on Environment and Development 1992

The UN Conference on Environment and Development in 1992 was held in Rio de Janeiro, with representatives from some 167 countries. The purpose of this third international environment conference was to review progress since the earlier conferences in 1972 and 1982 in safeguarding the human environment and promoting human welfare. The results were the Rio Declaration on Environment and Development; the adoption of Agenda 21 on environment and development (a programme for the 21st century); the creation of a Commission for Sustainable Development; the adoption of a Convention on Protecting Species and Habitats (the Convention on Biological Diversity); and the adoption of a Framework Convention on Climate Change. The secretary-general to the conference was Maurice F. Strong.

UN World Summit for Social Development 1995

The UN World Summit for Social Development held in Copenhagen in 1995 was essentially a follow-up conference to the UN World Summit for Children held in

Continued on page 74

Continued from page 73

1990. The achievements between the conferences have been outlined in the *State of the World's Children* (UNICEF 1995). Malnutrition has been reduced; poliomyelitis has been eradicated in much of the world; deaths from measles have more than halved; and immunization levels have been maintained or increased. There has also been progress in preventing vitamin A and iodine deficiencies, and in providing safe water. It was reported at the conference that as a result of these improvements, 2.5 million fewer children would die in 1996 compared with 1990; and at least 750 000 fewer children would be disabled, blinded, crippled or mentally retarded. However, there was little progress to report in India, Pakistan and Bangladesh, or in sub-Saharan Africa, where there was a lack of resources to make improvements. Diarrhoea still killed about three million children a year, with pneumonia still the biggest single killer of children.

UN World Food Summit 1996

The World Food Summit was held at the Rome headquarters of the FAO in November 1996, once again bringing together world leaders to discuss food security. At a similar conference in 1974, countries had pledged a goal of eradicating hunger within a decade. This had not happened. The FAO estimated that in 1996 about 14% of the world's population suffered from chronic malnutrition. More than 80 nations had been identified as low-income food-deficit countries (LIFDCs), more than half of them in sub-Saharan Africa. The world's population, expected to increase by 50% by 2020, faced a declining per-person supply of tillable land and water. The record low cereal stocks added to the urgency of the summit.

The summit finalized a 'Declaration on World Food Security' that identified the basic causes of the problem and the actions needed by governments. The summit's goal was to reduce the number of undernourished people to half their present numbers no later than 2015. Although reducing poverty was a major focus of the declaration, the future need for stable and expanding food supplies and effective emergency food assistance was also emphasized. A plan of action was adopted by the summit to achieve its objectives. No new international agency was to be created. The FAO Committee on World Food Security gained responsibility for monitoring the programme.

Box 3.2 summarizes the impressive range of UN agencies concerned with the environment, while the text of the chapter amplifies some of these, adding others of international importance.

3.2 UN Commission on Sustainable Development

The concept of a Commission on Sustainable Development was first debated at the UN Conference on Environment and Development 1992 and proposed subsequently in Agenda 21. A decision to create the Commission was taken by the UN in December 1992, with the Commission beginning work in early 1993. Countries are invited to report regularly to the Commission on their progress

Box 3.2 United Nations Agencies Concerned with the Environment

- *UN Capital Development Fund (UNCDF).* Established in 1966, UNCDF provides aid to developing countries by means of grants and loans. Its resources are primarily used to assist the 30 least-developed countries.
- *UN Centre for Human Settlements.* Established following the UN Conference on Human Settlements 1976, the centre addresses the problems of urban and rural communities. It is located in Nairobi, Kenya.
- *UN Children's Fund (UNICEF).* A programme devoted to supporting national efforts to improve the health, nutrition, education and general welfare of children.
- *UN Commission for Sustainable Development.* Created by the UN in 1992, following the UN Conference on Environment and Development, to promote international co-operation in the implementation of Agenda 21, with particular emphasis on policies to promote sustainable development.
- *UN Conference on Trade and Development (UNCTAD).* A UN standing conference set up in 1964 to assist the less-developed nations in attaining improved rates of economic growth; it arranges aid and finance while promoting trade
- *UN Development Program (UNDP).* The largest UN provider of grant funding for development and the main body for co-ordinating development assistance. UNDP aims to create more productive societies and economies in the low-income countries. It aims at a reduction of poverty.
- *UN Economic and Social Commission for Asia and the Pacific (ESCAP).* A regional economic commission created by the Economic and Social Council of the UN in 1978. In 1990, ESCAP produced a report on the state of the environment in Asia and the Pacific, with emphasis on widespread poverty and the lack of basic necessities.
- *UN Economic Commission for Europe (UN ECE).* A regional economic commission created by the Economic and Social Council of the UN in 1947 to raise the level of post-war economic activity in Europe. The environmental activities of the UN ECE have broadened considerably over the years, with the pioneering of important environmental conventions.
- *UN Educational, Scientific and Cultural Organization (UNESCO).* Established in 1946 to promote international collaboration in education, science and culture; major activities have been the Man and Biosphere Program and work on the World Heritage List.
- *UN Environment Program (UNEP).* Created by the UN Conference on the Human Environment 1972 to implement the recommendations of environment conferences and co-ordinate environmental activities within the UN system. It is based in Nairobi, Kenya.
- *UN Food and Agriculture Organization (FAO).* Established in 1945, a UN agency concerned with investment in agriculture and emergency food supplies.
- *UN Fund for Population Activities (UNFPA).* Known as the UN Population Fund, a trust fund under the jurisdiction of the UN Development Fund. UNFPA funds family planning and population control projects in more than 120 countries.
- *UN Global Environment Facility (GEF).* Established initially by the World Bank in 1991, a facility to provide additional grant and concessional funding for the achievement of global environmental objectives. The facility was subsequently endorsed by UNDP and UNEP. Nations contributing financially to the fund are known as 'participants'.

Continued on page 76

Continued from page 75

- *UN Industrial Development Organization (UNIDO).* Established in 1967, UNIDO is a UN agency that aims to assist in the industrialization of the developing countries by co-ordinating other UN agencies devoted to that end.
- *UN International Atomic Energy Agency (IAEA).* Established in 1957, the agency is based in Vienna, Austria. Its primary purpose is to promote the peaceful uses of atomic energy, and to ensure that nuclear material is not diverted to the manufacture of nuclear weapons.
- *UN International Labour Organization (ILO).* Established as a UN agency in 1946, the ILO has produced codes and guidance to improve occupational and working conditions. It co-operates closely with the World Health Organization.
- *UN International Maritime Organization (IMO).* The IMO is a UN body with far-reaching responsibilities for maritime safety and regulation, including pollution discharges from ships. The World Maritime University, as part of IMO, is based in Sweden.
- *UN University.* The Institute of Advanced Studies of the UN University, inaugurated in 1996, directs its research programmes towards issues of global importance. The centre is based in Tokyo. Research is undertaken in the multithematic areas of economic development and environmental management, mega-cities and urban development, multilateral diplomacy and environmental governance, and the policy aspects of science and technology in relation to sustainable development.
- *World Bank.* An agency of the UN, the World Bank comprises the International Bank for Reconstruction and Development, the International Development Association, the International Finance Corporation, and the Multilateral Investment Guarantee Agency. The common objective of these bodies has been to help to raise the standards of living in the developing countries, channelling resources from the richer to the poorer nations. The World Bank was established in 1945 and is based in Washington, DC.
- *World Food Council (WFC).* A UN organization created in 1974 upon the recommendation of the World Food Conference, held in the same year. Based in Rome, the WFC co-ordinates information and suggests strategies for food policies, reporting to the UN General Assembly. Attention is focused on the needs of developing countries and on the reduction of food-trade barriers between developing and developed countries.
- *World Food Program (WFP).* A UN organization established in 1961 as a joint project of the UN General Assembly and the UN Food and Agriculture Organization (FAO). It seeks to stimulate economic development through food aid and emergency relief. Its chief organs are the Committee of Food Aid Policies and Programs and the joint UN/FAO Administrative Unit. The objectives of WFP are specifically geared to eliminating emergency situations that arise from insufficient food supplies. It also seeks to further land reclamation and irrigation.
- *World Health Organization (WHO).* Created in 1948, the WHO is an agency of the UN. The work of WHO encompasses the control of communicable diseases, water supply and waste disposal, air and water pollution, nutrition, food hygiene, occupational health, and environmental health impact assessment.
- *World Meteorological Organization (WMO).* A UN agency since 1951, the WMO's primary role is to maintain a world-wide meteorological system and to grapple with problems of a global nature, particularly in respect of world climate change.

Box 3.3 Agenda 21

Adopted by the UN Conference on Environment and Development on 14 June 1992, Agenda 21 is a comprehensive document summing up the international consensus on actions necessary to move humanity towards the goal of sustainable development during the 21st century. The objective is to alleviate poverty, hunger, sickness and illiteracy world-wide, while at the same time arresting the deterioration of the ecosystems on which humanity depends to sustain life. Agenda 21 is divided into four principal sections:

1. *Social and economic dimensions*: the seven chapters which comprise the first section address central issues such as poverty, health, consumption patterns, population, trade and resource transfers between countries, human settlement patterns, and integration of environment and development in decision-making.
2. *Conservation and management of resources for development*: the 14 chapters of this section deal with a very wide range of issues such as the atmosphere, land resources, deforestation, desertification, mountain development, agriculture, biological diversity, biotechnology, oceans, freshwater resources, toxic chemicals, hazardous wastes, solid wastes and radioactive wastes.
3. *Strengthening the role of major groups*: the nine chapters of this section deal with women's affairs, youth and children, indigenous peoples, NGOs, local authorities, trade unions, business and industry, science and technology, and farming.
4. *Means of implementation*: the eight chapters of this section deal with finance and the funding of Agenda 21 programmes, technology transfer, science in sustainable development, education and training, capacity building, institutional arrangements, legal instruments, and information for decision-making.

in the implementation of Agenda 21 and with respect to environmental management generally. The Commission may make recommendations to the UN General Assembly and the UN Agencies (see Box 3.3).

The Commission comprises representatives from 53 countries: 13 from Africa, 11 from Asia, 10 from Latin America and the Caribbean, and 19 from East and West European states. In respect of funding, the Commission must monitor the promised official development assistance of 0.7% of GDP due from the developed nations. The Commission must also review progress in respect of technology transfer, and receive and analyse input from non-governmental organizations and bodies outside the UN. The implementation of UN Conventions must also be kept under review.

The Commission meets annually for up to three weeks, addressing various topics within Agenda 21 such as health, human settlements and fresh water, toxic chemicals and hazardous waste, land, desertification, forests and biodiversity, atmosphere, oceans and sustainable development.

At the Earth Summit II Conference in 1997, five years after Earth Summit I was held in Rio de Janeiro, the Commission tendered a report warning world leaders that humanity faces a global water crisis by 2025 caused by pollution and water overuse. This report contains the first UN warning about the risk of conflict caused by disputes over access to water. Current disputes involve more than 300 transboundary rivers world-wide, as well as many aquifers. Other outcomes regarding human welfare were also disappointing, although the rate of growth of the world's population has eased. However, by 1997 there were more than 450 million extra people on Earth than there had been at the time of the Rio Conference. Further, the promised financial support from the richer countries had dwindled.

The secretariat of the UN Conference on Environment and Development estimated that the average annual costs up to 2000 of implementing Agenda 21 in the developing countries would be over US$600 billion, including about US$125 billion on grant or concessional terms from the international community. These were order-of-magnitude estimates only, subject to actual programmes. Official aid, promised at the UN Conference, would be the main source of this funding. A further source of funding was to be the Global Environment Facility and the World Bank.

3.3 UN Environment Program (UNEP)

Created by the UN Conference on the Human Environment 1972, UNEP was charged with implementing its recommendations and those of subsequent environmental conferences. Based in Nairobi, Kenya, UNEP is subject to a governing council. Its activities are supported by an environment fund, to which nations contribute. UNEP's environment assessment programme, Earthwatch, has four closely linked components: evaluation and review, research, monitoring, and exchange of information. Earthwatch contains such important and operational elements as the Global Environmental Monitoring System (GEMS), the International Referral System (INFOTERRA), and the International Register of Potentially Toxic Chemicals (IRTC). There has been much successful activity in the area of regional seas, resulting in several conventions; for example, the Barcelona Convention for the Protection of the Mediterranean Sea Against Pollution. UNEP has promoted the Vienna Convention for the Protection of the Ozone Layer, and the Basel Convention on the Control of Transboundary Movement of Hazardous Wastes and their Disposal. UNEP also provided support for two conventions on wildlife protection: the Convention on International Trade in Endangered Species of Wild Fauna and Flora (CITES) and the Convention on the Conservation of Migratory Species of Wild Animals (CMS). In 1978, UNEP established an EIA division. Later, the Governing Council of UNEP adopted goals and principles for EIA, for adoption throughout the world. In 1982, the Governing Council adopted the Montevideo Program for the Development and Periodic Review of Environmental Law which has been a mainspring behind the development of more recent conventions and agreements.

In 1990, UNEP established the UNEP Collaborating Centre on Energy and Environment at the Riso National Laboratory, Denmark. Core funds come from UNEP, the Danish International Development Agency (DANIDA) and Riso National Laboratory. The Centre works on contract with other bilateral and multilateral agencies. The mandate of the Centre is to promote the incorporation of environmental considerations into energy policy and planning, especially in developing countries. The Centre collaborates with institutions and experts world-wide and provides direct support to UNEP headquarters. The Riso National Laboratory is the largest research institution in Denmark, the main research areas being energy, environment and materials. It considers projects at the national, regional and global levels. UNEP's Collaborating Centre has focused on climate change, mitigation analysis, environmental and developmental economics, national and international policy instruments, energy efficiency and energy sector reform.

3.4 UN Development Program (UNDP)

The UN agency UNDP was created in 1965 and provides assistance to low-income nations to develop their natural resources and human capabilities. It is the UN's largest provider of grant funding for development and the main body for co-ordinating UN development assistance. The primary aim is the reduction of poverty and the encouragement of sustainable activity. Its funds come from annual voluntary contributions from the members of the UN and its related agencies. In 1997, some 58% of UNDP's total resources were allocated to countries designated as 'least developed' by the UN General Assembly.

UNDP activities focus on four priority areas: poverty elimination; creation of jobs and sustainable livelihoods; advancement of women; and protection and regeneration of the environment. A *Human Development Report* has been published annually since 1990. Capacity-building programmes have been supported in the areas of natural resource management, food security, forestry, water and sanitation, energy and urban development. The UNDP assisted developing countries and NGOs to prepare for the 1992 UN Conference on Environment and Development. As a follow-up to that conference and the adoption of Agenda 21, UNDP launched Capacity 21, a plan which became fully operational in 1993. By 1995, Austria, Canada, Denmark, Finland, France, Germany, Italy, Japan, the Netherlands, Norway, Sweden, Switzerland, the United Kingdom and the USA had pledged substantial financial support. UNDP also supports the Global Environment Facility (GEF). The headquarters of UNDP are in New York.

3.5 The World Bank

An agency of the United Nations, the World Bank comprises the International Bank for Reconstruction and Development (IBRD) and its affiliates, the International Development Association (IDA), the International Finance Corporation

(IFC) and the Multilateral Investment Guarantee Agency (MIGA). The common objective of these bodies, known collectively as the World Bank, has been to help to raise the standards of living in developing countries, channelling resources from the richer to the poorer nations. The Bank was established in 1945. In 1970, the Bank created an Office of Environmental and Health Affairs, but with limited resources. Since 1979, investment and assistance began to extend to afforestation and reforestation, soil conservation, flood mitigation and control, range management, wildlife protection, and abatement of air and water pollution. However, in more recent years, the Bank has been accused of failing in its original role of reducing poverty and of disregarding in whole or part the environmental effects of major development projects. It has, it is claimed, financed roads for settlers who devastated the Amazon rain forest and financed dams that forced many thousands to be resettled in less satisfactory circumstances. Some of these criticisms were confirmed by the Bank itself (see Box 3.4). Since 1991, there has been much more attention to these problems; several projects have been modified as a result of environmental assessment. In 1992, the Bank published its three-volume sourcebook on EIA. In 1994, the Bank developed the Environmentally Sustainable Development (ESD) Triangle to indicate a continuing shift in its approach (see Figure 3.1).

In June 1997, the World Bank issued its Green Top 10 Plan, a list of proposed actions to address the world's most pressing environmental problems. The Plan pointed out that world-wide energy-related subsidies, amounting to US$800 billion annually, rarely benefited the poor and inevitably harmed the environment. Carbon dioxide emissions had increased by nearly 25% since the Rio Earth Conference in 1992, and 1.3 million people continued to be adversely affected by polluted air.

Among the proposed actions were the global phase-out of leaded gasoline (petrol) and a reduction in the manufacture and use of chlorofluorocarbons (CFCs). The Plan also encouraged the trading of greenhouse emissions, enabling countries which fail to reach their targets to buy pollution credits from countries that have achieved more than their allotted targets. This enables countries to exceed their targets agreed to internationally, by incurring a financial penalty.

Box 3.4 The World Bank

In its first report on the environment, the World Bank admitted that in the past it had financed projects that proved environmentally damaging. It singled out the Indira Sagar-Sardar Sarova Dam in India which would submerge 600 000 ha of land and displace 100 000 people, and the Carajas iron-ore mine in Brazil, the development of which destroyed a large area of forest. The report also added that half of the loans made during the preceding year had included measures to protect the environment, and other loans had been made specifically for environmental improvements.

Source: The World Bank and the Environment: First Annual Report, Fiscal 1990.

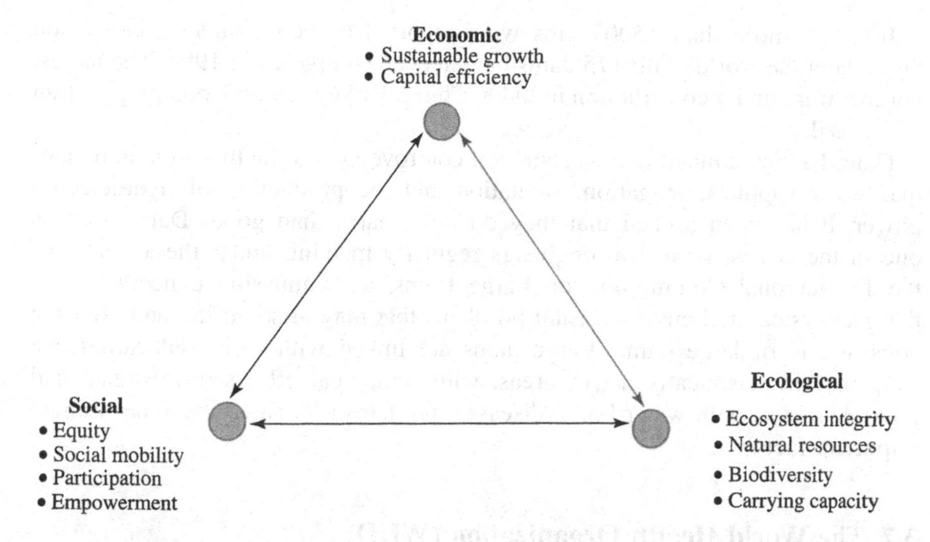

Figure 3.1 *The World Bank ESD Triangle promoting environmentally sustainable development. Reproduced by permission from Serageldin and Steer (1994)*

At one time, the World Bank was dominant in the developing world, being the main source of finance. However, the flow of funds from private sources has progressively increased. In 1996, World Bank lending to the developing world was US$21.4 billion, whereas the volume of private money reached US$170 billion. Whereas the World Bank had once been the only investment banker in the developing world of emerging markets, the scene has now been transformed.

3.6 The World Commission on Dams

Created in 1998, the World Commission on Dams aims to bring a more responsible approach to investment in large dam projects by conducting the first-ever independent global review of their costs and benefits. It will develop standards and guidelines for the benefit of countries faced with future dam-building decisions. Its report and recommendations will be delivered in the year 2000.

The 12-member Commission was created on the initiative of the World Bank and IUCN. It is chaired by Kader Asmal, South Africa'a Minister of Water Affairs and Forestry. The Commission works closely with advocates and opponents of large dams, in an effort to overcome the deadlock that has polarized the international debate on this subject. The commissioners include representatives from the Murray-Darling Commission, Oxfam International, the California Institute of Technology, the China Ministry of Water Resources, the Environmental Defence Fund, and university and appropriate private sector representatives.

In 1997, more than 1500 dams were reported as being under construction throughout the world, with 175 dams having been completed in 1996. The largest number were under construction in India, China, Turkey, South Korea, Japan, Iran and Brazil.

Dams have continued to be a source of controversy despite their role in municipal water supplies, irrigation, sanitation and the production of hydroelectric power. It has been argued that they do more harm than good. Dam safety is one of the issues, with dam engineers regularly meeting under the auspices of the International Commission on Large Dams, a commission concerned with the geophysical and environmental problems that may arise during and after the construction of large dams. Large dams are linked with increased earthquake frequency in seismically active areas, with ecological effects downstream and possible increases in water-borne diseases. As dams age, safety becomes a more important factor.

3.7 The World Health Organization (WHO)

The World Health Organization, an agency of the UN, came into being in 1948. Based in Geneva, WHO absorbed the health activities of the UN Relief and Rehabilitation Administration which had assisted the health departments of many countries with both advice and practical aid, as well as the Paris office of epidemic intelligence, and the health organizations of the former League of Nations.

WHO became the sole international health organization. It operates with a high degree of autonomy, though directed and guided by the World Health Assembly which meets annually. The work of WHO encompasses the control of communicable diseases; water supply and waste disposal; air and water pollution; standards for biological and chemical substances; nutrition; food hygiene and food standards; occupational health; the effects of radiation; psychosocial influences; carcinogenic risks; and environmental health impact assessment.

Health has been defined by WHO as 'a state of complete physical, mental and social well-being and not merely the absence of disease or infirmity'. However, most assessments of health still rely upon morbidity and mortality statistics, such as infant- and child-mortality rates and the average expectations of life in various countries.

Smallpox (variola), until 1977 one of the world's most dreaded plagues, has been declared eradicated. Routine smallpox vaccination has been discontinued in most countries. The battles against such diseases as malaria, typhoid, typhus, schistosomiasis (bilharziasis) and yellow fever, continue.

In May 1997, WHO published the results of an assessment of 12 toxic organic pollutants conducted by the International Program on Chemical Safety. The report found sufficient evidence to warrant international action to reduce or eliminate the discharge of the following chemicals: polychlorinated biphenyls (PCBs), dioxins, furans, aldrin, dieldrin, dichlorodiphenyl-trichloroethane (DDT), endrin, chlordane, hexachlorobenzene (HCB), mirex, toxaphene and heptachlor, all of which being substances that can be transported long distances from their sources by air and water.

In 1998, WHO entered a phase of restructuring, with greater emphasis on such issues as chronic non-communicable diseases, urban environmental health problems, rising rates of cancer, cardiovascular diseases, mental illness, and the promotion of national health insurance systems. The task of reforming WHO has fallen to Dr Gro Harlem Brundtland, former prime minister of Norway and chair of the World Commission on Environment and Development.

3.8 UN International Children's Emergency Fund (UNICEF)

Created by the UN in 1946, UNICEF is a body devoted to assisting national efforts to improve the health, nutrition, education and general welfare of children, particularly in less-developed countries. Other activities include the development of health services and the training of health personnel, the construction of educational facilities, and teacher training. These activities are financed by governments and private voluntary contributions.

In 1998, the UNICEF publication *The Progress of Nations* confirmed that immunization rates in children in the developing world have risen from 5% to 80% during the past 20 years. Indeed, the world-wide immunization of children against six major diseases – measles, tetanus, whooping cough, tuberculosis, poliomyelitis and diphtheria – has been one of the greatest triumphs of public health. Even so, some countries, such as Niger and Haiti, still fell short of the mark. Even the USA failed to immunize 11% of its children under one year old against measles. Measles still kills about one million children a year – deaths that could be prevented by vaccination. Cuba, a relatively poor country, managed to vaccinate 98% of its children, one of the best success rates in the world.

3.9 International Labour Organization (ILO)

Established in 1919, the ILO is an intergovernmental agency in which representatives of governments, employers and employees participate to improve occupational and working conditions. The ILO was recognized by the UN in 1946 as a specialized agency. Within the context of the working environment, the ILO has produced codes and guidance in respect of air pollution, noise and vibration, occupational cancer, chemical hazards and control of risks. It co-operates closely with the World Health Organization.

World Employment is an annual report published by the ILO. The 1996–97 report stated that about one billion people world-wide, or about one-third of the global labour force, were either unemployed or underemployed in 1995. This compared with about 820 million in 1993 and 1994. The ILO warned that the growing numbers of the 'working poor' would aggravate economic problems and social unrest.

The ILO argues that sustained economic growth is the best recipe for getting people back to work. The average level of unemployment in the European Union during 1995 had been 11.3% of the workforce, while unemployment in Central and Eastern Europe remained in double digits. On the other hand, unemployment in the USA had been below 6% for 26 straight months.

3.10 International Seabed Authority (ISA)

The Convention on the Law of the Sea (the Montego Bay Convention) came into force in 1994, after many years of debate. Among other things, the convention created an International Seabed Authority and established the principle of Exclusive Economic Zones.

The Authority is based in Kingston, Jamaica. Its function is to oversee and regulate the exploitation of the rich mineral resources of the oceans, through a licensing system. The Authority assesses and predicts the environmental impact of activities and proposed activities in the International Seabed Area; studies the effects of pollution; oversees exploration activities; develops rules and regulations for the conduct of seabed activities; approves plans for activities; and develops mechanisms for the enforcement of responsible behaviour.

Under the Exclusive Economic Zone (EEZ) arrangement, coastal nations assume jurisdiction over the exploration and exploitation of marine resources including fish and seabed minerals in their adjacent sections of continental shelf. An EEZ is defined arbitrarily as a band extending 320 km (200 miles) from the shore, with median lines agreed between nations separated by seas of lesser distance. The introduction of EEZs resulted in the exclusion from many areas of high-performance long-distance foreign fleets. The British fleet of 168 distant-water trawlers gradually disappeared, to be replaced by a fleet of compact coastal-type vessels.

The UN declared 1998 the International Year of the Ocean – recognition of our increasing reliance and impacts on the world's ocean resources, and the need to protect the oceans and to use their resources in a sustainable manner.

3.11 International Whaling Commission (IWC)

The International Whaling Commission is an international body formed in 1946 with the purpose of framing regulations for 'the conservation, development and optimum utilization of whale resources'. It operates under the terms of the Convention for the Regulation of Whaling. It meets each year to review the condition of whale stocks. Its original membership of 14 nations has gradually increased to include a number of non-whaling nations. In view of the virtual failure of the whale management programmes, the IWC in 1982 voted to discontinue commercial whaling world-wide, commencing in 1986. Some countries still persisted in some commercial whaling. In response, the IWC in 1994 suggested the creation of a vast oceanic sanctuary that would give whales permanent protection from commercial hunts. First proposed by France, the Southern Ocean Sanctuary covers an area more than five times the size of Australia. It embraces waters from Bass Strait to Antarctica and encircles the globe in a vast swathe. Japan and Russia objected to the sanctuary. The sanctuary is used by seven species of endangered whales.

Iceland left the IWC in 1992 to resume hunting minke whales. Norway stayed in the IWC, but resumed commercial whaling in 1993. Japan kills about 300 whales a year for research purposes. There are now 39 members of IWC, and with the exceptions of Japan and Norway, they all continue to oppose commercial whaling, confirming a total ban in October 1997. Japan and Norway argue that the purpose of the IWC is to manage whaling, not to ban it. Opponents to whaling argue that it is better to ban it, than to open the doors to the bloody and cruel excesses of the past.

In 1997, the Humane Society International, the world's largest animal welfare and conservation organization, started to campaign for a universal whale sanctuary covering all oceans to be established by the year 2000. Professor David Bellamy has been supporting this move.

3.12 World Trade Organization (WTO)

The World Trade Organization was created through the Uruguay Round of the General Agreement on Tariffs and Trade (GATT), to give expression to and co-ordinate the application of that agreement. It came into operation in 1995. The main aim of WTO is to reduce or abolish trade barriers. The first step was the opening up of the banking and insurance markets. A continuing responsibility is also to settle trade disputes between nations through adjudication, and to resolve issues involving labour standards and the environment.

The General Agreement on Tariffs and Trade was an international treaty which came into force in 1948 to promote world trade through the progressive reduction of trade barriers. Through rounds of talks over many years, tariff reductions have been progressively achieved in the face of many impediments. The Uruguay Round of talks ran for seven years between 1986 and 1993. The result was a commitment to a substantial reduction in tariff protection for industrial products, agricultural produce and services.

Environmentalists have objected to the 1993 GATT on the following grounds:

- Trade liberalization encourages economic growth and so damages the environment.

- By limiting national sovereignty, countries cannot apply any environmental measures they choose.
- GATT does not allow countries to keep out a product because of the way that product is produced or harvested.
- GATT disapproves of the use of trade measures to influence environmental policies abroad.

3.13 The International Institute for Environment and Development (IIED)

The International Institute for Environment and Development was founded in 1971, largely through the inspiration of Barbara Ward (Baroness Jackson). It is a global organization, set up to further the wise use of natural resources, with sustainable development as the guiding principle. The scope of the Institute ranges from forests and fisheries to the overcrowded living conditions in the cities of the developing world, from the Antarctic to the tropics, and working alongside the people of the developing world in Africa, Central and South America, and South-east Asia. The objective of the IIED process is to enable its partners to become self-reliant and to improve their livelihoods on a sustainable basis. IIED is funded by private and corporate foundations, international organizations, governments and concerned individuals.

IIED was one of the first organizations to recognize the importance of environmental economics, setting up the London Environmental Economics Centre at University College. Its publishing arm is Earthscan Publications. It published *Blueprint for a Green Economy* (1989), and *Blueprint 2: Greening the World Economy* (1991). The IIED has offices in London, Buenos Aires and Washington, DC.

3.14 The Beijer International Institute of Ecological Economics (BIIEE)

The Beijer International Institute of Ecological Economics was created in 1990, under the auspices of the Royal Swedish Academy of Sciences, with the object of achieving co-operation between ecologists and economists on issues concerning the natural environment. The Institute promotes research such as the Baltic Drainage Basin Program, the Resiliance Program, and the Role of Property Rights. It also promotes a teaching and training programme. In 1995, the Institute launched a new journal, *Environment and Development Economics*. The Institute is based in Stockholm.

While environmental economics is an extension of neoclassical economics, ecological economics begins from different foundations reflecting a new way of thinking about the economy and the environment, and the relationship of humanity with the natural world. It sees the economy as being within the global ecosystem with critical interdependence. It also recognizes that people are both citizens and consumers, and respond to ethical considerations as well as economic self-interest. Citizens tend to make decisions on the basis of what is seen to be right. The anthropocentrism of conventional economics may be contrasted with the ecocentrism of ecological economics. Ecocentrism takes the view that everything natural has intrinsic value, i.e. value in itself. Biocentrism goes one step further, emphasizing the intrinsic value of living things; its emphasis on the value of life provides it with a different philosophical standpoint.

3.15 The Organization for the Prohibition of Chemical Weapons (OPCW)

The Organization for the Prohibition of Chemical Weapons is an international organization that was created in 1997 with the objective of destroying all existing stocks of chemical weapons and prohibiting the use and development of this category of weapons of mass destruction through the regular inspection of chemical industries. Based in The Hague, the Netherlands, in 1998 OPCW had the support of 112 nations, including all permanent members of the UN Security Council.

3.16 Summary

This chapter has provided details of many international agencies which develop policies and implement programmes of benefit to the world's environment. Notable among these agencies is the UN Commission on Sustainable Development which carries a responsibility for the implementation of Agenda 21. Other sterling bodies include the UN Environment Program (UNEP), the UN Development Program (UNDP), the World Bank and its component agencies, the World Commission on Dams, the World Health Organization (WHO) under Gro Harlem Brundtland, the International Labour Organization (ILO), the International Seabed Authority, the International Whaling Commission, the World Trade Organization (WTO), and two international research institutions, mostly creations of recent years. Collectively, they reflect growing world concern for common resources and the need for improved environmental management, and also the current trend towards globalization. Benefits are to be sought through international co-operation and not solely through international competition. The level of concern reflects the fact that the world now seems so much smaller than it did before the Second World War. At the same time, the number of nations has vastly increased, now numbering some 200 or more. Nationalism, like patriotism, is no longer enough. We share a small planet which is in need of careful maintenance.

Further Reading

Gilpin, A. (1996) *Dictionary of Environment and Sustainable Development,* John Wiley, Chichester.

International Atomic Energy Commission (1991) *Electricity and the Environment,* IAEE, Vienna.

McDonald, J. (1996) The World Trade Organization and environment protection law and policy, a paper presented to the 2nd Environmental Outlook Conference, March 1995, Australian Centre for Environmental Law and the Australian Department of the Environment, Sport and Territories, Sydney. *Conference Proceedings,* The Federation Press, Sydney.

OECD (1997) *Economic Globalization and the Environment,* OECD, Paris.

UNDP (1998) *Human Development Report,* Oxford University Press, Oxford.

UNEP (1995) *Energy, Pollution, Environment and Health,* UNEP, Nairobi.

World Bank (1996) *Environment Matters and the World Bank,* World Bank, Washington, DC.

World Health Organization (1998) *World Health Report 1998,* Geneva.

All institutions and agencies publish annual reports.

4

Sustainable Development

4.1 Introduction

Sustainable development is economic development that meets the needs of the present without compromising the ability of future generations to meet their own needs. The concept has been widely embraced, but few have been able or willing to translate this noble concept into policies which are different to those already prevailing. Most may readily agree that cutting down the world's forests, harvesting all the fish in the oceans, indifference to the loss of biodiversity on an increasing scale, runaway global warming, radioactive contamination, accumulated toxic wastes, and a world population that has outstripped the means of adequately feeding itself, would lead to a disastrous future. However, most current policies aim at preventing such disasters.

Beyond food, clothing and accommodation, it is virtually impossible to predict the needs of people some eight generations ahead. Box 4.7 proves, beyond reasonable doubt, that in 1800 it would have been virtually impossible to predict the needs of people in the year 2000.

Likewise in 2000, it is virtually impossible to visualize the needs of people in the year 2200. Cities may be underground (as predicted by Wells), climates rigorously controlled, food genetically designed, all wastes productively recycled, transport so revolutionized as to eliminate death and injury, productivity so great that poverty has been eliminated, and the working day reduced to two hours. The challenge may be to combat boredom, and reflect on the barrenness of a falling population.

However, it is still possible today, as this chapter illustrates, to set out a path of development that will lead to many of the improvements described. Indeed, the outcome of the chapter is that the problems of intergenerational equity are much more readily addressed than the problems of intragenerational equity. The essential problem today is that material progress advances rapidly, yet the social and political background is beset by the baggage of tribalism, bigotry, hate-thy-neighbour, increasing inequality between nations and between individuals within the same society, and extremists in religion and politics.

The chapter closes with reflections on natural resources and natural resources accounting.

4.2 Sustainable Development: The Meaning

Development involves the application of human, physical, natural and financial resources to meet effective or prospective market demands and other human needs. The breadth of the concept is not always appreciated as it applies not only to industrial, commercial and financial institutions, but to the provision of infrastructure, sanitation, educational facilities, hospital and public health services, housing, roads, national parks, and tourist and recreational facilities. In all countries, development is a function of both the public and private sectors of the economy and voluntary activity. Activity of this nature may be profit-motivated, unpaid, or a matter of public policy. The general effect may be an improvement in per capita income, accompanied by benefit or detriment to the quality of life locally, regionally, nationally or globally. Development may be sustainable if the resources used are renewable, or non-sustainable if the resource base is exhaustible in the short term, or if substitutes cannot be found. Unsustainable activity may result in the exhaustion of fish stocks, massive soil erosion, salinization to an extent that causes degradation of soils and vegetation, the siltation of rivers, the destruction of wetlands and mangroves, the loss of rare and endangered species, the loss of areas of high genetic and conservation value, a legacy of abandoned factories and quarries, urban blight, increasing noise levels, and air and water pollution at an unacceptable level. The history of sustainable yield and sustainable development is a long one, being inseparable from the land in agricultural communities. Sustainable yield means living off the interest, rather than the capital or stock of a resource.

President Theodore Roosevelt in 1908 declared, 'The nation behaves well if it treats its natural resources as assets which it must turn over to the next generation increased and not impaired in value' (see Box 2.3). President Roosevelt thus struck a note echoed by others, from time to time, throughout history.

In 1980, the World Conservation Strategy (WCS) was launched by the World Conservation Union (WCU), the UN Environment Program, and the World Wide Fund for Nature (WWF). It demonstrated that development can only be sustained by conserving the living resources on which that development depends, and by the integration of development and conservation policies. It urged every country to prepare its own national conservation strategy, and many did. The chief successor to the WCS has been the document, *Caring for the Earth: A Strategy for Sustainable Living*, published by the same bodies in 1991. It included a wide range of recommendations for legal, institutional and administrative reform.

In 1987, the World Commission of Environment and Development (the Brundtland Commission), in *Our Common Future*, its report to the Governing Council of UNEP, defined sustainable development as 'development that meets the needs of the present without compromising the ability of future generations to meet their own needs'. Sustainable development considers both the living and non-living resource base with regard for conservation and the advantages and disadvantages of alternative courses of action for future generations. It allows the use of depletable resources in an efficient manner with an eye to the substitution of other resources in due course. Sustainable development, the report argued, called for much more emphasis on conserving the natural resource base on which all development depends; and a greater regard for equity within society and between rich and poor nations, with a planning horizon that extends further than the present generations alive today. It requires an integration of economic, social and environmental considerations in decision-making at both government and corporation level.

Subsequently, the World Bank (Serageldin 1996) advanced a more positive concept:

> Sustainability is to leave future generations as many opportunities as we ourselves have had, if not more . . . leaving future generations more capital per capita than we have had, although the composition of the capital we leave to the next generation will be different in terms of its constituent parts than the capital we have used in our generation.

Clearly, these enlightened concepts of sustainability inevitably clash with the short-term considerations of many enterprises. Furthermore, consideration of the needs of future generations is beset with problems. It would have been impossible at the beginning of the 18th century or the 19th century to predict the developments and hence needs of the coming 100 years. The question of equity within the present generation is beset with challenges, let alone the question of equity between generations. The actions of the present generation may reduce the endowment of resources to future generations, but there is no way of knowing whether future generations could be, or should be, compensated in some way, for their needs are unpredictable. To keep coal in the ground may be pointless when future generations may well not have any use for it. Future sources of energy may, for example, be solar or hot rock, both unleashing boundless quantities of energy. Through shifts in energy usage, global pollution may abate on a scale unimaginable today. Alterations in lifestyle and transportation could have similar effects. The main difficulty in implementing long-term policies, however beneficial, is that governments and other decision-makers respond to short-term pressures and demands, especially when politically attractive, at the expense of future considerations even those involving the present generation. Industry is likewise drawn to short-term considerations for this is where the future competitive edge frequently lies.

Box 4.1 is an important commentary on economic growth and environmental policy by a range of eminent writers, providing further insight into the nature of sustainability. Box 4.2 attempts to summarize the principles of a sustainable society, extracted from the update of the World Conservation Strategy.

Box 4.3 derives from a rare attempt to specify some of the ingredients of sustainable development. Few speakers or writers have attempted this. The reason

Box 4.1 Economic Growth and Environmental Policy

Economic growth is not a panacea for environmental quality; indeed it is not even the main issue. What matters is the content of growth – the composition of inputs (including environmental resources) and outputs (including waste products). This content is determined by, among other things, the economic institutions within which human activities are conducted. These institutions need to be designed so that they provide the right incentives for protecting the resilience of ecological systems. Such measures will not only promote greater efficiency in the allocation of environmental resources at all income levels, but they would also assure a sustainable scale of economic activity within the ecological life-support system. Protecting the capacity of ecological systems to sustain welfare is of as much importance to poor countries as it is to those that are rich.

Source: K. Arrow, B. Bolin, R. Costanza, P. Dasgupta, C. Folke et al. (1995) Economic growth, carrying capacity, and the environment, Science, 268 (28 April): 521.

Box 4.2 The Principles of a Sustainable Society

1. A duty of care for other people and other forms of life, now and in the future; to share fairly the benefits and costs of resource use and environmental conservation. This is an ethical principle.
2. The direction of development and economic growth towards the improvement of the quality of human life in the broadest sense.
3. The conservation of life-support systems and biodiversity, ensuring that the uses of renewable resources are sustainable, that is within the capacity of the resource for renewal.
4. Minimizing the depletion of non-renewable resources such as minerals, oil, gas and coal, through recycling, economical use, and switching to renewable substitutes.
5. Keeping within the carrying capacity of the Earth. Human numbers and lifestyles should be kept within the carrying capacity of the Earth, while that capacity is expanded by careful management.
6. A re-examination of values and adjustment of behaviour to support the new ethic; discouragement of values that are incompatible with a sustainable way of life.
7. Encourage local communities to care for their environments, taking part in decision-making processes and active discussions.
8. A national framework for integrating development and conservation, involving all interests, seeking to identify and prevent problems before they occur.
9. A firm alliance between all countries to achieve global sustainability; global and shared resources, especially the atmosphere, oceans, major ecosystems, and endangered species need to be managed on the basis of common purpose and resolve.

Source: Derived from Caring for the Earth: A Strategy for Sustainable Living (1991), IUCN, UNEP, WWF, Gland, Switzerland.

Box 4.3 The Priorities for Sustainable Development

- A transition to population stability, to reduce the need for growth everywhere.
- Poverty reduction will require considerable growth and development in developing countries, balanced by negative growth for the richer countries.
- The large-scale transfer of resources from the richer to the poorer countries.
- Higher prices for the exports of the poorer countries.
- Markets will have to learn to function without expansion, without wars, without waste, and without advertising.
- Economic policy will have to suppress certain activities in order to allow others to expand, so that the sum total remains within the biophysical budget.
- The transition to renewable energy will need to accelerate.
- The upgrading of human capital through education and training and adequate employment; job creation in preference to automation.
- The alleviation of poverty by direct amelioration and targeted aid.
- Major technological transfers from rich to poor.
- The limitation, arrestation and even reduction of economic activities in the richer countries.
- In conflicts between biophysical realities and political realities, the latter must eventually give ground.
- As economic stability in the richer countries may depress the terms of trade and lower economic activity in developing countries, additional large-scale transfers to the poorer countries will be needed.
- The primary need is for income redistribution and population stability.
- Exceptional political wisdom and leadership will be required; political resolve will be the scarcest resource.
- Other criteria include no net increase in the concentration of greenhouse gases in the atmosphere; no net increase in acidification in surface waters and soils; no net increases in toxic chemical and heavy metal concentrations in soil and water; no net topsoil erosion; no net depletion of aquifers; the preservation of all or most remaining natural forests, estuarine areas, coral reefs, and other ecologically critical biomes of ecosystems; no further humanly induced extinctions of other species; the reduction of the energy intensity of GDP; an increase in the proportion of energy from renewable resources; an increase in recycling; avoidance of overgrazing.
- Part of the proceeds from consuming non-renewables should be allocated to research in pursuit of sustainable substitutes.
- The harvest rates of renewable resource inputs should be within the regenerative capacity of the natural system that generates them.
- If a nonrenewable resource has no renewable substitute, its use may be shared over some ethically determined number of generations.

Source: Derived from R. Goodland, H. Daly and S. El Serafy (1992) The urgent need for environmental assessment and environmental accounting for sustainability, in Proceedings of the 12th Annual Meeting of the International Association for Impact Assessment, April 1992, pp. 106–27.

Box 4.4 The Hartwick Rule

The Hartwick rule states that consumption may be sustained, when using exhaustible resources, only when the net proceeds of these declining stocks are invested in reproducible capital. The rule requires sufficient substitutability between exhaustible and reproducible stocks. In this way, aggregate output can be maintained constant in per capita terms. The rule simply means that if consumption depends heavily on the depletion of assets, wisdom requires the building up of other assets that offer long-term, sustainable returns instead of squandering all net returns on current consumption. It is now argued that the depletion of natural assets should be reflected in the national accounts. Solow also previously considered this question, in the same environment.

Sources: J.M. Hartwick (1977) Intergenerational equity and the investing of rents from exhaustible resources, American Economic Review, 66: 972–4; R. Solow (1974) Intergenerational equity and exhaustible resources, Review of Economic Studies Symposium, 29–45.

is not hard to find. In essence, it involves a complete reversal of normal expectations: in the richer countries, children must expect less than their parents or grandparents; rationing of say fuel, cars and air transport; promotion of labour-intensive activities such as basket-weaving; larger payments for imports; larger contributions by way of financial aid; negative economic growth; and the limitless expansion of poorer nations. The general effect is that of depression in material and human terms, with growing massive unemployment. It has been well noted that political resolve will be the scarcest resource, followed no doubt by human endurance. Misery is not a readily saleable commodity. Box 4.4 provides insight into the Hartwick Rule, a route by which wasting assets may be converted into sustainable development.

4.3 Property Rights and Duties

Property rights and duties are the rights or entitlements that accompany the ownership or control of assets such as land or water, together with the associated duties that may be imposed by tradition, resource and environmental laws, planning restrictions, international agreements and licensing or leasing conditions. Rights short of freehold rights include riparian rights, rights of common access, fishery rights, native land rights, rights of way, rights of entry, mining rights, air rights and water rights. All these rights are subject to conditions.

Property may be held freehold by the public sector at various levels of government; it may be leased for special purposes or used directly by government for roads, harbours, airfields, defence purposes, forests, wilderness, parks and gardens, approved developments, recreation, bridges, heritage and conservation purposes, lighthouses, fauna and flora reserves, tunnels, railways, residential development, rain forest, wetlands, mangroves, national and marine parks, plantations, buffer zones, schools and universities, and a variety of other purposes.

The private sector, at levels ranging from the householder to the corporation, may own and utilize land and water for a variety of similar purposes, subject to statutory consent and conditions. The corporate sector is more likely to engage in large-scale manufacturing, power generation, major shopping centres, automobile manufacturing, steelworks, mining of coal and other minerals, commercial fishing, hotels and resorts, entertainment facilities, aircraft construction, restaurants, commercial and business centres, ship building, and servicing.

Other property remains as common property with collective or co-operative ownership, such as condominiums and apartment blocks, co-operative farming enterprises, co-operative shops as in Britain, mutual enterprises, or commons open to the public.

There is also open access property, without an owner, such as the Arctic and the Antarctic, the oceans and the atmosphere; these international commons, being free, are more likely to be abused and overused, unless subject to international agreements and conventions. Thus the whale was almost hunted to extinction.

It has been argued that the allocation of property rights is essential to the rational allocation of resources and the resolution of disputes. Certainly, ownership with responsibilities may be recognized as better than open access, with no responsibilities. So the argument has a certain validity. At the same time, many environmental catastrophes have occurred in situations with fully allocated property rights, such as the incidents at the Three-Mile Island and Chernobyl nuclear plants, the *Exxon Valdez* disaster, the pollution of the tidal Thames, salinity in the Murray-Darling river system which affected several states in Australia, or the dust bowls in the United States. It may be argued that property rights, properly exercised, may offer better management of the environment. In reality it may offer no better contribution than that someone may be held responsible and blamed. At best, it may be conceded that property rights offer a necessary, but not sufficient, requirement for efficient resource management and environment protection. Further, population pressures on local resources may destroy all effective and sustainable management, and break down the rules of custom and tradition.

As mentioned in Chapter 1, Coase (1960) argued that, as a rule, no form of government action was required to deal with externalities; property rights, not government action, would resolve the problems. The Coase theorem simply fails against experience. Others have continued, however, to carry the banner, as if there were something inherently virtuous about private property and private profit when it comes to the environment. If this contained even an element of truth, direct regulation by government would never have emerged or gained popular support.

Arrow (1996) goes beyond both private property and direct regulation:

> When private individual property fails, economists usually think of state intervention ... but ... private property rights, frequently hard to define, on the one hand, and the supervision of the state on the other, only begin to exhaust the list of social devices to balance individual initiative with prevention of injury to others.

Hanna *et al*. (1996) also concede that

> current property rights regimes over natural resources are often inadequate to prevent resource degradation and overuse. Even more important, property rights regimes fail when overwhelmed by increases on both absolute population growth and per capita demand for resources. ... Property rights regimes are a necessary but not sufficient condition for the sustainability of natural systems. Property rights regimes have failed in the past and are continuing to fail in the pursuit of short-term profits, rapid technological change, cultural change, high levels of absolute population growth, increased per capita demand for resources, and inappropriate government policies.

In respect of dispute resolution by negotiation, mentioned in Chapter 1, which may achieve economic efficiency in the narrower sense, but not social equity, Cropper and Oates (1992) remark,

> Since most cases of air and water pollution, for example, involve a large number of polluting agents and/or victims, the likelihood of a negotiated resolution of the problem is small – transaction costs are simply too large to permit a Coasian solution of most major environmental problems.

See Box 4.5.

4.4 Intragenerational Equity (Fairness within Today's Society)

Intragenerational equity is a concept of fairness between individuals and groups, within society, locally, regionally, nationally and globally. The concept of human rights and dignity was embodied in the UN Universal Declaration of Human Rights in 1948, and in a whole range of conventions and declarations since then. Such rights include equality before the law; protection against arbitrary arrest; the right to a fair trial; the right to own property; freedom of thought, conscience and religion; freedom of opinion and expression; freedom of peaceful assembly and association; the right to work; the right to equal pay for equal work; the right to form and join trade unions; the right to rest and leisure; the right to an adequate standard of living; and the right to education. These rights are compromised in all countries by reasons of birth, gender, race, property, class, caste, political division, territorial ambition, inequality of income, denial of rights, persecution, genocide, conquest, prejudice, bigotry, arrogance, maladministration, and non-democratic forms of government. These factors work to the detriment of effective environmental management and sustainable development.

Box 4.5 Property Rights and Duties in Relation to Natural Resources: Regimes

- *Public sector regime*: the State owns the land and controls the use of resources on it. Examples are the public ownership of harbours and waterways, national and marine parks, national forests and reserves, defence lands, air rights, public housing estates, and state-owned infrastructure.
- *Private sector regime*: land is in private ownership, either individual or corporate, and the resources on it are privately controlled subject to government planning and pollution-control requirements. Examples include private forests, private dwellings, many agricultural, pastoral and industrial enterprises, airports, mines, multimedia enterprises, company towns, private infrastructure and office buildings.
- *Common property regime*: land, dwellings and resources owned by a group of people. Condominiums, co-operatives, mutual companies and common land are examples.
- *Open or free access regime*: a situation in which there are no property rights, and in which all have free access to resources without limit. Examples are oceans and unallocated resources such as Antarctica, subject to any international agreements and conventions that may apply such as fishing rights, whaling restrictions, exclusive economic zones, oil exploration rights, pollution-control conventions, mining bans, seabed protection and anti-dumping measures. Control measures have tended to follow overexploitation, whether on land or sea.

Despite enormous material progress (as summarized in Box 4.7) the world at the social level remains substantially tribal and barbaric. Certainly, democratic forms of government have made substantial progress while a variety of dictatorships have collapsed. Society remains, however, relatively primitive; it is the area of least progress. Progress seems at times to come by chance, or through the actions of a small number of heroic people. People seem to find it easier to hate than to love, particularly in respect of people who look or sound different. Issues of race constantly intrude, and the author is reminded of his encounter with the Ku Klux Klan in 1968. Humanity remains at a low ebb in these matters, even though we can reach Mars.

These reflections should not detract from some regard for the increasing burden falling on the workforce of the present day; a workforce which tends to shrink while becoming increasingly productive. Those engaged in growing food, mining and manufacturing are yielding progressively to those in services; while the total force becomes a smaller part of the total population. Box 4.6 attempts to sum this up, before we look at the claims of generations yet to be born.

4.5 Intergenerational Equity (Fairness between Generations)

Intergenerational equity is the concept that those living today should not compromise or restrict the opportunities open to future generations. It envisages a

Box 4.6 The Burden on the Present Working Population

- The need for an ever-diminishing, though highly productive, workforce to provide an adequate flow of raw materials, foodstuffs, goods and services, to an ever-increasing and increasingly dependent population with an ever-improving standard of living.
- An ever-increasing school-leaving age with an increasing proportion of school-leavers proceeding to higher education.
- An ageing population that retires earlier and lives longer than ever before; a characteristic of all countries.
- A level of unemployment in all societies adopting market principles.
- A level of unemployment arising from hospitalization, sickness, accident, disability, sole parenthood, part-time and casual employment, victims, those on maternity and special leave, and those confined to penal and other institutions.
- A level of taxation on incomes and expenditure to support defence, health, education, local government services, insurance, law enforcement, roads and railways, public superannuation schemes, government, overseas aid, subsidies, social security and international activities.

partnership among all the generations that will expect to thrive on the world's resources. It implies, at the very least, that each generation should hand over to the rising generation a world in as good an order as possible and with the full benefits of sustainable development. The World Commission on Environment and Development, in *Our Common Future*, stated that 'Sustainable development is development that meets the needs of the present without compromising the ability of future generations to meet their own needs.' Problems arise with the application of the concept. Clearly, a treeless, barren planet, scorched with ultra-violet radiation and littered with radioactive wastes, would not meet the criteria. Yet, considering more likely outcomes, it is impracticable to envisage the circumstances of generations as yet unborn. One has only to ask whether anyone living in 1800 could predict the needs and aspirations of those living in 1900; and whether anyone living in 1900 could predict the needs and aspirations of those living in 2000. (See Box 4.7 which selects some of the developments between 1800 and 2000.) This does not take account of the general advance in the standard of living for most people, the huge fall in infant mortality, the improvements in diet, and the striking extensions in the span of life.

Some have suggested that compensation funds should be set up which, with compound interest, would serve to compensate future generations for the ravages of the present. Apart from the risks to such funds which could be overwhelmed by inflation or simply forcibly acquired by the wrong people, such proposals cannot be morally supported. If the past is any guide, future generations will be substantially better off than those living at present. The future will inherit the present with all its progress in electronics and advanced technologies. To hand over a planet in fairly good trim is all that is morally required from the present.

Box 4.7 *Events since 1800 Proving the Unpredictability of the Progress of Future Generations*

1800	Arc light invented
1801	Trevithick steam vehicle invented
1803	The Fourdrinier papermaking machine invented; Denmark banned the import of slaves into the Danish West Indies
1804–5	Lewis and Clark's expedition across America; Dalton's theory of atoms
1809	Canned foods introduced
1815	The Davy miner's lamp invented
1820	Gas lighting in Pall Mall, London
1829	Stephenson's *Rocket* locomotive invented
1833	Slavery abolished in the British Empire; American Anti-slavery Society established
1834	The Hunt sewing machine invented
1839	Grove invented the fuel cell; photography established
1840	Penny Post introduced in Britain
1842	Gorrie's air-conditioning apparatus invented; first steam hammer
1843	The *Great Britain* Atlantic liner launched; first facsimile transmission
1844	The Morse Code invented
1846	Anaesthetics introduced
1848	The Seneca Falls Convention on female suffrage; the Communist Manifesto
1849	Hunt invented the safety pin
1850	The Swan electric lamp invented
1851	Mechanical refrigeration introduced
1852	*Uncle Tom's Cabin* published; the safety match invented
1853	The Bunsen burner invented
1859	Darwin's theory of evolution; the first oil well
1860	Tramways introduced in Britain; first trans-Atlantic submarine telegraph; the Yale lock invented
1862	The four-stroke engine invented
1863	The Geneva Convention
1865	Slavery abolished in the USA; Lister's antiseptic surgical procedures introduced; genetics
1867	First typewriter; Nobel demonstrated dynamite
1868	First plastics introduced
1869	The US transcontinental railway; the periodic table of the elements; margarine patented in France
1870	The Swan and Edison incandescent lamp; application of pasteurization to milk, wine and beer
1872	Yellowstone National Park declared
1876	Bell invented the telephone; anthrax organisms identified
1879	First electric locomotive
1882	First public power stations in London and New York
1883	Koch identified the germs of cholera and tuberculosis
1884	Evaporated milk patented; diphtheria isolated
1886	The Benz petrol-driven car invented; commercial aluminium produced

Continued on page 100

Continued from page 99

1889	First motion picture; the Eiffel Tower opened
1890	First electric underground railway opened in London
1892	The diesel engine invented
1895	X-rays discovered; Ross identified the source of malaria; Marconi wireless invented
1896	Radiation treatment for breast cancer
1897	Aspirin; viruses identified
1898	The Curies discovered polonium and radium; psychiatry founded
1900	Agriculture accounted for 68% of the labour force in Japan, 44% in the USA and 20% in Britain: the Zepperlin airship; Planck quantum theory
1901	Nobel prizes first awarded
1903	Wright brothers took to the air; the motorbike invented
1904	The St Louis World Fair; the ice-cream cone invented
1905	The Trans-Siberian Railway; Einstein's special theory of relativity
1908	Model T Ford appeared
1909	Bleriot flew across the English Channel
1912	Vitamins discovered
1914–18	The First World War
1919	World's first supermarket, Memphis; Alcock and Brown made the first non-stop Atlantic flight
1920	Women's suffrage, USA; the league of Nations
1923	The zipper invented
1925	Diesel-electric locomotives introduced
1926	Baird's TV system invented
1928	The *Flying Scotsman* locomotive; women's suffrage in Britain; Fleming discovered penicillin
1929	The Wall Street crash
1936	The printed circuit board introduced; first radiotelescope
1938	*Queen Elizabeth* launched; Britannica Book of the Year
1939	DDT as insecticide
1939–45	The Second World War
1940s	Forerunners of today's computer systems; the Colossus computer
1945	Atomic bombs exploded; the United Nations
1947	The Kon-Tiki expedition; the transistor invented; the polaroid camera invented
1948	The world's largest telescope at Mount Palomer
1950	The photocopier invented
1952	First video recorder
1953	Colour TV introduced; heart–lung machine invented
1954	H-bomb tests; Salk vaccine for poliomyelitis developed
1956	First nuclear power stations in Britain and the USA
1957	The Jodrell Bank radiotelescope; *Sputnik I*
1958	The Boeing 707; NASA established
1959	The hovercraft invented
1960	The contraceptive pill
1961	The first soft contact lenses
1964	The word-processor invented
1966	Harrier jump-jet introduced

Continued on page 101

Continued from page 100

1967	First human heart transplant
1968	Concorde's first flight
1969	US landing on the moon; Sony videocassette recorder
1972	First UN Conference on the Human Environment; first Landsat satellite
1973	Opening of the Sydney Opera House
1977	Smallpox eradicated
1981	The US space shuttle launched
1986	Chernobyl nuclear power station disaster
1987	The World Commission on Environment and Development
1992	The UN Conference on Environment and Development
1993	The Channel Tunnel
1997	*Sojourner* explored the surface of Mars
1998	Agriculture accounts for only 7% of workers in Japan, 3% in the USA and 2% in Britain
1999	Eyesight correction and mobile telephones mark the end of the road for spectacles and telegraph wires; high-intensity focused ultrasound treatment in surgery

Intergenerational equity remains a noble concept which cannot be translated into useful and practical policies other than the pursuit of responsible management of resources to serve the interest of current generations. There is every reason to believe that problems with population growth, forestry, fisheries, biodiversity, global warming, the ozone layer and world food supplies, will be resolved, while material progress continues to surge forward throughout many parts of the world. There is no evidence to the contrary. Box 4.7 summarizes our gift to future generations; they could ask for no more.

4.6 The Polluter-Pays Principle and the Internalization of Environmental Costs

Two versions of the polluter-pays principle have evolved over time. The first equates the price charged for the use of environmental resources with the cost of damage inflicted on society by using them. The price charged may be levied directly on the process that generates pollution or as the purchase price of licenses that entitle the holder to generate specific quantities of pollutants. The polluter may find the charges greater than the cost of abating the pollution using pollution-control equipment or by changing the processes or fuels. The difficulty with this procedure is deciding on the right price to charge, when the damage to society cannot be assessed in monetary terms in any realistic way. Further, it bestows the right to pollute on any scale, i.e. the right to continue to damage the health of members of the community in known and unknown ways over time. In addition, there is the inherent problem that the penalty paid may never reach the actual community affected, or if it does, it may never be distributed in an equitable way. The effects of some pollutants can only be assessed with difficulty, such as lead in children; to suggest compensation seems a most inappropriate response when prohibition is to hand. An overriding problem is the problem of attributing atmospheric pollutants to a particular firm when there are several similar plants in the neighbourhood. The task of allocating charges to vehicles seems impracticable.

In a later version, the polluter-pays principle asserts that the full cost of controlling pollution by whatever means to an adequate degree shall be undertaken by the polluter, preferably without public subsidy or tax concession. Hence the potential cost of pollution to society at large is translated into pollution-control costs which are internalized and reflected in the costs of production. Depending on the elasticity of demand, these cost increases may be transferred to the purchasing public. In many cases, the polluter-pays principle becomes a public-pays principle, though the public benefits from the measures taken. While all appropriate measures appear in the accountancy costs of polluters, the adverse environmental effects on the public are minimized or abated.

The polluter-pays principle was affirmed at the UN Conference on the Human Environment held in Stockholm in 1972, and by the OECD in the same year. The principle was reaffirmed by the OECD in 1985. Since then a hybrid philosophy has emerged stressing the need for overall statutory pollution-control regulation and the setting of ceilings augmented by economic instruments to encourage abatement activities beyond the requirements of regulations, to yield a market in pollution credits.

4.7 Precautionary Principle

The precautionary principle was adopted by the UN Conference on Environment and Development (the Earth Summit) in 1992, where it was agreed that, in order to protect the environment, a precautionary approach should be widely applied. The Rio Declaration on Environment and Development (Principle 15), interprets the precautionary approach as meaning that where there are threats of serious or irreversible damage to the environment, lack of full scientific certainty should not be used as a reason for postponing cost-effective measures to prevent environmental degradation. Critics of this approach are concerned about large commitments of resources to deal with vaguely defined problems. It should be noted, however, that the reference to cost-effective measures implies a high degree of certainty about the nature of the problem.

The precautionary principle already had a long history. It first emerged in Germany in 1980 when it was considered by the European Council of Experts on Environmental Matters in relation to marine pollution. Within a few years, the principle appeared in a variety of documents and treaties, such as the Montreal

Protocol on Substances that Deplete the Ozone Layer of 1988; the Bergen Ministerial Declaration on Sustainable Development of 1990; the Convention on the Protection of the Marine Environment of the Baltic Sea of 1992; the UN Framework Convention on Climate Change and the UN Convention on Biodiversity, both of 1992. Since then the concept has been incorporated in the Maastricht Treaty of 1993. There have been few guides as to interpretation.

The British White Paper on the Environment of 1992 stated:

> Where the state of our planet is at stake, the risks can be so high and the costs of corrective action so great, that prevention is better and cheaper than cure . . . Where there are significant risks of damage to the environment, the Government will be prepared to take precautionary action to limit the use of potentially dangerous materials or the spread of potentially dangerous pollutants, even when scientific knowledge is not conclusive, if the balance of likely costs and benefits justifies it.

The Australian Intergovernmental Agreement on the Environment of 1992 offers the following guidance:

> In the application of the precautionary principle, public and private decisions should be guided by: careful evaluation to avoid, wherever practicable, serious or irreversible damage to the environment; and by an assessment of the risk-weighted consequences.

Clearly there are several different interpretations of the principle and there are many different ideas concerning its application. It may be simply regarded as a reversal of the principle of the burden of proof, that proponents of developments must offer proof that the proposal will not be harmful to the environment, subject to conditions. In which case we are not far from today's public inquiry, now well established in many countries.

A central question today, following the Kyoto Protocol, is whether enormously expensive measures should be adopted now to constrain global warming, when the scientific evidence is not yet conclusive. The precautionary principle will now be tested to the limit.

4.8 Natural Resources

A natural resource is any portion or aspect of the natural environment, such as the atmosphere, water, soil, land, minerals, wildlife, mangroves, forest, flora, fauna, radiation, beauty, the coast, mountains, and environmental assets generally. Natural resources are one of the factors of production, entering into all economic activities. Natural resources can be of several types:

1. non-renewable resources such as coal, petroleum, natural gas, uranium, and minerals of all sorts, which, once consumed, are gone forever in that form;

2. renewable resources such as water, fish, timber, crops and livestock;
3. non-expendable resources such as beautiful views and assets with an existence value, which are valued for their own sake;
4. rare and endangered species which may reach a threshold of irreversible depletion, and below that level are lost forever.

Technological change, recycling, innovation, exploration, discovery and substitution may ease the problems of some non-renewable resources for a time, and careful management and sustainable activities may enhance the production of renewable resources and protect rare and endangered wildlife.

There is a certain degree of elasticity of substitution between non-renewable and renewable resources, as when solar energy displaces the use of fossil fuels in home heating and the generation of electricity. The availability of natural resources has often been the key to economic success. Egyptian civilization flourished on the strength of the fertile Nile Valley; Spain flourished on gold from the Americas; Britain thrived on iron ore and coal; New Zealand and Argentina thrived on farming; and the Southern US States prospered on cotton and slaves.

However, many countries today showing striking economic success are notably poor in natural resources, such as Japan, Korea, Taiwan, Denmark and Switzerland. Australia has slipped relative to others in the OECD league notwithstanding impressive natural resources which would otherwise have ranked it as the wealthiest nation in the world.

Other nations, rich in natural resources, such as Mexico, Nigeria, Brazil, Russia and Papua New Guinea, have relatively poor standards of living for the masses (though not for the few). There are clearly many more factors involved in economic success than the immediate availability of natural resources, such as a prevailing political context of peace and stability, a climate of incentive, the development of skills and talents, the ease of movement of capital and labour, the financial infrastructure, the ease of trade and transportation over long distance, the accessibility of markets, and the energy released in countries that are not well-endowed but have an overarching determination to succeed.

Box 4.8 presents a classification of natural resources, while Box 4.9 quotes from Garrett Hardin's famous text on the *Tragedy of the Commons*. To illustrate the exhaustion of resources, Box 4.10 cites the case of Nauru which exhausted its resource base while maintaining a high GDP for many years.

Figure 4.1 illustrates the McKelvey scheme, a classification system for solid non-renewable resources, developed by the US Geological Survey and by the US Bureau of Mines. It is named after Dr V.F. McKelvey, director of the Geological Survey. It defines categories of resources in general terms of certainty of occurrence and economic feasibility of extraction. The McKelvey scheme illustrates in summary form the current status of our knowledge of a particular resource, and can serve as a general guide in determining which categories are most likely to satisfy future requirements. A major deficiency of the classification is that it conveys no impression of the availability or likely availability of resources

Box 4.8 A Classification of Natural Resources

Renewable resources	*Non-renewable resources*
Pasture grasses	Petroleum
Agricultural crops	Oil shale
Animal forage	Natural gas
Forest crops	Coal
Livestock	Peat
Harvested wild animals	Uranium
Trees and timber	Iron ore
Fish	Bauxite
Birds	Gold and silver
Flowers	Copper
Insects	Chromium
Air	Zinc
Water (surface water and groundwater)	Cobalt
Biomass	Lead
Wood fuel	Manganese
Solar energy	Mercury
Geothermal energy	Molybdenum
Hot rock systems	Nickel
Hydroelectricity	Platinum
Magma bodies	Tungsten
Ocean thermal energy	Tin
Tidal energy	Many other minerals
Wave energy	
Wind energy	

Box 4.9 The Tragedy of the Commons

The tragedy of the commons develops in this way. Picture a pasture open to all. It is to be expected that each herdsman will try to keep as many cattle as possible on the commons. Such an arrangement may work reasonably satisfactorily for centuries because tribal wars, poaching, and disease keep the numbers of both man and beast well below the carrying capacity of the land. Finally, however, comes the day of reckoning, that is, the day when the long-desired goal of social stability becomes a reality. At this point, the inherent logic of the commons remorselessly generates tragedy.

As a rational being, each herdsman seeks to maximize his gain ... the rational herdsman concludes that the only sensible course for him to pursue is to add another animal to his herd. And another, and another ... but this is the conclusion reached by each and every rational herdsman sharing a common. Therein is the tragedy. Each man is locked into a system that compels him to increase his herd without limit ... in a world that is limited.

Ruin is the destination toward which all men rush, each pursuing his own best interest in a society that believes in the freedom of the commons. Freedom in a commons brings ruin to all.

Source: G. Hardin (1968) The tragedy of the commons, Science, 162: 1243–8.

Box 4.10 Nauru: The Exhaustion of the Resource Base

Nauru is an island republic in the south-western Pacific Ocean, with an area of 21 km^2 and a population of a little more than 9000. Nauru's economy has been almost exclusively based on the mining, processing and export of phosphate, the island being covered with beds of phosphate rock that are derived from rich deposits of guano, the excrement of sea birds. The country's gross domestic product (GDP) has been the highest in the Pacific and among the highest world-wide. By the mid-1990s the phosphate deposits had been virtually exhausted, while the reclamation of land for agricultural purposes lagged behind. Nauru has been left as a moonscape after decades of mining. In 1993, Australia, Britain and New Zealand agreed on a compensation package, payment being made over 20 years. The Nauru Government has committed itself to a rehabilitation programme to 'recreate the Garden of Eden that was once Nauru'. Nauru launched a law suit against Australia in the International Court of Justice, seeking further, massive compensation. Australia settled out of court in 1993.

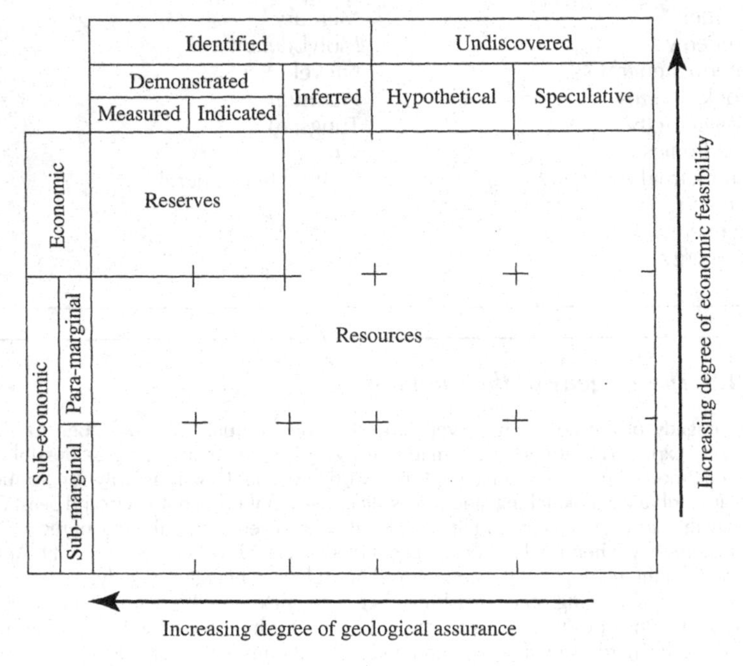

Figure 4.1 *The McKelvey classification*

through time; the figures represent a judgement at a particular time and in terms of current economics of a resource or reserve. A lack of appreciation of the nature of reserves and resources has led some writers into error. Box 4.11 gives the predictions of Jevons.

Box 4.11 *Jevons on the British Coal Crisis, 1865*

I draw the conclusion that I think anyone would draw, that we cannot long maintain our present rate of increase of consumption . . . the check to our progress must become perceptible considerably within a century from the present time; that the cost of fuel must rise, perhaps within a lifetime, to a rate threatening our commercial and manufacturing supremacy; and the conclusion is inevitable, that our happy progressive condition is a thing of limited duration.

Britain's supremacy did not last; but the price of coal had nothing to do with it, while energy consumption increased exponentially.

Source: William Stanley Jevons (1835–82) in his book The Coal Question (1865).

4.9 National Income Accounts and Natural Resource Accounting

The national income accounts, published by all countries under the guidance of the United Nations, represent an attempt to measure the income and production of each country. The most commonly used indicator of national output has been the gross domestic product (GDP), which is a measure of the total market value of currently produced finished goods and the value of services rendered. While there are statistical difficulties, the result is a fair representation of the output of a country. Some kinds of activities do fall beyond its scope, such as many domestic services, the informal economy, losses to the resource base, and any adverse effects on the environment or the prospects for sustainable development. Still less does it measure at all accurately that more subtle concept, the quality of life. However, the rule that only finished or final goods are to be counted, avoids double or treble counting of raw materials, and intermediate products. For example, the final value of automobiles already includes the value of the steel, glass, rubber, plastics, electronics and other components used to make them. National income accounting remains an inexact science, yet it remains an indispensable tool for national economic planning and a basis of continuous debate. It remains one of the most significant inventions of the 20th century.

While the aim of national income accounting is to provide a framework for the analysis of the underlying economic system and its performance, there is a marked asymmetry in respect of the value of natural resources. Human-made assets such as buildings and equipment are valued as productive assets and are written off against the value of production as they depreciate. Natural resources are not so valued, and their loss entails no charges against income that reflect a decrease in potential future production.

The failure of national income accounts to measure the depletion of natural resources became a central issue at the UN Conference on Environment and Development 1992. Agenda 21, adopted by that conference, included a programme to establish systems for integrated environmental and economic accounting. In effect, the UN System of National Accounts (SNA), dating back to 1968, was to be modified in some way to take account of losses in the resource base on which ultimately all economic activity depended. A range of workshops, jointly sponsored by the World Bank and UNEP, have revealed a great range of opinions as to how this integration might be achieved.

During recent years, proposals have been made to modify the national accounting system with regard to environmental factors. However, the majority of national accounts experts have rejected the possibilities of substantial changes to the conventional national accounts. Instead, it was decided to establish a satellite system outside the traditional framework of the national accounts for the purpose of describing environmental–economic relations (United Nations 1993). A satellite system would enable a broader range of data to be incorporated. Close connections would remain between the traditional economic accounting system and the new satellite system. The focus would be on national data.

Natural resource accounting as now envisaged deals with stocks and changes in stocks of natural assets. These natural assets comprise biological assets such as living plants and animals of economic importance, subsoil assets such as proved reserves, air, land, soil and water, including territorial and aquatic ecosystems. In natural resource accounting, measurements in both physical and monetary units are necessary, comprising flow as well as asset accounts.

Practical experience on how to integrate existing national accounts with environmental accounts has been gained through pilot studies carried out by the UN Secretariat and the World Bank in Mexico, Papua New Guinea and Thailand.

In addition, a number of developed countries, including France, Norway, Canada, Japan, the Netherlands, Australia and the USA, have proposed or set systems of environmental accounts. Natural resources take priority in the accounts of France and Norway, while pollution and environmental quality are the central issues in Japan and the USA. The approaches of Canada and the Netherlands lie somewhat in between. See Box 4.14 for comments on the Australian approach, in which caution is urged.

In both France and Norway, systems of resource accounting have been adopted as a supplement to, not a replacement for, the system of national accounts. The French experience has been documented by J. Theys in Ahmad *et al.* (1989), and by P. Cornière (1986). The Norwegian system has been described by the Norwegian Central Bureau of Statistics (1987). Both the French and Norwegian systems are also described in Pearce *et al.* (1989).

The French natural patrimony accounts are set up in physical units appropriate to each type of resource, generally without attempting to assign values. The Norwegian system includes natural resource accounts for the most important energy sources, such as coal, petroleum, natural gas, timber and hydroelectricity, measured in terms of their physical units. No attempt has been made to assign values to these physical resources.

While a purely physical accounting process has considerable shortcomings, it does enable decision-makers to consider the impact of major policy decisions on the national stock of natural resources. However, as Repetto *et al.* (1989) argue, it remains one crucial step away from adjusting national product with a depletion allowance reflecting the use of, or extraction of, natural resources previously accounted for as the free gifts of nature. It is acknowledged, however, that there are limits to monetary valuation, set mainly by the remoteness of the resource in question from the market economy. Some resources such as minerals and subsurface water may be readily evaluated, but others, such as wild animals, do not contribute directly to production and can be valued in monetary terms only through quite roundabout methods. While the faults are obvious, the proposed solutions are many; there has been little consensus.

Eventually, a credible standard valuation technique to measure depletions should be adopted and applied to different resources, across a variety of countries. The work undertaken by the UN (1993) represents a preliminary but vital step in this direction.

Despite the progress made, many environmentalists still want to go further with an estimated single measure of the effect of environmental damage on economic growth. This goal of a green GDP remains a beacon, a light on the hill, but many statisticians are concluding that it is virtually unattainable in the sense of allocating values to such things as fine public parks, rare and endangered species, and vanishing forests. Marginal values are even more difficult to assess, for the loss of a total asset may be calamitous, but the loss of a small amount may be negligible. Putting monetary values on environmental assets remains something of a black art (*The Economist* 18 April 1998).

The general trend then has been to relate economic activity, measured in monetary terms, to environmental assets measured in physical units. This does not meet the objectives of environmentalists who have the support of the World Bank.

Box 4.12, quoting from *Our Common Future*, the report of the World Commission on Environment and Development (1987), restates the case for green accounting, and Box 4.13 reinforces the argument (Repetto *et al.* 1992). In Box 4.14, McCarthy (1993) reflects a sense of reservation in Australia, while Box 4.15 states the current UN position on integrated environmental and economic accounting.

4.10 Summary

This chapter emphasizes the breadth of true development, as distinct from growth alone which can be solely materialistic, narrow, distorting, and damaging to social and environmental considerations. True development spreads out the benefits of progress, reducing its more harmful effects. It brings the good news to all.

Defining 'sustainable development' is, however, much more difficult, even taking into account the efforts of the World Commission on Environment and

Box 4.12 The World Commission on Natural Resources

... income from forestry operations is conventionally measured in terms of the value of timber and other products extracted, minus the costs of extraction. The costs of regenerating the forest are not taken into account, unless money is actually spent on the work. Thus figuring profits from logging rarely takes full account of the losses in future revenue incurred through degradation of the forest ... similarly incomplete accounting occurs in the exploitation of other natural resources, especially in the case of resources that are not capitalized in enterprise or national accounts: air, water and soil. In all countries, rich and poor, economic development must take full account in its measurements of growth of the improvement or deterioration in the stock of natural resources.

Source: World Commission on Environment and Development (1987) Our Common Future, Oxford University Press, Oxford, p. 52.

Box 4.13 Natural Resources Ignored in Most National Accounts

A country could exhaust its mineral resources, cut down its forests, erode its soils, pollute its aquifers, and hunt its wildlife and fisheries to extinction, but measured income would not be affected as these assets disappeared. Ironically, low-income countries, which are typically most dependent on natural resources for employment, revenues, and foreign exchange earnings are instructed to use a system for national accounting and macroeconomic analysis that almost completely ignores their principal assets ... Codified in the UN system of national accounts closely followed by most countries, this difference in the treatment of natural resources and other tangible assets provides false signals to policymakers ... it confuses the depreciation of valuable assets with the generation of income ... the results can be illusory gains in income and permanent losses in wealth.

Source: R. Repetto, W. Macgrath, M. Wells, C. Beer and F. Rossini (1992) Wasting assets: natural resources in the national income accounts, The Earthscan Reader in Environmental Economics (eds A. Markandya and J. Richardson), Earthscan, London.

Box 4.14 The Australian National Accounts

Ultimately, the Australian Bureau of Statistics intends to move down the path suggested by the Revised System of National Accounts, that is, to develop satellite accounts enabling the derivation of environmentally adjusted GDP ... However we are concerned that unrealistic expectations are held in some quarters regarding the efficacy of the measures proposed, the extent of the data base required, the accuracy of any values derived given that they cannot be based on observed market values, and the timetable that can be realistically achieved.

Source: P. McCarthy (1993) Problems and Prospects for Green Gross Domestic Product (GDP) in the Australian National Accounts, Environmental Economics Conference, Department of the Environment, Sports and Territories, Canberra.

Box 4.15 Integrated Environmental and Economic Accounting

In the past few years, an increasing number of scientists have called for a synthesis of the ecological and anthropocentric points of view. The exploitation of nature for economic purposes has reached its limit . . . the benefits of some types of environmental use have resulted in disbenefits with respect to other competing uses . . . the exploitation of nature has reached a point where human beings are impairing their own living conditions.

The short-term exploitation of environmental resources should be replaced by a long-term concept of preserving the environment for both human and natural needs . . . Not only the functions of the environment for human use, but the environment itself, should be kept intact, even if there might not be an apparent human use for it. An intact environment is considered to be part of the natural patrimony and might very well prove to be a precondition of human survival. These considerations affect the design of an integrated system for environmental and economic accounting.

Source: UN Department of Economic and Social Information and Policy Analysis Handbook of National Accounting (1993) Integrated Environmental and Economic Accounting, Series F, No. 61, UN New York, pp. 2–3.

Development and the World Bank. Sustainability, like apple pie and motherhood, is everyone's favourite. Business people are very keen on sustainable profits, sustainable growth, and sustainable investments; academics may search for sustainable arguments; designers may seek sustainable performances. In the end, sustainability simply means keeping things going for a long period of time, perhaps indefinitely.

The concepts of sustainable agriculture, sustainable forestry and sustainable fisheries also reflect the idea of permanence. A sustainable society is one that lives on the interest of its activities and not on the underlying capital. The chapter reflects on the characteristics of a sustainable society, and introduces the Hartwick rule.

The chapter examines the role of property rights in the efficient allocation of resources, but does not find the magic often attributed to them by writers. Property rights are only of social benefit when tied to responsibilities.

The thorny issue of intragenerational equity must be approached, albeit cautiously, for it stands at the core of history and turns on the essential unfairness of most things human. However, it has a role in environmental impact assessment.

However, intergenerational equity receives rougher treatment. It has been argued by many writers that we of this generation should compensate future generations for the resources we use today. Investments of funds, growing at compound rates of interest, have been suggested; while deliberately keeping resources in the ground has been a proferred alternative. This chapter rejects these proposals, arguing that the present owes the future nothing. First, the needs of future generations cannot be predicted. Secondly, the future automatically inherits the scientific and technical achievements of the present. What we achieve now adds to the inheritance of future generations. They need nothing more. However, for the benefit of the three generations alive today we need to manage this small planet better.

The chapter concludes by examining the polluter-pays principle, the internalization of environmental costs, the precautionary approach, the nature of natural resources and natural resource accounting.

Further Reading

Ahmad, Y.J., El Serafy, S. and Lutz, E. (1989) *Environmental Accounting for Sustainable Development*, A UNEP–World Bank Symposium, World Bank, Washington, DC.

Barrett, S. (1993) *The Theory of Property Rights: Transboundary Resources*, Beijer International Institute of Ecological Economics, Beijer Discussion Paper Series No. 44, Stockholm.

Bartelmus, P., Stahmer, C. and Tongeren, J. van (1991) Integrated environmental accounting framework for an SNA satellite system, *The Review of Income and Wealth*, **37**(2): 111–48.

Cameron, J. and Aboucher, J. (1991) The precautionary principle: a fundamental principle of law and policy for the protection of the global environment, *Boston College International and Comparative Law Review*, **14**: 1–27.

Common, M.S. (1995) *Sustainability and Policy: Limits to Economics*, Cambridge University Press, Cambridge.

Hanna, S. and Munasinghe, M. (eds) (1995) *Property Rights and the Environment: Social and Ecological Issues*, The Beijer International Institute of Ecological Economics and the World Bank, Washington, DC.

Kirkby, J., O'Keefe, P. and Timberlake, L. (eds) (1995) *The Earthscan Reader in Sustainable Development*, Earthscan, London.

Lutz, E. and El Sarafy, S. (1988) *Environmental and Resource Accounting: An Overview*, Working Paper No. 6, World Bank, Washington, DC.

Munasinghe, M. and Shearer, W. (eds) (1995) *Defining and Measuring Sustainability: The Biogeophysical Foundations*, UN University and the World Bank, Washington, DC.

OECD (1997) *Sustainable Development: OECD Policy Approaches for the 21st Century*, OECD, Paris.

Pearce, D.W. and Warford, J.J. (1993) World without End: Economics, Environment and Sustainable Development, Oxford University Press, New York.

Pejovich, S. (ed.) (1997) *The Economic Foundations of Property Rights: Selected Readings*, A and M University, Texas.

Repetto, R., Magrath, W., Wells, M., Beer, C. and Rossini, F. (1989) *Wasting Assets: Natural Resources in the National Income Accounts*, World Resources Institute, Washington, DC.

Rowley, C.K. (ed.) (1993) *Property Rights and the Limits to Democracy*, The Locke Institute and George Mason University, US.

UN (1993) *Handbook of National Accounting: Integrated and Environmental and Economic Accounting*, Series F, No. 61, UN, New York.

World Conservation Union, UNEP and WWF for Nature (1991) *Caring for the Earth: A Strategy for Sustainable Living*, WCU, Gland, Switzerland.

5
Direct Regulation

5.1 Introduction

Direct regulation through statutes and subordinate rules has been the normal approach to pollution control and environment protection throughout history. This approach has been known in the USA as 'command and control'. All legislation, regulations and ordinances in the Western world have been of this type. This approach embraces many aspects of policy, touching on environmental management, such as pollution control (air and water pollution control and noise abatement), land management and contaminated sites, town planning and zoning, pesticides and hazardous chemicals, product performance standards, river and catchment management, development consents, environmental impact assessment, the translation of international conventions, marine management, health and safety at work, conservation and heritage programmes, radioactive substances, habitat protection, local and regional structure plans, toxic substances control, ocean dumping, sewage treatment, drinking water quality, and management of wetlands, mangroves, rain forest, reefs, swamps, areas of high conservation value, sand dunes, endangered species, coastal zones and recreational values. Few writers have direct experience of more than one of these subjects, and some writers have no experience of environmental management whatever.

Box 5.1 outlines the characteristics of a direct control system, while Box 5.2 stipulates the duties and functions of the US Environmental Protection Agency. Figure 5.1 illustrates some of the OECD achievements in waste-water treatment and air pollution emissions over the last 15 years, using direct regulation.

Direct regulation involves the setting of standards or objectives as indicated in Box 5.1. Ambient quality standards are concerned with the protection of the receiving environment and may be regarded as social or community aims; they do not indicate what the performance should be for individual emitters. This is the task of emission standards, which restrict emissions from point sources by weight or concentration. Other standards relate to processes, products and environmental management systems (see Box 5.3).

In respect of pollution control, best-practice concepts have emerged over time, the most famous of which is the adoption of the *best practicable means*, a concept which may be found in British legislation of the 19th century. Alternative specifications are to be found in Box 5.4.

Box 5.1 Characteristics of a Direct Control (Command and Control) System within Government

- The passing of environmental laws and regulations by government at national, state and local level
- The creation of an Environmental Protection Agency to implement the legislation
- The distribution of various duties among other branches of government
- The drafting and introduction of codes and regulations relating to pollution control
- The drafting and introduction of codes relating to environmental planning
- The drafting and introduction of codes implementing international conventions
- The formulation of procedures for environmental impact assessment at policy, programme and project levels
- Procedures for the establishment and maintenance of national and marine parks, nature reserves and public open space
- The protection of rare and endangered species
- The identification and preservation of places, buildings and assets of heritage value
- The establishment of procedures for the conduct of public inquiries in appropriate instances
- The enforcement of laws and regulations relating to pollution of air, water and land; excessive noise; environmentally hazardous chemicals; contaminated sites; nuisances; noxious accumulations; and odours
- The creation of systems for the prior approval of proposed developments, and the licensing of discharges
- The creation and maintenance of adequate monitoring and recording systems
- The creation of environmental plans at national or state, regional and local level to promote good town and regional planning and contribute to the strategic control of pollution
- The promotion of the aims of natural resource conservation, recycling, cleaner production, efficient waste disposal, litter reduction, and beautification
- The establishment of policies and plans to integrate economic, social and environmental progress and further the objectives of sustainable development

Box 5.2 Duties of the US Environment Protection Agency

- To protect and enhance the environment to the fullest extent possible under the laws enacted by the US Congress
- To mount an integrated, co-ordinated attack on environmental pollution in collaboration with state and local governments
- The development of national programmes and regulations for air pollution control; national standards for air quality; emission standards for stationary sources; emission standards for hazardous pollutants; the Acid Rain Program; the Emissions Trading Program; and field training
- The study, identification and regulation of noise sources; and radiation protection programmes

Continued on page 115

Continued from page 114

- The development of national programmes and regulations for water pollution control and water supply; water supply quality standards and effluent guidelines; and field training
- Guidelines and standards for the land disposal of hazardous wastes; assistance in the operation of waste management facilities; the recovery of energy from solid waste; the administration of Superfund; and ocean dumping
- The development of national strategies for the control of toxic substances; assessing the impact of new chemicals and chemicals with new uses
- The control and regulation of pesticides; tolerance levels for pesticides in relation to food, fish and wildlife; and the investigation of pesticide accidents
- Enforcement procedures in relation to control programmes in air, water (including groundwater), toxic substances, solid waste management, radiation and noise
- The conduct of conferences, hearings and other legal proceedings
- The conduct of national research programmes
- Environmental impact assessment under the National Environmental Policy Act 1969
- Advice to the US Council on Environmental Quality

The Agency was established in 1970, following the amalgamation of 15 separate agencies. It has a staff of some 7000 people.

The costs of pollution control vary considerably according to the choice of plant, the source of energy, and other process characteristics. Typically, pollution control costs as a percentage of total plant and equipment costs, for industries in Europe and North America, have been as follows: iron and steel processing,

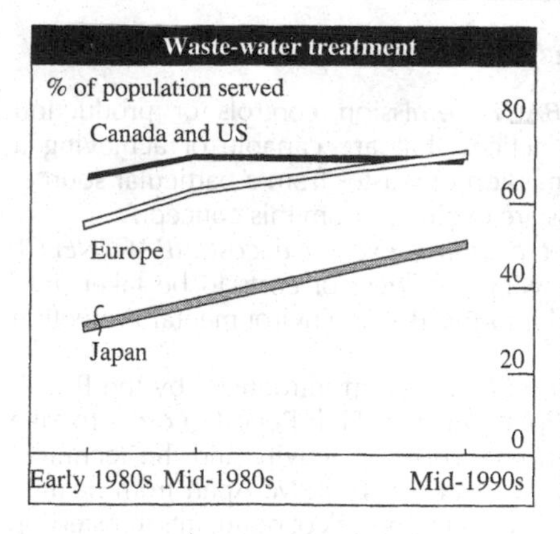

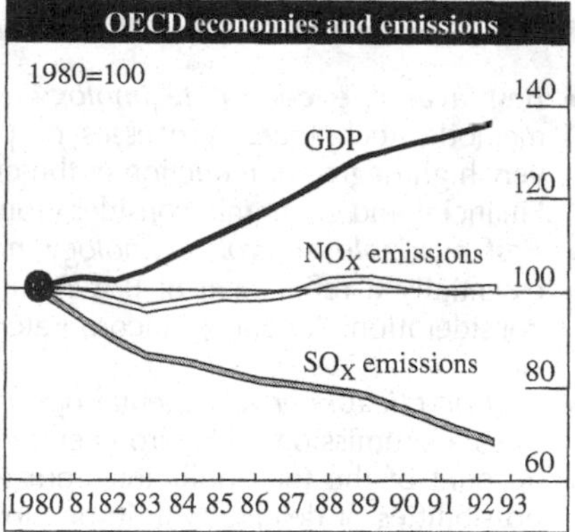

Figure 5.1 *OECD achievements in waste-water treatment and air pollution emissions. Reproduced by permission from OECD (1995)*

Box 5.3 Environmental Standards

- *Ambient quality standards*: environmental objectives in respect of the receiving environment, e.g. the maximum concentration of sulphur dioxide in the atmosphere, the maximum concentration of nitrates in drinking water, or the maximum noise levels in residential districts.
- *Emission standards*: the maximum allowable discharges of pollutants into the environment from fixed points, e.g. the maximum emissions that may be discharged into the atmosphere from industrial stacks measured in weight per unit time or concentration, or the maximum biochemical oxygen demand (BOD) that may be discharged into water from a sewer or waste pipe.
- *Process standards*: standards that govern processes of production restricting emissions in general within industrial premises, e.g. restrictions on the emission of fluorides from aluminium pot plants, or the specification of stages in effluent treatment.
- *Product standards*: standards specifying the characteristics of potentially polluting products, e.g. detergents, fertilizers, chemicals, insecticides, fuels, motor vehicles, refrigerators, fire suppressants, lawn-mowers, pneumatic drills, air conditioners, aircraft, paints and containers.
- *International standards*: environmental standards progressively introduced at an international level promoting satisfactory environmental management, e.g. in 1994 the International Organization for Standardization (ISO) issued ISO 14000, a set of environmental management guidelines. Concurrently, the British Standards Institution issued BS 7750, also on environmental management systems; while the European Union has issued an eco-management and audit regulation. Environmental audits are common to these standards. The World Trade Organization promotes the adoption of international environmental standards.

Box 5.4 Pollution Control: Best-Practice Concepts

- *Best available control technology (BACT)*: emission controls or production methods, techniques, processes or practices that are capable of achieving a very high degree of reduction in the emission of wastes from a particular source. Financial and economic considerations are excluded from this concept.
- *Best available control technology not entailing excessive costs (BACTNEEC)*: essentially a refinement of BACT, allowing the level of costs to be taken into consideration. A concept incorporated into the British Environmental Protection Act.
- *Best practicable environmental option (BPEO)*: a term introduced by the British Royal Commission on Environmental Pollution in its Fifth Report in order to take account of the total pollution from an enterprise or activity and the technical possibilities of dealing with it; possibly a successor to BPM. Apart from dealing with all pollutants, BPEO was to take into account the risk of pollutants transferring from one medium to another.

Continued on page 117

Continued from page 116

- *Best practicable means (BPM)*: a commonly used approach to pollution control requirements from industrial and other premises. The word 'practicable' is taken to mean 'reasonably practicable having regard, among other things, to the state of technology, to local conditions and circumstances, and the financial implications of the measures'. The concept is much easier to apply than the ambient quality approach.
- *Good control practice (GCP)*: the application of established pollution control methods, or practices, that a control agency considers to be sufficient in order to achieve an adequate degree of reduction and subsequent dispersion of wastes emitted from a particular source. GCP might be appropriate in a relatively isolated situation.
- *Maximum achievable control technology (MACT)*: emission control requirements that are more exacting than those imposed by BACT. This higher standard may involve the application of new, original or innovative control technology to emission sources, almost regardless of cost.

Box 5.5 A Judgement on 'Command and Control' (CAC)

The term "command-and-control" encompasses a very broad and diverse set of regulatory techniques–some admittedly quite crude and excessively costly. But others are far more sophisticated and cost sensitive ... it can be quite misleading to lump together in a cavalier manner CAC methods of regulatory control and to contrast them as a class with systems of economic incentives ... at most points in the area, environmental quality will be higher under a CAC system than under the least-cost solution.

Source: M.L. Cropper and W.E. Oates (1992) Environmental economics: a survey, Journal of Economic Literature, ***XXX****: 675–740.*

20%; non-ferrous metals, 12%; electricity generating plant, 10%, more with gas scrubbing. Other costs are also incurred in the acquisition of buffer zones, landscaping, environmental impact statements, delays and operating requirements and maintenance (Gilpin 1990).

In Box 5.5, Cropper and Oates (1992) express an opinion on direct regulation, countering the claim that economic incentives produce least-cost solutions for equal environmental achievement.

5.2 Direct Regulation (Command and Control)

Boxes 5.6–5.8 outline the history of environmental legislation in Britain, the USA and Australia. At first it was fragmentary and limited in scope; it was not until after the Second World War that more far-reaching measures were adopted in respect of air and water pollution, insecticides, town planning, environmental

Box 5.6 Britain: Evolution of Environmental Policy

1273 Use of coal prohibited in London as being 'prejudicial to health'
1306 Royal Proclamation prohibits use of sea-coal (coal by sea from Newcastle) in furnaces
1648 Londoners present petition to Parliament to prohibit importation of coal from Newcastle, because of its ill-effects
1661 John Evelyn presents his thesis on London smog to King Charles II
1798 Malthus' famous *Essay on the Principles of Population*
1817 Birmingham introduces first smoke-control measure
1848 Public Health Act prohibits black smoke
1863 Alkali Works Act to control industrial pollution from certain industries
1875 Public Health Act: the Great Charter
1903 Letchworth, the first garden city founded
1905 Lincoln: first use of chlorination to check typhoid epidemic
1906 Alkali Works Act, to extend the 1863 Act
1909 Town Planning Act
1922 Carr-Saunders report on *The Population Problem*
1932 Town and Country Planning Act
1936 Public Health Act; Housing Act
1937 Food and Drugs Act
1940 The Barlow Report on the geographical distribution of population
1942 The Beveridge Report on Social Insurance; the Uthwatt Report on compensation and betterment
1943 The Abercrombie Plans for London
1945 Water Act; Distribution of Industry Act
1946 New Towns Act; new towns programme begins
1947 Town and Country Planning Act
1948 Rivers Boards Act
1949 National Parks Act; Nature Conservancy created
1951 Rivers (Prevention of Pollution) Act; Dartmoor National Park and Lake District National Park established
1952 Town Development Act; London smog disaster, 4000 died
1954 The Beaver Report on the London smog disaster
1956 The Clean Air Act; the first smoke control order, West Bromwich
1957 Windscale nuclear reactor incident
1960 Clean Rivers Act; Noise Abatement Act; Radioactive Substances Act
1961 Pippard Report on pollution of the tidal Thames; River Thames clean-up begins; Holme Pierrepont public inquiry, following which a development application for a 2000 MW power station was rejected
1962 Ratcliffe-on-Soar power station public inquiry
1964 Harbours Act
1965 National Environmental Research Council established
1966 The Aberfan disaster, 144 died, of which 116 were children
1967 Civic Amenities Act; *Torrey Canyon* oil spill
1968 Town and Country Planning Act; Clean Air Act

Continued on page 119

Continued from page 118

1970	Royal Commission on Environmental Pollution created; Department of Environment established
1971	Town and Country Planning Act; Oil Pollution Act; Roskill Commission reports on future London airpórt; International Institute of Environment and Development established
1972	Poisonous Waste Act; Local Government Act
1973	Land Compensation Act; Water Act; Nature Conservancy Council established
1974	Control of Pollution Act; Health and Safety at Work Act; Dumping at Sea Act
1975	Dobrey Report on the British development control system
1976	Greater London Development Plan
1978	Inner Urban Areas Act; Parker Windscale inquiry; Commission on Energy and the Environment
1980	Highways Act
1983	Conservation and Development Programme
1984	Survey of derelict land
1985	Britain adopts European Directive on Environmental Impact Assessment
1988	Town and Country Planning (Assessment of Environmental Effects) Regulations
1989	Electricity Act; London Conference on Climatic Change
1990	Town and Country Planning Act
1991	Environmental Protection Act; Planning and Compensation Act; *Guide to the Environmental Implications of Government Policies* published by the Department of the Environment, UK; integrated pollution control introduced
1992	Town and Country Planning (Assessment of Environmental Effects) Regulations; Harbour Works (Assessment of Environmental Effects) Regulations; EIA guides published by Kent and Essex County Councils, the Passenger Transport Executive Group, and the Department of the Environment; British Report to the UN Conference on Environment and Development
1993	Clean Air Act; Radioactive Substances Act; British Standard 7750 on Environmental Management Systems
1994	Environment Agency for England and Wales established; National Environmental Technology Centre created; Conservation (Natural Habitats) Regulations; Strategy on Sustainable Development
1996	Environment Agency took over responsibilities formerly exercised by the National Rivers Authority, the Inspectorate of Pollution, and local authority waste inspectors; Landfill Tax, imposing charges for material dumped into landfill sites; *Sea Empress* runs aground at the entrance to Milford Haven harbour, Wales, creating the worst oil spill since the *Torrey Canyon* disaster of 1967.
1997	Under the Kyoto Convention on Climate Change, Britain is committed to reduce its production of six greenhouse gases by 12.5% of 1990 emissions by 2010; the Royal Commission on Environmental Pollution suggested dramatic increases in energy prices and significant improvements in the efficiencies of car engines
1998	The Advisory Committee on Business and the Environment urges the British Government to consider a levy on carbon dioxide emissions. Energy taxes and emissions trading being considered

Box 5.7 USA: Evolution of Environmental Policy

1872 Creation of Yellowstone National Park
1890 Creation of Yosemite National Park
1899 Refuse Act
1908 State of the Union Message by President Theodore Roosevelt on the need for the conservation of natural resources
1935 Historic Sites, Buildings and Antiquities Act
1940 Bald and Golden Eagle Protection Act
1946 Start of Pittsburgh clean-up
1947 Los Angeles anti-smog programme launched; Federal Insecticide, Fungicide and Rodenticide Act
1948 Federal Water Pollution Control Act
1956 Water Pollution Control Act re-enacted as a permanent measure; Fish and Wildlife Act
1962 *Silent Spring* published by Rachel Carson; White House conservation conference
1963 Federal Clean Air Act
1964 Wilderness Act; national wilderness preservation system
1965 Federal Water Quality Act; Solid Waste Disposal Act; Further Clean Air Act; Anadromous Fish Conservation Act
1966 Environmental impact assessment policy established; Clean Water Restoration Act; National Historic Preservation Act
1967 Federal Air Quality Act
1968 National Trails System Act; Wild and Scenic Rivers Act
1969 National Environmental Policy Act; Council on Environmental Quality appointed
1970 US Environmental Protection Agency (EPA) created; Environmental Quality Improvement Act; Water Quality Improvement Act; Clean Air Amendment Act; Mining and Minerals Policy Act; President's Message on the Environment
1972 Noise Control Act; Coastal Zone Management Act; Clean Water Act; Marine Mammal Protection Act
1973 Endangered Species Act
1974 Safe Drinking Water Act; Solar Energy Research, Development and Demonstration Act
1975 Energy Policy and Conservation Act
1976 Toxic Substances Control Act; Resource Conservation and Recovery Act; Magnusan Fishery Conservation and Management Act; National Forest Management Act; Federal Land Policy and Management Act
1977 Environmental impact procedures strengthened; Clean Water Act; Clean Air Amendment Act; Surface Mining Control and Reclamation Act; Soil and Water Resources Conservation Act; President's environmental message
1978 Environmental impact assessment regulations promulgated; Renewable Resources Extension Act; Public Rangelands Improvement Act; Surface Mining Control and Reclamation Act; Cooperative Forestry Assistance Act; National Energy Conservation Policy Act; Solar Photovoltaic Research, Development and Demonstration Act; Uranium Mill Tailings Radiation Control Act; National Ocean Pollution Planning Act

Continued on page 121

Continued from page 120

1979 Introduction by the US Environmental Protection Agency of the 'bubble concept' for the management of pollution
1980 Alaska National Interest Lands Conservation Act; Comprehensive Environmental Response, Compensation and Liability Act (Superfund); Wind Energy Systems Act; Low-level Radioactive Waste Policy Act; Act to Prevent Pollution from Ships
1982 Coastal Barrier Resources Act; Reclamation Reform Act; asbestos-in-schools rule; Nuclear Waste Policy Act
1983 Times Beach found too contaminated with dioxins for human habitation
1985 International Security and Development Act
1986 Emergency Wetlands Resources Act; Right-to-Know Act
1987 Water Quality Act; Driftnet Impact Monitoring, Assessment and Control Act
1988 Ocean Dumping Ban Act
1989 First nationwide survey of more than 320 toxic chemicals released to air by industry; tanker *Exxon Valdez* goes aground in Alaska; North American Wetlands Conservation Act; Marine Pollution and Research and Control Act
1990 Clean Air Act to reduce air emissions substantially; California Air Resources Board introduces strictest vehicle-emission controls ever; Coastal Wetlands Planning, Protection and Restoration Act; Coastal Barrier Improvement Act; Oil Pollution Act; Food Security Act; Pollution Prevention Act; Antarctic Protection Act; Global Change Research Act; Food, Agriculture, Conservation and Trade Act; National Environmental Education Act
1991 US signs UNECE convention on environmental impact assessment in a transboundary context
1992 US Congress requires the analysis of the environmental effects of major US federal activities abroad; National Geologic Mapping Act
1993 California Desert Protection Act
1997 Regulatory Improvement Act; new ambient air quality standards; US Senate unanimously passes the Byrd-Hagel Resolution asking that the US not sign a protocol with legally binding emissions limitations unless developing countries agree to binding limitations within the same time frame
1998 The US President announces a five-year multi-billion dollar package of tax incentives and research programmes to reduce emissions and address climate change

Source: Adapted from Gilpin (1996).

impact assessment, public inquiries, catchments and endangered species, culminating in a range of UN conferences such as the UN Conference on the Human Environment in 1972 and the UN Conference on Environment and Development in 1992. This period saw the setting up of many national environmental agencies and a proliferation of legislation. At the same time, graduate training in the environmental sciences and environmental law came into being in many universities throughout the world. Media attention on oil spills, greenhouse gases, nuclear incidents, woodchipping, smogs, threats to natural resources, population growth,

Box 5.8 Australia: Evolution of environmental policy

1879	Royal National Park established, Sydney
1958–64	Australian states introduce clean air legislation
1964–79	Australian states introduce solid- and liquid-waste management and anti-litter legislation
1970	New South Wales (NSW) establishes the State Pollution Control Commission; Victoria (Vic) establishes the Environment Protection Authority and the Land Conservation Council
1970–71	Australian states introduce water pollution control legislation
1971	Australian Environment Council established; Western Australia (WA) establishes its Environment Protection Authority
1972	South Australia (SA) establishes its Environment Protection Council
1972–78	Australian states introduce noise control legislation
1973	Tasmania establishes a Department of the Environment
1974	Australian (federal) government introduces EIA legislation
1975	Australian Government creates the Great Barrier Reef Marine Park Authority and the Australian Heritage Commission
1978	Victoria introduces EIA legislation
1979	NSW introduces EIA legislation and creates a Department of Environment and Planning
1980	The Northern Territory establishes a Conservation Commission
1981	SA establishes a Department of Environment and Planning
1983	National conservation strategy adopted by most states; World Heritage Properties Conservation Act
1985	National unleaded petrol (gasoline) programme adopted
1986	WA Environment Protection Act
1987	Murray-Darling Basin Commission created; Ningaloo Marine Park created off WA coast; Queensland creates a Department of Environment and Heritage; Environment Institute of Australia established
1992	Intergovernmental agreement on the environment; national strategy adopted for ecologically sustainable development; NSW creates its Environment Protection Authority
1994	National strategy for the conservation of Australia's biological diversity
1996	National Environment Protection Council formed
1997	NSW passes the following: the Native Vegetation Conservation Act; the Contaminated Land Management Act; the Protection of the Environment Operations Act; and the Environmental Planning and Assessment Amendment Act. Under the Kyoto Convention on Climate Change, Australia negotiates an increase in its carbon dioxide emissions of 8% by 2010. No regrets measures were introduced
1998	Federal government passes Environment Protection and Biodiversity Conservation Act; WA and NSW introduce new environmental laws; Great Australian Bight Marine Park proclaimed; Blue Mountains national park created

threatened wildlife, coastal erosion, marine pollution, transboundary pollution and environmental economics intensified.

Until quite recent times, the whole programme for the protection of the environment, pollution control and town planning rested on a system of legislation, regulations, monitoring, prior approval of plant, prosecutions and fines. In the USA this was known as 'command and control' and elsewhere simply as statutory regulation. The essential characteristics of a direct control system are set out in Boxes 5.1 and 5.2. The ramifications of such a system are much wider than is generally recognized by many writers. Even within the narrower confines of pollution control, the ingredients are not always appreciated.

5.3 Environmental Impact Assessment (EIA)

EIA is the critical appraisal of the likely effects of a policy, plan, programme, project or activity, on the environment. To assist the decision-making authority, assessments are carried out independently of the proponent, who may have prepared an environmental impact statement. The decision-making authority may be a level of government (local, state or federal) or a government agency (at local, state or federal level). Assessments take account of any adverse environmental effects on the community; any environmental impact on the ecosystems of the locality; any diminution of the aesthetic, recreational, aesthetic, scientific or other environmental values of a locality; the endangering of any species of fauna or flora; any adverse effects on any place or building having aesthetic, anthropological, archaeological, cultural, historical, scientific or social significance; any long-term or cumulative effects on the environment; any curtailing of the range of beneficial uses; any environmental problems associated with the disposal of wastes; any implications for natural resources; and the implications for the concept of sustainable development. EIA extends to the entire process from the inception of a proposal to environmental auditing and post-project analysis.

The outcome of an EIA is a recommendation to the decision-maker to either (1) approve the project in its entirety, subject to a range of recommended statutory conditions or requirements; or (2) reject the proposal, outright and completely. Outright rejection is rare, but it does occur, while some proponents withdraw their applications in the face of growing public objection.

The general effect of EIA procedures has been to improve the quality of proposals, rather than to prohibit development. To those committed to zero-economic growth this has been a profound disappointment. EIA is presented as the servant of economic growth and the avenue to resource degradation.

Boxes 5.9–5.12 provide windows into the process of EIA. Box 5.9 summarizes the matters to be taken into account when making a decision on whether a development application is to be accepted. Boxes 5.10 and 5.11 provide examples of the rare instances in which development applications have been rejected. Box 5.10 describes the failure of the Central Electricity Generating Board for

Box 5.9 EIA: Matters for Consideration in Deciding a Development Application

In considering and determining a development application, whether from the public or private sector, a decision-making or approving authority should take into consideration such of the following matters as are of relevance to the proposed development or activity:

1. provisions of the following:
   (a) any environmental planning instrument, national, state, regional, provincial or local plan relevant to any proposed preferred and alternative sites, together with any draft instruments or policies, conservation or management plans;
   (b) any conservation plan creating national parks or reserves, protecting rare or endangered species of flora or fauna, safeguarding elements of biodiversity, preserving heritage assets or natural resources, including mangroves, wetlands, bushland, waterways, coral and marine life;
   (c) any government strategies for the development of roads and transport systems, energy facilities, transmission systems, mining or extracting natural resources and minerals, forestry activities and logging, agriculture, aquaculture and infrastructure;
2. the results of any specific studies such as ecological, social, hazard, health, urban, resource or economic impact assessments;
3. the nature and character of the existing environment and something of its natural or industrial history, the nature and character of alternative sites or techniques which might be available;
4. the nature and character of the proposed development or activity;
5. the certain, likely or possible impact of the proposed development or activity on the environment in the short, medium or long term; the means proposed to control pollution and protect the environment; and the means of monitoring and auditing such measures within an environmental management plan;
6. the environmental consequences of the construction or development phase; and the decommissioning phase;
7. the effect of the proposed development or activity on the landscape or scenic quality of the locality; on any wilderness area or other natural asset;
8. the social effect and the economic effect of the proposal;
9. the physical character, location, siting, bulk, scale, shape, height, density, design and external appearance of the proposal;
10. the size and shape of the land to which the development application relates, its zoning, the siting of any building or facility thereon and the area to be used;
11. whether the land involved is subject to flooding, tidal inundation, subsidence, slip, soil erosion, or has been contaminated by previous activity;
12. the relationship of the proposed site to developments on adjoining land, existing or already approved;
13. whether the proposed means of entrance and exit from the subject land are adequate, and whether ample provision has been made for the loading, unloading, manoeuvring and parking of vehicles within that development;

Continued on page 125

Continued from page 124

14. the amount of traffic likely to be generated by the development or activity, particularly in relation to the capacity of the road system in the locality, and the effect on traffic generally in the regional road system; the alternative of rail transport;
15. whether additional public transport services are needed;
16. whether other infrastructure such as water and sewerage facilities, housing and workforce accommodation, are also needed;
17. whether the developer or proponent has agreed to make financial contributions towards the improvement of public infrastructure requirements;
18. whether adequate provision has been made for the landscaping of the subject land and whether any trees or other vegetation already on the site should be preserved;
19. whether the proposal is likely to cause soil erosion, or chemical contamination;
20. the contribution of the proposal to clean technology, waste minimization, recycling and reuse of materials, and acceptable disposal of unavoidable residues;
21. any regional cumulative effects that may occur;
22. the likely contribution to the greenhouse effect;
23. in respect of mining, the adoption of progressive rehabilitation;
24. the environmental implications of the importation of natural resources to serve the proposed development or activity; and the environmental implications of the use of the final product;
25. any representations by public authorities, corporations, private bodies, commercial interests, local councils, non-government organizations, and individuals or groups of residents, in relation to the proposal, either favouring or opposing;
26. any special circumstances of the case, particularly in the context of the public interest;
27. the proposal in relation to the precautionary principle, polluter-pays principle, intragenerational and intergenerational equity;
28. the proposal within the context of sustainable development.

Box 5.10 Holme Pierrepont, England, 1961

In 1961, in the face of severe electricity shortages, the Central Electricity Generating Board (CEGB) for England and Wales had selected a site at Holm Pierrepont, on the east side of the City of Nottingham, in the East Midlands, for a major power station, the first of 10 such installations. The plant was to be latest design, comprising four 500 MW units. The coal-fired plant would be connected to a single multi-flue stack some 200 m in height, ensuring adequate dispersal of the flue gases during all meteorological conditions. Some 600 tonnes of sulphur dioxide would be dispersed daily, produced from the daily consumption of 20 000 tonnes of coal. Dust-arresting equipment would remove over 99.3% of the fly ash carried forward to the stack. High efflux velocities would be maintained. The development application was placed before the Minister for Power who ordered a public inquiry. The public inquiry was held in Nottingham, and was attended by many objectors and

Continued on page 126

Continued from page 125

representatives of the CEGB. The procedures allowed full legal representation by all parties, with extensive cross-examination of witnesses, a normal British procedure.

Witnesses for the CEGB argued that this excellent plant would have minimal adverse environmental impacts. However, searching questions remained unanswered. What was the ultimate fate of the 600 tonnes of sulphur dioxide released each day? Of the 40 tonnes of dust emitted per day, how much was in the respirable range reaching the lungs of residents? What was the effect of 20 trains a day bringing coal to the power station on the level crossings of the city? How would 4000 tonnes of ash produced per day be adequately handled? How could the CEGB disguise the presence of huge steam generators and eight immense cooling towers, on a flat site open to the city? Witnesses stumbled on some of these questions.

The decision of the Minister of Power stunned the CEGB. The development application had been rejected in its entirety, leaving a 2000 MW hole in the programme of an industry beset with electricity shortages, black-outs and brown-outs. The Board was stunned as the reasons for the rejection were entirely environmental.

The report of the commission of inquiry stressed the openness of the site in relation to the city, the wind directions for most of the year, the uncertainties regarding plume dispersal, the untested claims in respect of dust-arresting equipment, the absence of knowledge regarding the fate of large amounts of sulphur dioxide, the possible effects of warmed water on the river, and the poor record of CEGB in respect of pollution control generally. All these factors mitigated against this development application, notwithstanding the technical merits of the station, being the first power station in Britain to have a high single stack with high-efficiency electrostatic precipitators for dust removal. A dramatic revision of site selection procedures was undertaken. It became essential to select sites more carefully with regard to possible resident reactions, and to research some of the more pressing questions, not yet fully answered. In the end it was concluded that much of Britain's sulphur pollution was reaching Scandinavia.

The CEGB went back to Nottingham quite quickly. The new site at Ratcliffe-on-Soar was to the south of the city, further away and screened by hills. Being less intrusive, it attracted less opposition. The sulphur in the fuel was restricted to 1%, and the coal trains would no longer run through the city of Nottingham. Notwithstanding a long and exhausting public inquiry, the development application was successful without modification. Success rolled on to Fiddlers Ferry, Didcot, and a chain of others.

Box 5.11 Kurnell Peninsula, Sydney, Australia, 1987

In 1987, Bayer Australia Ltd put forward a proposal to develop facilities for the formulation and storage of agricultural and veterinary products at Kurnell Peninsula, Sydney, Australia. A summary of the Commission of Inquiry findings and recommendations to the NSW Minister for Planning is given below.

Continued on page 127

Continued from page 126

This Commission of Inquiry concerns a highly controversial development application by Bayer Australia Limited to develop its land at Kurnell and establish facilities for the formulation of agricultural and veterinary products, the filling of aerosols for house and garden products and the storage of certain pharmaceuticals, imported chemicals and finished products awaiting distribution to customers. Products include insecticides, fungicides, herbicides, cattle dips and sprays, and sheep drench, many of which contain toxic chemicals. It should be noted that this proposal is for the formulation (mixing) of chemicals and not for the manufacture of chemicals, although the prospects of future manufacturing were foreshadowed by Bayer at the Inquiry. The Bayer site has been unused for some time. It was formerly owned by Phillips Australia Chemicals Pty Ltd, who produced carbon black and synthetic rubber there.

The Commissioners have completed a review of all the submissions presented to the Inquiry since conclusion of the public hearings together with subsequent written submissions from parties on conditions that might be attached to consent should it be granted. These submissions on conditions included Bayer's submission setting out its strong objections to a number of environmental controls proposed by the Department of Environment and Planning as a result of the Department's environmental assessment of the project.

As a result of the review of submissions the Commissioners recommend to the Minister that consent to the proposed Bayer project at Kurnell be refused.

In summary, the Commissioners find as follows:

1. We are not satisfied on the evidence put to the Inquiry that the Bayer project would operate without harmful effect on the local environment of the Kurnell Peninsula, in particular, nature reserves, oyster leases, fishing and prawn breeding grounds and wetlands of Botany Bay generally.
2. On the evidence put to the Inquiry considerable uncertainties in present knowledge exist about the impact of toxic chemicals on sensitive marine environments of the kind found at the Botany Bay wetlands.
3. The evidence does not establish with certainty that emissions of toxic chemicals even at low levels from all sources would not have detrimental environmental impact on the Kurnell Peninsula particularly the aquatic environment.
4. The proposals for the containment of surface water run-off and collection on-site do not establish with certainty that contaminated waters would not enter the aquatic environment.
5. Spillages of toxic chemicals are likely to occur on Captain Cook Drive through accidents. Captain Cook Drive is of sub-standard design and construction and in poor condition. It carries considerable traffic, residential, recreational and heavy trucks. Captain Cook Drive runs along the foreshore fringe of the wetlands. Toxic chemical spillages are incapable of containment on Captain Cook Drive and could have disastrous long-term impact on the wetlands of Botany Bay. The Department of Environment and Planning's recommended road improvements for Bayer to undertake would not eliminate this risk on Captain Cook Drive. Bayer objected to undertaking these works. No means have been identified to adequately mitigate such damage, or to reverse the long-term harm to the environment that may occur.
6. Bayer in its submissions to the Inquiry objected strongly to a number of environmental controls on the project which the Department of Environment and Planning considered essential if the plant was to operate in a satisfactory environmental manner. The Commissioners are not confident the stringent controls proposed are capable of practical application and strict enforcement on the Bayer site, particularly in view of Bayer's strong objection and submissions by the State Pollution Control Commission on difficulties of adequate, independent and constant supervision of the proposed high temperature incinerator.

Continued on page 128

Continued from page 127

7. Conservation of the wetlands should have priority at Kurnell Peninsula and those wetlands should be protected from environmental damage likely to be caused by location of new industries at Kurnell.
8. Alternative sites have not been adequately explored.
9. Our findings do not preclude other types of industry locating within the industrial zone at Kurnell provided it can be established, with certainty, on the merits of each case, that hazardous industry is likely to operate without damage to the environment of the Peninsula and the surrounding waterways. Some anomalies exist in present and proposed zonings and land use controls on the Peninsula in relation to industrial zoning, the lands of Besmaw Pty Ltd, proposed regional open space, marine and nature reserves.
10. It is regrettable that Bayer Australia Limited have spent in the order of $4 million on acquisition and upgrading of buildings and facilities on the site prior to determination of their development application. The Commissioners find this surprising in view of the controversial nature of the Bayer pesticide and veterinary products plant and its potential for adverse impact on the local environment. However, should the project be refused consent, as we recommend, Bayer should be able to recoup, at least, a substantial amount of funds already spent on acquisition and improvement of the site through sale.

For these reasons, which are commented upon in more detail in the Commissioners' report, we recommend that the Bayer Australia Limited application for Kurnell, be refused.

NB The decision of the NSW Minister for Planning was to immediately refuse the development application of Bayer Australia.

Source: Report to the Honourable Bob Carr, Minister for Planning and the Environment, New South Wales, by Commissioners John Woodward and Dr Alan Gilpin on the Proposed Bayer Australia Ltd Facilities for the Formulation and Storage of Agricultural and Veterinary Products at Kurnell, Sydney, January 1987, being an Inquiry Pursuant to Section 119 of the NSW Environmental Planning and Assessment Act 1979.

Box 5.12 New International Airport, Hong Kong

The international airport at Kai Tak, located on the eastern fringe of Kowloon, has been replaced by a new international airport with two independent runways at Chek Lap Kok to the west of Hong Kong. Land for the airport was formed by levelling the islands of Chek Lap Kok and Lam Chau, just off Lantau Island and by using excavated material from marine borrow areas. The entire programme comprised 10 interlinked infrastructure projects, with a completion year of 1997. Apart from the airport itself, these projects included 1669 ha of land reclamation, a harbour crossing and land tunnel, the longest road/rail suspension bridge in the world, a new town in Tung Chung, a 34 km airport railway and more than 30 km of expressway. The whole

Continued on page 129

Continued from page 128

was known as the Airport Core Program, being one of the largest infrastructure programmes in the world. Moving the airport from its original location provided relief to the nearby 350 000 residents affected by severe aircraft noise and eased the congestion of air traffic.

The airport master plan study began in 1990, establishing a basic airport configuration. The configuration selected involved the dredging and disposal of more than 70 million m^3 of marine mud and a requirement for over 150 million m^3 of fill. An early decision was made to retain a sea channel between the airport island and the coast of North Lantau. This enabled the natural coastline west of Tung Chun to be largely preserved, allowing also for tidal flushing of a potential area to the east of Chek Lap Kok.

An early EIA study of construction impacts identified a number of significant effects. About 70 dwellings would be adversely affected by noise, the construction programme commencing in 1992, on a 24-hour basis. The Hong Kong Executive Council granted exemption from the Noise Control Ordinances for the construction programme. However, this exemption was subject to conditions, including provision for the installation of air conditioners in affected dwellings, allowing windows and doors to be closed. In addition, a temporary 10 m earth berm was to be installed along the southern edge of the airport as a barrier to construction noise. Attention was to be given to noise minimization from the completed operational airport. Aircraft noise impacts were predicted for various assumed aircraft fleet mixes using Noise Exposure Forecast contours. It was concluded that by the year 2000 an acceptable level of noise would only be exceeded for a small number of dwellings at Sha Lo Wan on North Lantau, west of Tung Chung. These villagers were to be relocated, and all new noise-sensitive land uses were to be excluded from the airport vicinity.

Blasting, loading, transport and placement of fill material were the primary sources of dust. Concrete/asphalt plants were also a source of pollutants. In order to control the amount of dust, permanent monitoring stations were set up.

The destruction of the two islands, along with their terrestrial and marine life, necessitated a number of compensatory conservation measures, among them the rescue, study and possible re-establishment of the rare Romer's tree frog.

The excavation works required stripping the existing vegetation, marine mud being disposed of by marine dumping. Potentially hazardous material and chemical wastes would require special measures. However, the EIA study confirmed that the airport island would have an insignificant effect on water quality, bulk, and tidal regimes, subject to the preservation of a sea channel some 200 m wide, to assist tidal flushing.

Apart from incorporating environmental measures in the construction projects, the Hong Kong Government forged links with the community to provide quick responses to their problems. An Environmental Project Office was created to provide a continuous overview of environmental effects, and to co-ordinate and monitor remedial action. The management of the airport development represented a further step in the implementation of EIA procedures, which became mandatory for all major public projects in 1997. The Hong Kong Environmental Impact Assessment Ordinance was enacted in January 1997, together with a Technical Memorandum on the EIA process.

The new international airport was officially opened in July 1998 by President Jiang Zemin.

Continued on page 130

Continued from page 129

Sources: Hong Kong Environment Protection Department (1992) Environment Hong Kong 1992, Government Printer, Hong Kong; A.M.M. Liu and L.Y. Ng (1995) Environmental Issues in the New Airport Development in Hong Kong, a paper to the Catalyst '95 Conference, University of Canberra, Australia; Elvis W.K. Au (1998) Status and progress of environmental assessments in Hong Kong: facing the challenges in the 21st century, Impact Assessment and Project Appraisal, vol. 16(2), Beech Tree Publishing, Surrey, England.

England and Wales to gain consent for the construction and operation of a major new power station at Holme Pierrepont in the East Midlands of England. Box 5.11 describes the failure of Bayer Australia to gain consent for a formulation plant on the Kurnell Peninsula, near Sydney. Finally, Box 5.12 describes the matters taken into account in the development and environmental management of the new international airport in Hong Kong.

5.3.1 The Cost of Environmental Impact Assessment

5.3.1.1 Direct Costs

The direct compliance costs of the assessment process do not appear to have proved a significant problem for large companies, especially if environmental assessment is integrated with feasibility studies. These costs usually represent only a small proportion of total project costs.

The Bureau of Industry Economics (1990) in Australia has estimated these direct costs to be generally less than 1% of the total project costs. The Environmental Protection Administration in Taiwan has estimated the costs in that country to range from about 0.1% to 1.5% of a project's total cost (Gilpin and Lewis 1990). It does not appear that these costs are of particular concern.

5.3.1.2 Indirect Costs

A corporation's planning horizon needs to be extended to allow for the whole of the planning, environmental, public inquiry, and licensing procedures. In some instances this involves planning 5–10 years ahead, or even longer in the case of electricity generation. By forward planning, much of what might be described as delays can be avoided. However, unexpected problems and delays may arise that could not reasonably have been anticipated.

These delays could arise from the following:

1. a lack of co-ordination between the responsible authorities or levels of government involved;
2. conflicting demands between agencies and levels of government;
3. the failure of agencies and governments to observe time limits;

4. an increase in the number of authorities and agencies involved;
5. an unexpectedly large volume of public opposition;
6. a public inquiry not reasonably expected;
7. parliamentary opposition not reasonably anticipated;
8. significant deficiencies being identified in the EIS by a discerning public and by agencies requiring further studies to be undertaken and involving perhaps adjournments during a public inquiry.

These indirect costs can be considerable, amounting to perhaps 10% of the total project costs. They are particularly onerous in the case of a large electricity generating system and transmission network, where delay incurs the continuing operation of less efficient plant beyond the planned time, with more efficient plant with lower generating costs being temporarily excluded from the system.

Some of these unexpected delays and interruptions are beyond the control of the company and its officers; others, perhaps through better public and agency consultation, might have been avoided or minimized. The ultimate rejection of a development application inflicts the maximum burden of costs, both direct and indirect.

5.3.2 Other Impact Studies

A variety of other impact assessments are carried out from time to time for specific purposes and they need to be clearly identified:

- *Environmental health impact assessment (EHIA)*: an assessment of the effects on people of aspects of a policy, project or programme recognized as having potentially adverse health impacts. In 1982, the WHO recommended that EHIA studies should be conducted for all major development projects. EHIA embraces the following: risks and hazards (direct and indirect) involving explosion, shock, heat blast, vibration and destruction of property; biological factors such as parasites, helminths, protozoa, bacteria, mycobacteria, rickettsia and viruses; toxic, carcinogenic or mutagenic chemicals and heavy metals; ionizing and non-ionizing radiation; noise; dust and other irritants; and excessive temperature or humidity. The possible implications for human health may be measured in morbidity and mortality rates.
- *Ecological impact assessment*: an assessment of the potential ecological impacts of a proposed development, involving the following minimum steps: (1) the identification of the ways in which a proposed development will directly or indirectly affect the ecology of the proposed site and adjoining areas; (2) the quantification wherever possible of the value of these disturbed ecosystem elements in a local, regional and national context, both in the short and long term; (3) the modifications flowing from specified mitigation measures; and (4) a proposed monitoring programme should development proceed. Many of

the above stages of an ecological assessment rely on accurate data gathering and the correct identification of species in the field.

- *Fauna impact statement (FIS)*: a specialized impact statement concerned solely with the possible effects and implications of a proposed policy, plan or programme or activity of faunal or animal life. It is a subset of an ecological impact statement.
- *Social impact assessment (SIA)*: a subset of environmental impact assessment, an appraisal of the effect on people of major policies, plans, programmes, activities and developments. Social impacts or effects are those changes in social relations between members of a community, society or institution, resulting from external change. The changes may be physical or psychological, involving social cohesion, general lifestyle, cultural life, attitudes and values, social tranquillity, relocation of residents, and severance or separation. For example, in the construction of large hydroelectric dams, large populations may be relocated into unfamiliar environments. The consequences may include social discontent, unhappiness, increased illness, and a loss of productivity and income.
- *Hazard and risk assessment*: an essential component of many an environmental impact statement, embracing the potentially adverse effects of a project involving fire, heat, blast, explosion or flood arising from a manufacturing plant, transportation system, or other major item of infrastructure. An assessment reveals hazards to life and limb and property, and is expressed in the form of risk probability. Safety depends both on the location of a plant and the safety precautions and back-up arrangements adopted, together with the degree of training and alertness in the plant. Buffer zones and correct routeing of vehicles are also essential.
- *Regulatory impact statement (RIS)*: as distinct from an environmental impact statement, a RIS is an assessment of the likely costs and benefits of regulations created under a wide range of legislation. A RIS may not be a major document, perhaps only three or four pages. The concept is that both parliament and the public should be given greater economic insight into draft regulations before the legislature. In this way a proposal for a carbon tax or other environmental levy would be open to the public for comment. It would be accompanied by a brief presentation of the likely costs and benefits.

5.3.3 Conclusion

Over the last 30 years, the principle of environmental impact assessment (EIA) has emerged in many countries. Signal events have been the passing of the 1969 National Environmental Policy Act in the USA and the directive of the European Union on environmental impact assessment in 1985. It has now been widely adopted in Asia also.

The reason for the success of the EIA principle, and the associated principle of the public inquiry, has been the increasing complexity of public decision-making. Only a systematic approach enables decision-makers to receive balanced advice on whether a project, programme or policy should proceed or not, and in particular what conditions should be imposed on the construction, operation and decommissioning of the plant with provisions for site rehabilitation. Quite often, such decisions are beyond local capability and require attention at state, provincial or national level. Occasionally, development applications have been refused outright, while in other cases applicants have withdrawn their applications in the presence of strong public opposition. The central problem in most cases is to establish a range of conditions with which the successful applicant must comply. Such conditions relate to timing, air and water pollution control, noise management, transportation issues, working hours, workforce facilities, monitoring, risks and hazards, buffer zones, emergency procedures, visual amenity, infrastructure, auditing arrangements, compliance with laws, regulations and conventions, and relationships with the local community.

Public inquiries take time, though often no more than three months. While the costs of delays are expensive, an EIA procedure carefully implemented tends to minimize long-term problems and yields a better outcome for the general public than might otherwise have been the case. Industry has progressively improved its EIA skills, producing environmental impact statements of quality, and addressing fundamental issues head-on. Proponents often contribute at public inquiries to the framing of statutory conditions with which they will need to comply. Annual reporting to government agencies is often required in more controversial cases.

EIA has emerged as an important environmental planning tool for the 21st century.

5.4 Town Planning

Another matter often ignored in the literature on economic incentives is the role of town planning or environmental planning in influencing the character of settlements and setting a strategic framework for environmental management over a wider area. The location of industry, commerce and residential development within airsheds or catchments, and the planning of transportation systems and infrastructure investments, set the scene within which pollution control and environmental control measures may be adopted for particular developments. The two approaches – environmental control and pollution control – cannot be separated. Town planning instruments as outlined in Box 5.13 are another example of direct regulation or command and control achievements.

5.5 International Conventions

International agreements or conventions are often negotiated by the UN or regional governments when dealing with a common problem. It is important to note that such agreements, though acceded to by the representatives of a

Box 5.13 The Objectives of Town Planning

- The pursuit of social improvement through changes in the physical infrastructure of urban centres, maximizing opportunities for personal choice and protecting the individual and the family from the adverse actions of others
- The orderly arrangement of the metropolitan area into residential, commercial, industrial, port and recreational areas, each part performing its function at least cost and with minimum conflict
- The promotion of an efficient system of transportation and communication
- The achievement of optimal standards in respect of infrastructure, lot sizes, building spacing and alignment, sunlight, open space and parks, parking facilities, fire and emergency facilities, road pavement, access and aesthetics
- The promotion of a safe, clean and adequate public water supply, sewerage system, energy supply system, schools and other public services
- The promotion of good quality housing in a variety of types
- The minimization of air pollution, water pollution, noise and vibration
- The promotion of a satisfactory solid waste management system with recycling arrangements and well-located landfill sites
- The progressive development of tree planting, landscaping, anti-litter programmes, bicycle paths, nature walks, national and marine parks and reserves, community access to natural features, and recreational schemes
- The identification, preservation and restoration of items of heritage value
- The identification of natural resources with measures against the alienation of the most valuable of them
- To achieve consistency between local plans and any metropolitan, regional, provincial, state, federal or national plans and relevant international conventions
- To promote environmental impact assessment
- The provision of ample opportunities for public participation in the decision-making process, including participation in public inquiries
- Finally, the promotion of economic prosperity and 'the good life' for all

Box 5.14 Some International Conventions Relating to Resources and the Environment

Convention for the Creation of a Joint Arctic Council (the Ottawa Convention)
Signed in Ottawa in September 1996, an agreement to create a joint council for the Arctic, with the aim of protecting the environment while providing for the long-term development of the region. The signatories were Canada, Denmark (on behalf of Greenland), Finland, Iceland, Norway, Russia, Sweden and the USA.

Convention for the Prevention of Marine Pollution by Dumping from Ships and Aircraft (the Oslo Convention)
A convention signed in Oslo in 1972 by 12 European nations pledging to take all possible steps to prevent the pollution of the north-east Atlantic. The signatory

Continued on page 135

Continued from page 134

nations were Belgium, Britain, Denmark, Finland, France, Germany, Iceland, the Netherlands, Norway, Portugal, Spain and Sweden. A commission was created to administer the convention.

Convention for the Prevention of Pollution from Ships (the Marpol Convention)
A convention concluded in London in 1973 at the end of a conference convened by the International Maritime Organization. The convention effectively superseded a previous convention for the prevention of the pollution of the sea by oil, but does not deal with dumping from ships and aircraft which is covered by the Oslo Convention. The Marpol Convention aims to eliminate pollution of the sea by both oil and noxious substances, from ships of any type.

Convention for the Protection of the Natural Resources and the Environment of the South Pacific Region (the Noumia Convention)
Adopted in 1986, the Noumia Convention is an agreement to take appropriate measures to prevent, reduce and control pollution in the South Pacific region, including the Australian east coast. In the same year, two protocols were adopted relating to co-operation in pollution emergencies, and the prevention of dumping.

Convention for the Protection of the Ozone Layer (the Vienna Convention)
A convention adopted by 21 countries together with the whole of the European Union in 1985 as a first step in protecting human health and the environment from the adverse effects of activities likely to modify the ozone layer. The convention was reinforced by the Montreal Protocol of 1988 which aimed at protecting the ozone layer by controlling the emission of chlorofluorocarbons and halons, with a specific programme. The 163 nations now participating have kept to the time-table, with the USA leading the way. CFCs are being replaced by greener alternatives.

Convention for the Protection of the World Cultural and Natural Heritage (the Paris Convention)
A convention initiated by UNESCO and adopted in 1972. It established a World Heritage Committee and a World Heritage Fund to allow financing of conservation projects. A World Heritage List was also created.

Convention for the Regulation of Antarctic Mineral Resource Activity (CRAMRA)
Concluded initially in 1988 by the parties to the Antarctic Treaty, CRAMRA is an agreement to manage the potential development of the Antarctic continent, with the creation of a Mineral Resource Commission. Australia opposed the convention, objecting to all mineral exploitation in Antarctica. In 1991, a protocol (the Madrid Protocol) on the protection of Antarctica was adopted; under this protocol mining was banned for a period of 50 years.

Convention for the Regulation of Whaling
A convention creating the International Whaling Commission, with responsibility for the conservation and optimal utilization of whale resources. It was signed in 1946. However, the management of whale stocks remained unsatisfactory for many years, with some species being hunted to virtual extinction. In 1982, the commission voted to ban commercial whaling world-wide, beginning in 1986. A few countries have defied the ban, at least in part.

Continued on page 136

Continued from page 135

Convention on Biological Diversity
Also known as the Convention on Protecting Species and Habitats, this convention was endorsed by the UN Conference on Environment and Development in 1992 to conserve world biological diversity, or variability among living organisms. Initially it was treated with reserve by the USA, but was subsequently endorsed in 1993, the US establishing a National Biodiversity Center. The convention requires nations to develop and implement strategies for the sustainable use and protection of biodiversity, and provides for annual conferences of the parties involved. The 1995 meeting dealt with marine biodiversity issues. The third conference, in 1996, discussed such issues as access to genetic materials, and the impact of intellectual property rights on the conservation and sustainable use of biological diversity, and equitable sharing of benefits derived from its use.

Convention on Climate Change
A framework convention endorsed by the UN Conference on Environment and Development in 1992, to protect the world's climate system, most notably against the effects of greenhouse gases and their warming influence. Developed nations are required to reduce their emissions of carbon dioxide and other greenhouse gases to 1990 levels or below. A succession of international conferences have followed in Berlin, Geneva, Bonn and Kyoto. The Intergovernmental Panel on Climate Change remains the principal advisory body.

Convention on International Trade in Endangered Species of Wild Fauna and Flora (CITES) (the Washington Convention)
A convention adopted in 1973 providing for the regulation of trade in plants and animals, dead or alive, and their parts and derivatives. It has been recognized that excessive international trade has contributed directly to the decline of many wild populations of rare and endangered species. The convention states that a listed protected species may not be the subject of international trade without a permit granted under the CITES procedures. The convention became effective in 1975; by 1997 it had 128 participating member countries. Timber species were not listed under CITES until 1992, when trade in Brazilian rosewood was banned; trade in other timbers is now banned or restricted. A Timber Working Group continually examines the relationship between CITES and the international timber trade.

Convention on Straddling Stocks and Highly Migratory Fish Stocks
An accord signed in 1995 regulating the catch of deep-water and migratory species, including tuna, swordfish and cod stocks. The convention requires that special efforts be made by participating countries to monitor fish stocks, while strengthening inspection and reporting requirements for vessels. Regulations under the agreement apply both to international waters and to national exclusive economic zones (EEZs).

Convention on the Control of Transboundary Movement of Hazardous Wastes and Their Disposal (the Basel Convention)
A convention finalized in 1992. Its aim is to encourage countries to minimize the generation of hazardous wastes and the transboundary movement of such

Continued on page 137

Continued from page 136

wastes. The convention was negotiated following concern about the dumping by industrialized nations of hazardous wastes on to African and other developing nations. In 1994, it was decided to prohibit immediately the export of hazardous wastes for final disposal from OECD to non-OECD countries, and to phase out by 1997 all exports of hazardous wastes destined for recycling operations from OECD to non-OECD countries.

Convention on the Law of the Sea (the Montego Bay Convention)
A UN convention adopted in 1982, which came into operation in 1984. It was the outcome of international discussions which began in the early 1970s, and aimed to establish a new legal regime for the oceans and their vast resources. The convention created the 200 mile (320 km) exclusive economic zones (EEZs) and an International Seabed Authority to administer the new regime, conserving minerals and food supplies.

Convention on the Prohibition of Fishing with Long Drift Nets in the South Pacific (the Wellington Convention)
Adopted in 1989, this was a convention emerging from the Tarawa Declaration of the 20th South Pacific Forum aimed at banning the practice of ecologically damaging drift-net fishing in the South Pacific.

Convention on Wetlands of International Importance, Especially as Waterfowl Habitat (the Ramsar Convention)
Adopted in 1971 and coming into force in 1975, the Ramsar Convention is an agreement by the parties to take action to create reserves and otherwise protect wetlands that are internationally important for reasons including their habitat value for rare or migratory birds. Japan, for example, has concluded agreements for migratory bird protection with the USA, Russia and Australia to protect migratory birds facing extinction. Australia has designated three areas in the Northern Territory, including Kakadu National Park, as wetlands of international significance. A protocol, adopted in 1982, strengthened the convention.

Convention to Combat Desertification
A UN treaty signed in Paris in 1994 and ratified later by individual countries. Its aim is to tackle the spread of deserts globally, and it involves the voluntary implementation of a national action plan to deal with the deterioration of farmland into desert. About one-fifth of the world's farmlands suffer from the effects of desertification.

nation, have no effect until adopted by the national parliaments involved, with appropriate legislation. National agencies can then implement the provisions of such treaties, and governments may proceed to ratify the convention, enabling it to come into force. These conventions join the armoury of direct controls. Box 5.14 summarizes some of the most relevant conventions relating to resources and environment.

5.6 Summary

This chapter attempted to outline the scope and scale of direct regulations (or command and control) as statutory instruments controlling or influencing the quality of the environment. Even economic instruments as described in the next chapter do not operate in an independent manner, but within a framework of legislation and regulation. There is no escape from this approach.

The main aims have been discussed: pollution control, environmental protection, environmental impact assessment, town or environmental planning and international conventions. Even this list is by no means exhaustive; controls reach out into many aspects of our lives with an environmental bearing.

In some important respects, the challenge has changed over the years. At one time, in Britain and elsewhere, it was enough to equip power stations with high stacks of over 200 m and high-efficiency dust-arresting equipment that would collect over 97% of the dust reaching the stack. Attempts might be made to limit the sulphur content of the coal to no more than 1%. All this would be enough to protect the locality and the region. However, complaints from Scandinavia and other parts of Europe regarding acid rain modified this approach, so that flue gas scrubbing became more urgent. However, with the displacement of coal with natural gas, this problem has eased. The continuous pressures for improvement came from the competing nuclear power industry, the world's growing sensitivity to global pollution, the need to gain development consents for new power stations within a reasonable time-scale, and to avoid the risks of brown-outs and black-outs. Close co-operation with the British central pollution inspectorate was clearly necessary, to achieve all this.

The problems of direct regulation have been compounded by the issues of hazardous wastes, radioactivity, solid waste disposal, recycling, product labelling, the need for better environmental planning at local, regional and national scales, with difficult planning decisions in respect of highways, airports, residential development, infrastructure, national parks and reserves, and now the Kyoto Convention on global warming. As the next chapter reveals, there is now increasing opportunity for economic incentives of various kinds, as well as or instead of penalties. These will tend to augment and not replace direct controls, providing a more equitable distribution of costs and benefits.

Environmental regulatory reform is under way in all OECD countries. The main types of reform include the simplification of bureaucratic procedures and paper work, greater flexibility in choosing the best means of achieving environmental standards, the progressive adoption of integrated permit and licensing schemes, the integration of environmental impact assessment (EIA) procedures into planning procedures, the setting of time limits for consultation and public participation, and the greater use of permit and emission charges. Other effective instruments include voluntary agreements, information schemes, and environmental management and audit schemes, all of which are experiencing substantial growth (OECD 1997a).

Further Reading

Clark, R. and Canter, L.W. (eds) (1996) *Environmental Policy and NEPA: Past, Present and Future,* St Lucie Press, New York.
Committee on Air Pollution (1954) *Report* (the Beaver Report) Cmnd. 9322, London.
Gilpin, A. (1986) *Environmental Planning: A Condensed Encyclopedia,* Noyes, New Jersey.
Gilpin, A. (1995) *Environmental Impact Assessment: Cutting Edge for the Twenty-First Century,* Cambridge University Press, Cambridge.
Gilpin, A. (1996) *Dictionary of Environment and Sustainable Development,* John Wiley, Chichester.
Kormondy, E.J. (ed.) (1989) *International Handbook of Pollution Control,* Greenwood Press, New York.
Ministry of Housing and Local Government (1961) *Pollution of the Tidal Thames: The Report of a Departmental Committee* (the Pippard Committee), HMSO, London.
OECD (1975) *The Polluter Pays Principle: Definition, Analysis, Implementation,* OECD, Paris.
The White House (1992) *National Report to the UN Conference on Environment and Development 1992, Rio de Janeiro,* The White House, Washington, DC.
UNEP (1991) *Urban Air Pollution,* UNEP, Nairobi.
UNEP (1992) *Chemical Pollution,* UNEP, Nairobi.
UNEP (1993) *Global Biodiversity,* UNEP, Nairobi.
UNEP (1994) *The Pollution of Lakes and Reservoirs,* UNEP, Nairobi.
UNEP (1994) *Pollution in the World's Megacities,* UNEP, Nairobi.
UNEP (1995) *Energy, Pollution, Environment and Health,* UNEP, Nairobi.
US Department of Commerce (1996) *Pollution Abatement Costs and Expenditures,* Government Printing Office, Washington, DC.
US EPA (1990) *Environmental Investments: The Cost of a Clean Environment,* EPA, Washington, DC.
US EPA (1997) *Pollution Prevention: 1997: A National Progress Report,* EPA, Washington, DC.

6

Economic Instruments

6.1 Introduction

The Organization for Economic Cooperation and Development (OECD) has played a leading role in the evolution of economic instruments to augment direct regulation practices, in the belief that this approach achieves more efficient outcomes. As this chapter indicates, examples of economic instruments are to be found in a number of countries, including schemes for the transfer of pollution rights, development rights and quotas, suggesting an increasing role for such instruments. Most notable is the US Emission Trading Program, with direct regulation progressively bringing down the ceiling on emission limits.

Established in 1961, the OECD has as its primary aim the achievement of the highest sustainable level of economic growth in its member countries with the expansion of world trade on a multilateral basis. The current membership of OECD comprises some 29 democratic countries with market economies: Australia, Austria, Belgium, Britain, Canada, the Czech Republic, Denmark, Finland, France, Germany, Greece, Hungary, Iceland, Ireland, Italy, Japan, South Korea, Luxembourg, Mexico, the Netherlands, New Zealand, Norway, Poland, Portugal, Spain, Sweden, Switzerland, Turkey and the USA.

In 1970, the governing council of the OECD created an environment committee to advise on patterns of growth and development that would be in harmony with protection of the environment. As a consequence of the work of this committee, the OECD took a lead in promoting the polluter-pays principle as a means of assigning the costs of pollution control, and a whole range of environment protection measures, including environmental impact assessment (EIA). In 1974, a Declaration on Environmental Policy was adopted, tending to mirror the declaration adopted at the UN Conference on the Human Environment held in Stockholm in 1972. In 1985, the OECD extended the principles of EIA to development assistance projects and programmes, in order to mitigate potentially adverse environmental effects, the council adopting a schedule of those projects and programmes most in need of environmental assessment.

Since then, the OECD has also encouraged the use of economic instruments in environmental policy, particularly since environmental policies have moved in the direction of prevention, as distinct from cure. Subsequent work has resulted in recommendations on the use of economic instruments (OECD 1991) as quoted in Box 6.1, and agreed guidelines for the application of economic instruments in

Box 6.1 Recommendations of the Council on the Use of Economic Instruments in Environmental Policy (OECD January 1991)

Inter alia, member countries should:

- make greater and more consistent use of economic instruments as a complement to, or as a substitute for, other policy instruments such as regulations, taking into account national socio-economic conditions;
- work towards improving the allocation and efficient use of natural and environmental resources by means of economic instruments so as to better reflect the social cost of using those resources;
- make efforts to reach further agreements at an international level on the use of environmental policy instruments with respect to solving regional and global environmental problems as well as ensuring sustainable development;
- develop better modelling, forecasting and monitoring techniques to provide information on the environmental consequences of alternative policies and their economic effects.

Source: OECD (1991), quoted in OECD (1997) Evaluating Economic Instruments for Environmental Policy, OECD, Paris, p.15.

member countries (OECD 1994). The most recent survey of the use of economic instruments in member nations is entitled *Evaluating Economic Instruments for Environmental Policy* (OECD 1997), findings of which are taken into account in this chapter.

Box 6.2 quotes the views of Baumol and Oates on incentives; Box 6.3 the views of Cropper and Oates on the role of economists; Box 6.4 the views of the Friedmans on market discipline; and Box 6.5 the views of Allan V. Kneese on incentives.

Box 6.2 Baumol and Oates on Incentives

In our view, the appropriate response to environmental problems is not to bring a complete halt to the expansion of the economy, but rather to build into it a powerful set of incentives to reduce those activities that degrade the environment. With such incentives as a basic part of our economic structure, we do not foresee the inevitable catastrophe envisioned by the neo-Malthusians. Continued growth and the associated increases in standards of living are consistent with improvement in environmental quality, if we adopt the measures needed to induce individual producers and consumers to economize on their use of environmental resources.

Source: W.J. Baumol and W.E. Oates (1979) Economics, Environmental Policy, and the Quality of Life, Prentice-Hall, Englewood Cliffs, NJ, p. 144.

Box 6.3 Cropper and Oates on the Views of Economists

When the environmental revolution arrived in the late 1960s, the economics profession was ready and waiting. Economists had what they saw as a coherent and compelling view of the nature of pollution with a straightforward set of policy implications... Economists saw pollution as the consequence of an absence of prices for certain scarce environmental resources such as clean air and water, and they prescribed the introduction of surrogate prices in the form of unit taxes or *effluent fees* to provide the needed signals to economize on the use of these resources.

Source: M.L. Cropper and W.E. Oates (1992) Environmental economics: a survey, Journal of Economic Literature, ***XXX****: 675–740.*

Box 6.4 The Friedmans on Market Discipline

Most economists agree that a far better way to control pollution than the present method of specific regulation and supervision is to introduce market discipline by imposing effluent charges. For example, instead of requiring firms to erect specific kinds of waste disposal plants or to achieve a specified level of water quality in water discharged into a lake or river, impose a tax of a specified amount per unit of effluent discharged. That way, the firm would have an incentive to use the cheapest way to keep down the effluent... the tax rate itself could be varied as experience yielded information on costs and gains.

Source: M. Friedman and R. Friedman (1980) Free to Choose, Macmillan, Melbourne, p. 217.

Box 6.5 Allen V. Kneese on Incentives

America's legal and regulatory systems address issues related to resource conservation and environmental improvement almost entirely through direct regulation of particular environmental media. Such an orientation continues to produce the familar field days for lawyers, heavy costs, a huge bureaucracy, ad-hoc and capricious impacts, and far-reaching intrusion by government into decisions about the design of industrial processes.

An incentive-oriented approach could not by itself deal with all the sticky problems that arise in achieving environmental management objectives but would, however, be based on a sound conception that prices should reflect all costs.

Source: A.V. Kneese (1998) Resources (newsletter of Resources for the Future), 130 (Winter), Washington, DC.

6.2 Fines, Penalties and Damages

The enforcement of pollution control legislation and other environmental control measures often results in the imposition of fines and penalties by the courts, or even the award of damages to victims. When the tanker *Exxon Valdez* ran aground in Prince William Sound, Alaska, in 1989, causing extensive damage to 2400 km of beach, a jury decided that the owners and the captain were reckless. Plaintiffs in the US Federal Court case included more than 34 000 commercial fishermen, native Alaskans and property owners, who claimed they had suffered harm from the spill. The jury, in one of the largest awards in legal history, ordered Exxon to pay US$5 billion in punitive damages to the plaintiffs. The captain was ordered to pay US$5000 in damages.

In New South Wales, Australia, the most serious offences under the pollution control legislation, subject to proof of wilfulness or negligence and harm to the environment, may result in a maximum penalty of A$1 000 000 for corporations, with fines of up to A$250 000 for individuals and/or the possibility of imprisonment for up to seven years. By 1998, in Australia, two company directors had served sentences for pollution offences. Other offences, attracting lower fines, are generally categorized as strict liability offences, i.e. the prosecution does not have to prove intent. Penalties for littering remain low at A$300. Apart from fines and terms of imprisonment, the courts may impose a wide variety of orders to mitigate pollution, award damages to victims, and restrain possible offenders.

If an environmental control authority wishes to avoid entanglement with other economic instruments, it may choose to rely on enforcement and penalties for breaches of prescribed standards and procedures, such penalties having a similar effect to a Pigovian tax. Strict liability tends to internalize, in a similar way to an appropriate tax. Legal liability may also provide compensation to the victim, unlike a Pigovian tax. Either way, pressure is imposed on the potential polluter. However, it must be recognized that most writers do not include fines and penalties as an economic instrument, largely because no doubt it belongs to the subject of direct control. This is often described in US literature, in a somewhat derogatory way, as the 'command and control' system.

Experience in several OECD countries, including Britain, over the past half-century suggests a generally high level of compliance with regulations and policies; visible pollution, once commonplace, is now a rarity. Experience shows that companies do not like adverse publicity, which could be damaging to often expensive programmes; they prefer to portray a good-neighbour image.

6.3 Carbon Taxes

A carbon tax is a tax on the consumption of fossil fuels (coal, oil and natural gas) to discourage consumption, reduce carbon dioxide emissions, provide funds to alleviate say payroll taxes and other labour charges, and promote other measures against the greenhouse effect. As the carbon content of fuels varies, fossil fuels would be taxed at different rates: coal would attract the highest rate, followed by oil, and then natural gas. Renewable, non-carbon-emitting energy sources, such as solar, wind, hot rock and geothermal power would not be taxed. A carbon tax would provide an incentive for all consumers to find the optimal mix of fossil and non-fossil fuels and energy for their individual circumstances. In Sweden, during 1990, environmental charges were imposed on fossil fuels, and a value-added tax on energy. The taxes embrace carbon dioxide, sulphur dioxide, and oxides of nitrogen.

In the USA it has been calculated that a carbon tax of about US$30 per ton phased in over five years would stabilize US emissions at 1990 levels by the year 2000, generating revenues of US$36 billion by the fifth year (Repetto 1993). However, the overall effect might not be all that great, depending on how the revenue was used to relieve taxes on labour and capital income. The effect might be revenue neutral for the government, and perhaps neutral also for the individual enterprise. In addition, energy costs make up only 2.6% of US industrial production costs, and a carbon tax might not raise average manufacturing costs by more than 0.1%. These findings refute the argument that a carbon tax can be imposed only at the cost of lost jobs and income. Indeed, it can be argued that there is a *double dividend* in protecting the environment while alleviating the incidence of payroll taxes on industry. However, it should be kept in mind that what industry does and does not do depends very much on the general economic climate, rather than the effects of particular taxes.

Instead of a carbon tax, governments could introduce carbon dioxide permits, and by controlling the total quantity of such permits governments could limit carbon dioxide emissions directly. This would cause a reduction in pollution, production, employment and investment more promptly than a severe carbon tax (Parry 1997). Any revenue-recycling effects would then depend on whether the permits were issued free, though rationed, or put up for auction. If the permits were auctioned off the revenue could be used by government to alleviate taxes in the economy. If given out free, there would be no revenue and no beneficial tax reliefs. A carbon dioxide emissions reduction policy, of either form, provides a benefit to society only if it raises revenue for the government and if one ignores the benefits in global terms, though these benefits are speculative at this stage (Nordhaus 1994).

6.4 Sulphur Taxes

Sulphur dioxide (SO_2) is a colourless highly pungent gas released when sulphur burns in air. The emission of sulphur dioxide in urban areas comes from industrial, commercial and domestic sources during the combustion of sulphur-containing fossil fuels, mainly coal and oil. During combustion, all the sulphur in fuel oil, and most of the sulphur in coal, is emitted from stacks into the general atmosphere as sulphur dioxide, with a small proportion as sulphur trioxide (SO_3). The discharge into the atmosphere of oxides of sulphur has led to the problem known as acid rain; oxides of nitrogen also contribute to this problem. Geographical areas significantly affected by acid rain include the north-east of the USA and Ontario, Canada; parts of Scandinavia, notably Sweden, and southern Norway. Other parts of Europe are affected by acid rain, and it has possibly contributed to the dieback of the Black Forest in Germany. While sulphur comes mainly from the combustion of coal and heavy oil in industrial plant and power stations, oxides of nitrogen which aggravate the problem arise mainly from stationary sources and transportation, most notably the automobile. Sulphur dioxide has also been a primary air pollutant in industrial towns and has resulted in restrictions in the sulphur contents of fuels, both coal and oil. The electricity authorities in Britain, for example, have placed restrictions on the sulphur content of coal to some power stations to no more than 1%, in order to facilitate development consents.

In 1979, a Convention on Long-range Transboundary Air Pollution was initiated by the UN Economic Commission for Europe, for the purpose of curbing the incidence of acid rain arising from the long-range transportation of pollutants, most notably sulphur dioxide. The convention came into effect in 1983. Regulations began to be introduced requiring power plants to be equipped with high-performance gas scrubbers. A protocol was introduced in 1985 (the Helsinki Protocol) requiring the reduction of sulphur emissions by at least 30%. A later protocol further restricted nitrogen oxides.

Soil and water acidification, forest damage, and fish deaths have affected large areas of Scandinavia. Transboundary air pollution has contributed up to 90% of total sulphur deposition in Sweden (OECD 1997b). This has been due to the meteorological patterns which influence the import of acid rain from emitting areas in Britain and the European mainland.

Sweden has traditionally used statutory measures to regulate sulphur emissions. Under the Environment Protection Act 1969, as amended, the National Licensing Board for Environment Protection issues permits for major stationary sources setting emission limits. There are also limits for the sulphur contents of various fuels. The sulphur tax introduced in 1991 did not replace those standards and procedures, but has occupied a complementary role. The tax aims simply at a faster and more cost-effective reduction of Swedish sulphur emissions to comply with the ambitious national target of an 80% reduction in sulphur emissions between 1980 and 2000. Cuts have already been substantial. The gradual acceptance of economic instruments for environment control has coincided with reforms to the highly distorted Swedish tax system. A reduction in public revenues from traditional sources was offset by a variety of environmental taxes and a broadening of the tax base. Despite a reduction in energy taxes, the sulphur tax remains in place.

The Swedish sulphur tax is a tax on the sulphur content of coal, oil (i.e. fuel oil, diesel and domestic heating fuel) and peat used for energy generation. The tax, in effect, is levied on the estimated emissions from a known fuel consumption. Fuels containing less than 0.1% sulphur are not taxed. Similar successes have been achieved in Norway, where sulphur taxes have been operating since 1970.

6.5 The Nitrogen Oxides (NO_X) Charge in Sweden

In 1985, the Swedish Parliament set a target for the reduction of annual emissions of nitrogen oxides. Emissions in 1980 were some 425 000 tonnes and this figure was to be dramatically reduced to 300 000 tonnes by the year 1995, i.e. a reduction of 30% from the 1980 figure. By 1997, this figure was substantially achieved. A large part of the reduction was achieved through controls on motor transport and industrial processes. Power stations accounted for little more than 10% of the total, as much output was hydropowered or nuclear. As future additional power would probably be drawn from fossil-fueled power stations, a nitrogen oxides tax has been directed towards power plant and large combustion plant in industry; the charge is intended to accelerate measures to reduce emissions, supplementing normal licensing requirements.

The instrument is a direct charge imposed on the measured emissions of this limited group of large emitters. The charge is imposed as a rate per kilogram of nitrogen dioxide. The choice of tax rate has been governed by estimates of the marginal cost of emissions abatement. These vary widely, so the charge is actually an approximation likely to achieve the desired results. Coupled with normal licensing requirements, it was estimated that the charge would lead to a substantial reduction in annual emissions. There were three principal methods by which the annual reductions were achieved: combustion controls, selective non-catalytic reduction and selective catalytic reduction. Almost all of the charge revenues are returned to the participating parties, but in proportion to their final energy output. Thus sources with high emissions relative to energy output are net payers to the scheme, while sources with low emissions relative to energy output are net recipients. The tax return arrangements do not detract from the incentives to reduce pollution (OECD 1997c).

6.6 Water Pollution Charge-Schemes in the Netherlands, France and Germany

The well-established water effluent charging systems in the Netherlands, France and Germany have attracted much attention in the literature (Anderson 1994), largely because of evidence not only of substantial revenue-raising but also some incentive characteristics.

In the Netherlands, the system of water pollution charges was created by the Pollution of Surface Water Act 1969. This established a system of discharge licences and a scheme of waste-water charges to help finance the costs of water treatment. Charges are levied at both national and provincial levels, being relatively high for both households and firms. The charging scheme is a mixed effluent/user system closely linked with direct regulation. The scheme strongly reflects the polluters-pays principle and is widely regarded as being the most successful application of effluent charges (Bresser 1988), with a major impact on polluting behaviour.

In France, six water basin authorities are responsible for water pollution control. These authorities have levied a water pollution charge on domestic and industrial water discharges since 1968. In respect of industrial and other non-domestic discharges, a water quality permit must be obtained from a basin authority, the permit governing the volume and concentration of effluent. The charges apply to both public and private sources, a charge being based upon the level of emissions in the effluent. Relevant characteristics are suspended matter, oxidizable matter, soluble salts, inhibitory matter and toxic substances. Charges were raised considerably following an adverse OECD report in 1989 arguing that the charges were too low to provide an incentive effect.

In Germany, water pollution charges were introduced under the Waste Water Charges Act 1976. The charges are levied on direct discharges into rivers, lakes, the sea and groundwater, by both industrial and municipal sources. The effluents from municipal treatment plants are not charged. The charges, actually introduced in 1981, have been progressively increased. The amount of the charge depends on the limits set down in the operating licence. There are arrangements for charge reductions where marked improvements are achieved. The system also allows dischargers to offset the costs of pollution control equipment against the charges. Over recent years, total revenues have fallen, notwithstanding increases in charge rates. A significant effect of the German effluent charging system has been the provision of financial resources to increase the number of capabilities of staff engaged in assessing and issuing water pollution permits, to improve monitoring, and to provide polluters with an incentive to review their discharges and consider technical improvements (Kraemer 1995).

6.7 Landfill Levies and Taxes

Charges are often imposed for the use of landfills, the effect being to minimize the amounts of waste disposed of in that way, while encouraging recycling. A landfill tax was introduced throughout Britain in 1996, with a standard rate per tonne and a reduced rate for inactive waste. The tax applies to all waste going to licensed landfill sites. Some of the income can be streamed by site operators into approved environmental trusts, with appropriate tax relief. Trust funds are used for projects such as the remediation of old landfill sites. The UK Environment Agency gives high priority to the prevention of fly-tipping, to minimize tax avoidance. The introduction of the tax was part of the British Government's commitment to extend the use of economic instruments for environment protection more widely.

British landfill levies appear modest compared with those charged in the Netherlands, Germany, Italy, the USA, Sweden, France, Finland and New Zealand, though higher than those charged in Spain, South-east Asia, Latin America and Australia (Riley 1996).

Most of the Australian states now have landfill levies. The most recent, introduced in Western Australia in 1998, allocates the levy to a trust fund solely for the purpose of developing and implementing waste management and recycling programmes, with particular reference to regional recycling plans.

6.8 User-pays Principle

The user-pays principle entails charging the consumers of services the whole long-run marginal cost of providing that service, covering both capital and running costs. Subsidizing is thus avoided; each consumer pays the difference in the costs to the system incurred with and without that consumer. The concept has considerable merit in relation to the main body of consumers in ensuring a rational, efficient and undistorted use of resources in respect of any service. It avoids distortions in demand, for if prices are artificially low, an inflated demand for community resources results.

On the other hand, the concept may at times conflict with certain other social objectives, e.g. that all homes should, if at all practicable, be connected to an electricity supply, to a telephone service, to sewerage, and be close to public transport. The provision of universal services in all or most locations, with high marginal costs beyond the reach of most consumers, must conflict with the user-pays principle. The resolution of these conflicting principles is a matter for public policy.

However, the principle is more easily applied to the consumers of public services involving the collection and treatment of effluents. Here, user charges are based on the costs of providing a collection and treatment service delivering a final effluent of acceptable quality. These user charges may be applied at a flat rate, but more often vary with the volume and nature of the pollutants discharged into a sewer. The primary aim of these charges is cost recovery, though depending on the severity of the charges, there is an incentive to reduce the volume and improve the quality of effluents, through internal recycling and improved technical efficiency. These charges are often known as trade waste charges, or sewer charges. The charges should just cover the full costs of treating that particular effluent.

6.9 Load-based Licensing

Many industrial emission licensing systems regulate the concentration of pollutants in the waste streams. While restricting the concentrations, these arrangements do not restrain the total load on the atmosphere. Applied to a power station, a licence may stipulate the maximum concentrations for particulates and sulphur dioxide, without restraining the total amount of either. While still limiting concentration limits where there is likely to be harm to human health or the environment, some licensing systems are now emphasizing the control of loads of pollutants discharged, with maximum load limits. Pollution load fees are related to the relative harmfulness of the pollutants, the state of the receiving environment and, in some cases, to the manner of discharge. The smaller the load of pollutants discharged, the lower the fee payable. Load-based licensing aims at achieving environmental objectives more effectively than traditional concentration-based licenses, while offering financial rewards for reducing pollution loads. Emissions of pollutants to air and water should be reduced to harmless levels at the lowest possible cost to business and the community.

6.10 Product Charges or Ecotaxes

Product charges are imposed on inputs to economic activities, or on products and services themselves, as a means of indirectly controlling environmental impacts. In some European countries, charges are levied on fuels according to their sulphur or carbon content, as an incentive to reduce emissions of sulphur dioxide or carbon dioxide. Differential taxes may be applied to recycled paper to encourage the reuse of paper, conserve timber supplies, and reduce waste disposal and litter. Product charges act as substitutes for emission/effluent charges; they alter the relative prices of products so influencing consumers and purchasers.

Belgium has imposed ecotaxes on products since 1993, applying them to goods considered to be harmful to the environment (Pittevils 1996). Ecotaxes are linked closely to existing indirect taxes and apply only to products consumed in Belgium. Product taxes in Belgium relate to all drink containers, throw-away razors, throw-away cameras, industrial packaging, batteries, pesticides, pharmaceutical products, and paper (Clercq 1996). The taxes are kept under review by a special commission, reporting to the federal government. The introduction of these taxes in Belgium proved to be much more complicated than the parliament had anticipated. It was not easy to combine environmental objectives on the one hand, with the free movement of goods and services within the European Union (of which Belgium is a member) on the other. This delayed the process of implementation. However, the effect on industry was marked, with an energetic search for solutions.

6.11 Subsidies

Most OECD countries provide some financial assistance for environmental investments by the private sector in the form of grants, soft loans, or tax concessions and allowances. The main objectives of these subsidies are (1) to assist the implementation of direct regulations, particularly where financial difficulties are being encountered; and (2) to support research, development, and the introduction of pollution control equipment and cleaner technologies (Panayotou 1994; OECD 1997c). It has been estimated that environmental subsidies in Europe range between 5% and 20% of total environmental investments. Pollution control costs in various industries have reportedly ranged from 10% to 20% of the initial capital cost of the plant. In Britain, air pollution control measures traditionally comprise some 10% of the cost of a new power station, with much higher costs for special gas scrubbers.

In 1997, countries providing subsidies for environmental investments and accelerated depreciation (the writing-down of a capital investment for tax purposes over a much shorter period) included Australia, Austria, Belgium, Canada, the Czech Republic, Denmark, Finland, France, Hungary, Japan, Mexico, the Netherlands, Norway, Poland, Spain and the USA (OECD 1997c).

In France, most environmental subsidies are drawn from emission charges; while polluters pay for their emissions, some 90% of the revenue is returned to them as assistance for environmental investments, while the remaining 10% goes to financing research into and development of new technologies.

In Germany, subsidies are provided mainly from the general budget; they are given in the form of soft loans (loans offered at rates of interest below normal commercial rates) to polluters facing strict environmental standards.

Italy has offered incentives to scrap older, more polluting cars, in favour of newer cleaner ones, boosting car sales while promoting a cleaner atmosphere.

In the USA, subsidies are less general and are applied mainly to waste treatment and noise abatement. Government subsidies for waste-water treatment were initiated in 1956 under the Water Pollution Control Act, at which time they varied between 30% and 75%.

In Australia, concessions are limited to heritage conservation works. However, under a Natural Heritage Trust created from the part-sale of Telstra, a capital injection into environmental works of A$1.15 billion was realized in the late 1990s. The Trust was primarily aimed at tackling the environmental and sustainable agricultural problems facing Australia, such as problems involving salinity, vegetation, rivers, biodiversity, landcare, coasts and clean seas, farm forestry and property management.

Another source of subsidy is the Global Environment Facility, created by the UN (discussed in Section 6.12). The granting of tax breaks and concessions on the environmental front is a clear breach of the polluter-pays principle but may be tolerated in limited circumstances.

The questions of subsidies and special concessions in the area of the environment needs to be viewed against much more significant subsidy issues. Many countries subsidize agricultural activities. In OECD countries, subsidies to farming by way of direct subsidies, cheap loans and guaranteed prices were worth 36% of agricultural production in 1996 (*The Economist* 14 June 1997). Fossil-fuel subsidies are world-wide, while electricity is typically subsidized in both developed and developing countries. Larsen and Shah (1995) estimated that global carbon dioxide emissions would be reduced by 4–5% if all energy subsidies were removed. Subsidies lead to over-investment in new plant, excessive energy consumption, and significant additional environmental impacts from power generation.

To avoid the impression that subsidies are a relatively new phenomenon, it should be remembered that subsidies for fireplace conversions in smoke control areas were introduced by the British Clean Air Act in 1956. Local authorities in Britain were obliged to make financial grants to the owners and occupiers of domestic dwellings in respect of the reasonable costs of converting heating installations to meet smoke control requirements. The national government contributed 40% of the cost, while the local authorities contributed 30% of the cost; the remaining 30% was borne by the owner or occupier. In other words, the costs of meeting smoke control requirements were subsidized to the extent of 70% of the total reasonable costs. The effect of smoke control areas was to eliminate smoke emission from the urban areas of Britain, at a reasonable cost, ensuring that only smokeless fuels were consumed.

6.12 Global Environment Facility (GEF)

In 1990, the World Bank, the UN Development Program and the UN Environment Program established the Global Environment Facility (GEF) to provide concessional financial assistance to the developing world for investments that would (1) protect the ozone layer; (2) protect international water resources; (3) protect biological diversity; and (4) reduce greenhouse gas emissions.

Measures to protect the ozone layer relate to steps required by the Montreal Protocol, an agreement reached in 1988 by more than 30 countries to control the emission of chlorofluorocarbons (CFCs) and halons to the atmosphere, thought to damage the ozone layer. By the mid-1990s, the developing countries continued to provide a substantial market for ozone-damaging chemicals made, but not sold, elsewhere.

International waters would clearly benefit from investments to prevent oil spills and toxic waste pollution. To protect biological diversity, the GEF may encourage debt-for-nature and debt-for-environment swaps and the conservation of tropical forest areas. To reduce greenhouse gas emissions, the emphasis is on investments in cleaner fuels and technologies in the energy sectors. An increase in the use of natural gas would qualify for assistance, as would investment in forests to create sinks for carbon dioxide.

Nations that have contributed financially to the GEF are known as 'participants', and at each of their regular meetings, the participants review and endorse the proposed work programme. The GEF participants are supported by a Scientific and Technical Advisory Panel composed of 16 independent members. The GEF represents a highly innovative approach to environmental protection, though problems have arisen over the recurrent financing of GEF projects. A lack of agreement emerged between industrialized and less-developed countries on the purpose and strategy of the GEF, and the linking of projects to development schemes run by dominant institutions, groups and companies. At a meeting in Cartagena, Colombia, at the end of 1993, these differences were debated; negotiations were completed in March 1994, when GEF funds were replenished with US$2 billion.

6.13 Superfund

The US Comprehensive Environmental Response, Compensation and Liability Act 1980, created a 'Superfund' to meet the costs of the clean-up of abandoned hazardous waste sites. The goal of this US Congress legislation was to eliminate the most serious threats to public health and the environment presented by uncontrolled hazardous waste sites, and to respond to such hazards in a cost-effective manner. The federal contribution to the Superfund was financed largely from a special tax imposed on feedstocks used by the petroleum refining and chemical manufacturing industries; the remainder from federal appropriations. Superfund was supplemented by 10%, matching grants from the states. Though some 600 companies pay superfund taxes, around 10 major petroleum and chemical companies meet about half the cost.

The Act created a Hazardous Substance Response Trust Fund. Taxes were also imposed on the owners and operators of hazardous waste disposal facilities in order to establish a second fund known as the Post-Closure Liability Trust Fund. In the event of the release of a hazardous substance, the procedures and methods to be followed are set out in a National Contingency Plan. To deal with the immediate problems, the Act required the preparation of a national priorities list. The US EPA estimated that no more than 2500 sites out of 20 000 identified would need priority treatment, taking some 10 years to complete the work.

In 1986, the programme was reauthorized with additional funds coming from a broader-based tax on corporate income, in addition to the taxes on chemical and petroleum feedstocks with federal appropriations. Funds were also provided for research into and development of new hazardous waste disposal technologies. The programme accelerated.

6.14 Environmental Levies, Special Charges and Rate (Local Tax) Relief

Local elected councils in many countries are becoming increasingly involved in environmental management, facilitating the protection of catchments, residential areas, public open spaces, agricultural activities, natural vegetation, and assets of heritage value. Various financial and economic instruments are being used to support these initiatives. Several cities have adopted differential taxation to penalize or encourage the use of land for particular purposes. Environmental levies are charged by many local governments in Australia (James 1997). In New South Wales, catchment management trusts raise funds via a catchment levy.

Some Australian city councils have introduced rate rebate schemes to fund the revegetation of groundwater recharge areas, to combat salination, prevent land degradation, or preserve structures or places of cultural, environmental, historic, heritage or scientific significance. Rate rebates may be granted to those entering into conservation agreements with local councils, or to those who undertake to adopt sustainable farm management practices.

6.15 Tradable Permits or Quotas: Marketable Instruments

During the past decade, solutions to environmental problems attracting government regulation or prescription have sometimes been augmented by market-based arrangements. Tradable quotas or permits have been used to help control and manage a number of environmental problems, including water allocation, water pollution, overfishing, lead pollution, ozone protection, and sulphur dioxide. A tradable quota regime addresses environmental externalities by allocating property rights in the form of a quota over the externality. The concept of allocating property rights over externalities was introduced by Coase (1960) and has found a limited range of application in the area of emission trading to control pollution. The approach may yet find a place in the control of greenhouse gas emissions.

6.16 US Emission Trading Program

The US Emission Trading Program was introduced into the USA in the early 1980s, in an attempt to introduce some free market principles into the use of environmental resources. The amount of pollution allowable from individual firms within a region or catchment is fixed by the US EPA, though the limits may be reduced progressively over time. At each stage, the region as a whole must meet the specified restrictions. Under a trading programme, there is an incentive to achieve reductions in emissions below the legal requirements, enabling a firm to expand, or to sell the resultant credits to other firms needing them. Sulphur credits are traded on the Chicago Board of Trade (*Environment Reporter* 28 August 1992) (See Box 6.6).

In order to reduce acid rain in the USA and Canada, Title IV of the Clean Air Act Amendments of 1990 established the US Acid Rain Program. Title IV of the Act sets as its primary goal the reduction of annual sulphur dioxide emissions by 10 million tons below 1980 levels by the year 2000. In other words, the programme aims to cut sulphur dioxide emissions by half. It will also substantially reduce nitrogen oxides emissions from electric utility plants.

To achieve these reductions, the scheme requires a two-phase tightening of the restrictions on fossil-fuel-fired power plants. Phase 1 began in 1995 and affects

Box 6.6 Terminology in the US Air Emissions Trading Program

- *Emission reduction credit (ERC)*: a credit given by the US EPA in respect of a reduction of emission from a stationary source below the permit or regulatory level. The credit may be traded internally or externally, involving other firms.
- *Netting*: introduced in 1974, some ERCs may be transferred between plants within the same premises. A reduction of pollutants from certain sources may allow an increase of similar pollutants from new sources within the same plant, achieving the same overall targets.
- *Offsets*: introduced in 1976, a rule that allows firms to enter non-attainment areas (i.e. areas where ambient standards are not being met), providing they have acquired ERCs from sources in the same area, to offset the new emissions, perhaps by a factor greater than one.
- *Banking*: the saving of ERCs for future sale or use.
- *Bubbles*: the enclosing of complete works within imaginary bubbles, with aggregate limits for pollutants imposed on each bubble; trading in ERCs takes place within the bubble. Managers may impose severe limitations on emissions that can be reduced relatively inexpensively, in return for some relaxation on emissions that are more expensive to control.
- *State implementation plans (SIPs)*: since 1970, plans prepared by individual US States to meet national ambient air quality standards. Annual sulphur dioxide emissions have since declined significantly, being halved by the year 2000.
- *Allowance*: the unit used in ERCs. An allowance authorizes a plant within a utility or industrial source to emit one ton of sulphur dioxide during a given year. At the end of each year, a unit or plant must be credited with allowances at least equal to its annual emissions, i.e. a unit that emits 5000 tons of sulphur dioxide must hold at least 5000 allowances that are usable in that year. Allowances are fully marketable commodities that may be bought, sold or banked. Trading is on the the Chicago Board of Trade.
- *Continuous Emission Monitoring (CEM)*: under the US Acid Rain Program, CEM equipment must be installed in all plant for the continuous monitoring of sulphur dioxide, oxides of nitrogen and carbon dioxide, along with other basic data.

110 mostly coal-burning power stations located in 21 eastern and midwestern US states. Phase 2 begins in the year 2000; it tightens the annual emission limits imposed on these larger plants, and also sets restrictions on smaller, cleaner plants, fired by coal, oil and gas. The Act also calls for a 2 million ton reduction in nitrogen oxides emissions by the year 2000. Much of this will be achieved by coal-fired utility boilers that will be required to use low NO_x burner technologies and to meet new emission standards. In 1998, the details of Phase 2 were announced (McClean 1998) involving a reduction in sulphur emissions from 8.7 million tons to 4.5 million tons by 2012.

The market-based sulphur dioxide (SO_2) allowance trading component of the Acid Rain Program allows utilities to adopt the most cost-effective strategies to reduce SO_2 emissions at units under their control. Affected utilities are also required to install systems that continuously monitor emissions of SO_2 and NO_2 and other related pollutants to ensure compliance and give credibility to the trading component of the programme. A variety of penalties apply for non-compliance with the programme.

Allowance trading is the core of the EPA's Acid Rain Program. One allowance authorizes a unit within a utility to emit one ton of SO_2 during or following a given year. At the end of each year, a unit must hold an amount of allowances at least equal to its annual emissions. A plant that emits 5000 tons of SO_2 must hold at least 5000 allowances that are usable in that year. Regardless of this, a plant must never exceed the limits under Title I of the Act to protect human health. Allowances are fully marketable commodities. Once allocated or acquired, allowances may be sold, bought, traded or banked for use in future years.

Utilities can reduce emissions by adopting energy conservation measures, increasing reliance on renewable energy, reducing usage, employing pollution control technologies, switching to lower-sulphur fuel, or adopting other alternative strategies. Allowances are allocated annually by the US EPA. The Act effectively places a cap of 8.95 million tons on the whole geographical area. Allowances are also available from three EPA reserves, subject to a range of restrictions. New units are not allocated allowances. Instead, they have to purchase allowances from the market, or from the EPA auctions and direct sales from the reserves. The EPA maintains an Allowance Tracking System (ATS); every account has a unique identification number and every allowance a unique serial number. Appeals may be directed to the Environmental Appeals Board.

The Acid Rain Program represents a dramatic departure from traditional regulatory (command and control) methods, harnessing the incentives of the free market to reduce pollution.

6.16.1 The Bubble Concept

The bubble concept was introduced as an early basis for emission trading in the USA. While an entire industrial facility must comply with the overall emission ceiling set by say a State Implementation Plan, the bubble concept enables managers to develop their own strategies for different levels of control at different sources. Managers may impose more severe requirements on plants that can be restricted fairly inexpensively, in exchange for reduced controls on outlets that are more expensive to control. There is no incentive within the bubble concept to reduce total emissions below the required limits, but it provides a measure of flexibility. It should be noted that the emphasis is on emissions and not on ground level concentrations; hence, the plant that is more expensive to control may be the plant that contributes most to ground level concentrations. Also, stack height is an important element in control – an element ignored in the bubble concept. Short stacks make a disproportionate contribution to ground level concentrations. Consequently, the characteristics of emissions may be just as important as the total mass (see Figure 6.1).

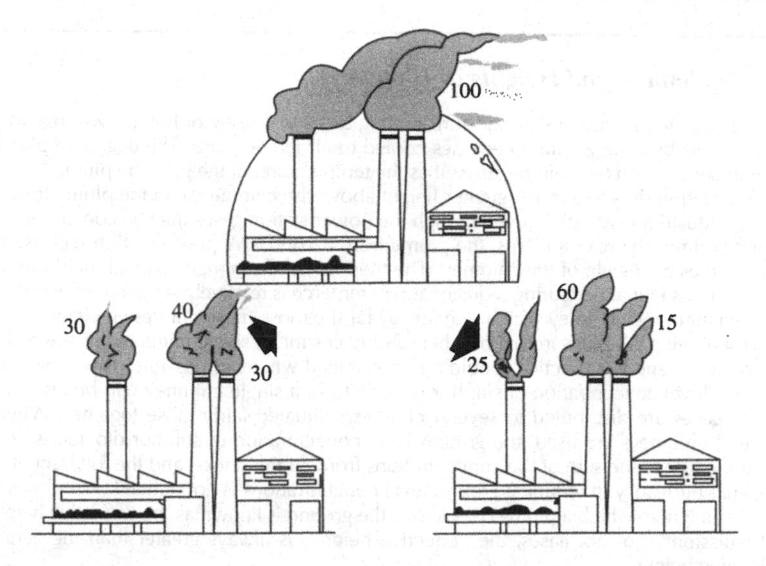

Figure 6.1 *The bubble concept: arrangements (percentages) that meet the target for emissions from the installation as a whole. (Source: US EPA)*

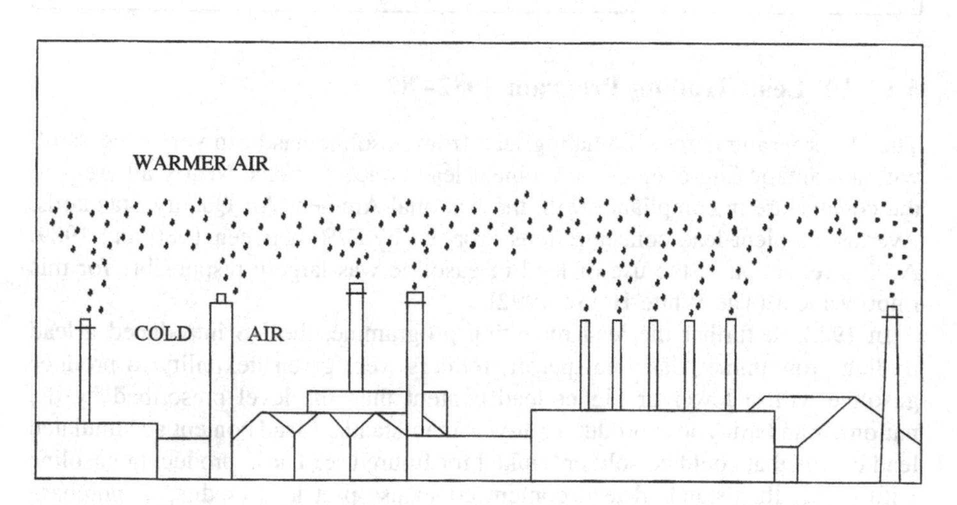

Figure 6.2 *Temperature inversions commonly trap pollution from short stacks – another factor often ignored by theorizers*

Box 6.7 sets out the relevant points about stack heights, while Figure 6.2 illustrates another factor limiting the dispersion of gases quite frequently, i.e. the temperature inversion. Both these factors tend to be ignored by commentators. The bubble concept appears, therefore, to be seriously flawed.

Box 6.7 Number and Heights of Chimneys

Hot gases are less dense than the surrounding air, and a body of hot gas will rise as a result of its buoyancy until it becomes cooled to air temperature. The degree of plume rise depends upon the volume, as well as the temperature, of the gas. The plume from a power station rises to a much greater height above the chimney than the plume from a small industrial installation, even though the power station gases may be cooler. Under normal atmospheric conditions, the plume from a 2000 MW power station will rise to three times the height of the chimney. The thermal rise from many typical fuel-burning installations in manufacturing industry and commerce is relatively small compared with the height of the chimney which remains by far the most important design element.

Discharging the gases from a number of furnaces into a single chimney increases the thermal lift, and this practice should be encouraged wherever possible. The maximum ground level concentration of sulphur dioxide from a single chimney will be less than if the gases are distributed to several chimneys situated fairly close together. Where several chimneys are used, the ground level concentration of sulphur dioxide is very nearly equal to the sum of the concentrations from all the stacks, and the full benefit of thermal buoyancy in reducing ground level concentrations is not utilized.

The height to which a plume rises above the ground is known as the "effective height of emission". For hot gases, the "effective height" is always greater than the actual chimney height.

Source: A. Gilpin (1963) Control of Air Pollution, Butterworths, London, pp. 331–2.

6.17 US Lead Trading Program 1982–87

The US programme for eliminating lead from gasoline has been very successful, with a dramatic improvement in ambient lead concentrations. Nearly all areas of the country are in compliance with the National Ambient Air Quality Standards. Average ambient lead concentrations dropped by 87% between 1980 and 1989. A 99% reduction in the use of lead in gasoline was largely responsible for this improvement (The White House 1992).

In 1982, to further the lead reduction programme, the US introduced a lead trading programme. Gasoline (petrol) refiners were given flexibility to produce gasoline with a lower or higher lead content than the level prescribed by the national standard. Those producing lower-than-standard lead content accumulated lead credits that could be sold or banked for future use. Those producing gasoline with higher-than-standard lead content could use past lead credits, or purchase them from other firms. About 15% of total lead rights were traded and 35% were banked and traded, or used later on (Panayotou 1994). The US EPA estimated the annual savings from lead trading to be US$200 million. This means that the lead standard mandated by regulation was achieved at a cost 20% lower with trading than without trading. The merits of the lead trading scheme were as follows:

1. there was consensus about the objective of phasing out lead in gasoline;
2. lead in gasoline can be easily monitored;

3. only a small number of refineries were involved; and
4. no complex approval process was involved.

The trading of lead rights (interrefinery averaging) resulted in a very active market. The programme allowed refiners to trade the severely limited rights to lead additives. While the programme officially expired in 1986, refiners were permitted to make use of banked rights throughout 1987. The success of the scheme stemmed largely from the absence of a large body of restrictions on trades; a quarterly report to the EPA on gasoline production and lead usage was all that was required (Cropper and Oates 1992).

6.18 Chlorofluorocarbons (CFCs) Phase-out Market Mechanisms

The Montreal Protocol on Substances that Deplete the Ozone Layer was an international agreement reached in 1988, by more than 30 countries, aimed at protecting the ozone layer by controlling the emission of chlorofluorocarbons (CFCs) and halons. Under the agreement, initiated by UNEP, CFC consumption was to be progressively reduced world-wide, being phased out completely by 1996. The consumption of halons was not to be allowed to increase above the levels of 1988. This protocol implemented the Convention for the Protection of the Ozone Layer (the Vienna Convention) of 1985. The CFCs and halons had been found to be harmful in the upper atmosphere as ozone-depleting agents. Such impairment of the ozone layer would allow more ultraviolet light to reach the Earth, with detrimental effects such as an increasing incidence of skin cancer.

CFCs had been widely used in a range of products and industrial processes, such as refrigerators, polystyrene packaging, in the electronics industry and as aerosol propellants. The US, among others, banned production of CFCs by 1996. US measures also included a tradable permit regime covering CFC manufacturers and importers, an excise tax on ozone-depleting chemicals, the development of alternatives, and changes to defence procurement rules.

Alternatives to CFCs were found more readily than had been anticipated, and the change-over proved much cheaper than expected. Consumption of CFCs fell rapidly; reductions occurred ahead of the legally mandated timetable. The two main market instruments in the package of measures – the tradable permits scheme and the ozone-depleting chemical tax – stimulated the progress made, with low administrative costs.

The marketable permit system allocated CFC allowances to existing importers and producers of CFCs and halons, based on each firm's market share in 1986; the initial allocation of allowances involved less that 30 firms. Firms could trade permits among themselves, while the system capped and progressively reduced the aggregate use of CFCs. The system was operated by a staff of four at the US EPA. As the market for CFCs declined, the system allowed firms to rationalize production between different production facilities according to the least-cost pattern of supply. The ozone tax was progressively increased, and the scope of the tax was broadened to include additional compounds. This added substantially to the prices of CFCs, and was another reason for the rapid fall in CFC demand.

6.19 Three Interesting Cases

The Wisconsin system of transferable discharge permits (TDP), trading of rights for phosphorus discharges into the Dillon Reservoir in Colorado, and a nutrient management system for the Tar-Pamlico River, North Carolina provide three interesting case studies of discharge control.

The Wisconsin TDP system established a framework under which the rights to BOD discharges could be traded among sources. However, since the scheme's inception in 1981 on the Fox River there has been only one trade, when a paper mill transferring its treatment activities to a municipal waste-water facility also transferred its discharge rights to that facility. A set of quite severe restrictions appears to have discouraged transfers. First, trades must be justified on the basis of need, and this does not include cost savings. Secondly, trades are effective for periods only between one and five years. Thirdly, trading is limited to avoid hot-spots, i.e. locations of high toxicity. Other uncertainties also discouraged trading.

According to a US EPA study in 1983, the Dillon Reservoir, Colorado, would have become eutrophic if waste-loads had continued to increase. In 1984, Colorado introduced a scheme requiring advanced treatment methods for all point discharges, special measures for non-point sources, and a plan that allowed the trading of rights to discharge phosphorus into the reservoir. The aim was to restrict pollution to 1982 levels, despite economic growth. The EPA reviews and approves all trades recommended by the State. Every three years, the Colorado Water Quality Control Commission is required to review and set waste-load allocations. In this case, few restrictions were imposed on trading and potential annual savings have been significant. Although only one trade has been completed, trading is expected to increase.

In 1989, the Tar-Pamlico River, North Carolina, was designated a Nutrient Sensitive Water. Following this a nutrient management strategy was devised to ease eutrophication problems and outbreaks of fish diseases. Water quality standards have been established using chlorophyll as a direct measure of algal growth and an indicator of eutrophication. During Phase 1 of the programme (1990–1994) there were provisions for a nutrient-reduction trading programme. In Phase 2 (1995 onwards) a long-term nutrient-reduction programme was to be implemented. The trading programme was designed to allow the waste-water treatment plants in the river basin to fund less-expensive non-point source controls to avoid the high compliance costs associated with the upgrading of major facilities.

A coalition of waste-water treatment plants (the Tar-Pamlico Basin Association) is held responsible for achieving a nutrient loading target for the entire association. Each member must monitor and report on loadings. It is anticipated that trading will achieve equivalent or better water quality than would have been achieved under the originally proposed effluent limits.

6.20 NSW Nitrogen Oxide (NO_x) Trading Scheme

In 1999, New South Wales, the largest state in Australia, with a population of five million, introduced a tradable permits scheme for nitrogen oxides together with several other measures as part of its Air Quality Management Plan. A licensing scheme requires heavy industries, including power stations, to reduce NO_x emissions to help the capping of 1998 levels by 2005. The plan allows companies to buy credits to permit an increase in emissions at a given plant, but only when there has been an equivalent reduction at another plant. The total scheme is administered by the NSW EPA, which adopts particular measures to prevent the development of hot-spots and the hoarding of credits by particular companies. Companies may also acquire credits by buying and scrapping thousands of old and inefficient vehicles. The total pollution load over Sydney will gradually be constrained.

The plan to contain car use will involve a measure known as Vehicle Kilometres Travelled (VKT). Unless constrained, the VKT in Sydney is expected to jump from 30 billion to 40 billion kilometres by 2020. The aim is to achieve zero growth over that period by a variety of measures including a substantial increase in public transport, e.g. the light rail networks already being developed. The public transport strategy for Greater Sydney may take 10 years to complete, during which time it is hoped that public transport usage in western Sydney will double. The population of Sydney is about 3.5 million.

There is also a supplementary scheme to reduce emissions from private homes, including solid fuel heaters, petrol lawn mowers, and various powered garden tools.

6.21 Deposit-Refund Systems

In deposit-refund systems, a surcharge is imposed on the price of potentially polluting products, and when these products are returned to approved collection points, the surcharge is refunded. The surcharge is intended to avoid pollution and promote recycling. Deposit-refunds have been applied to beverage containers, car batteries, tyres, car bodies and paint cans. Beverage containers include glass and plastic bottles and cans. Mandatory or voluntary deposit-refund systems are to be found in parts of the USA such as Michigan and Oregon, as well as in Sweden, Norway, Denmark, Germany, Belgium, Finland, Hungary, Poland, Austria, Canada, the Netherlands and South Australia.

South Australia is the only state in Australia that has introduced specific legislation for deposit-refunds. A High Court challenge by a major brewery indicated that the system had arguably created market impediments for beer producers in South Australia compared with producers in other states. The South Australian scheme has nevertheless enjoyed wide public acceptance since its introduction in 1975, with high return rates for beverage containers.

6.22 Performance Bonds

Performance bonds involve the placing with an environmental agency of sufficient funds to cover the cost of rehabilitation of a project in the event of the failure of the enterprise concerned to discharge its conditions of consent. Thus a guarantee is provided to government in advance against the risk of default by the developer, particularly in respect of rehabilitation requirements. Bonds may also be used to implement other environment protection measures, such as security against environmental damage, and for the use of a wide range of natural resources where proper rehabilitation and restoration is required.

In a specific application, mine bonding is the depositing of a sum of money (the posting of a bond) by a mining company before the commencement of operations, sufficient to guarantee the reclamation and rehabilitation of the area to be mined. If a miner goes bankrupt or fails to comply with the conditions imposed, the mining company forfeits the bond to the consent authority which then becomes responsible for the rehabilitation of the mine site. If conditions are met, the bond is refunded with interest.

In the state of Queensland, Australia, under the Mineral Resources Act 1989 bonds are routinely required. The higher the risk of poor environmental performance, the greater the bond required. A category system has been adopted to determine the risk of lease-holder non-performance on a particular lease; there are six categories. A security may be lodged as a cash payment (on which interest will be paid), a guarantee or indemnity from a financial institution, or a written guarantee from a company.

6.23 Environmental Assurance Bonds

The environmental assurance bond is a variant of the performance bond, to suit situations in which there are no precedents. Agents undertaking projects for which there are no precedents are required to post a bond with the environmental authority equal to the expected or possible worst case losses. It matches the value placed by the environmental authority on allowing the activity to proceed given the current state of knowledge about its wider and longer-term effects. The bond may be reviewed from time to time as experimental data become available on the user or external costs of the activity.

Such bonds should be required from resource users only where the future environmental costs of present activities are commercially uninsurable, because the actuarial risks cannot be calculated from historical data. The purpose of a bond is to indemnify society against the potential environmental costs of unprecedented activities, and provide an incentive for research activity to minimize losses that could arise out of the use of the resource. The bond is a precautionary instrument in the sense that it imposes the cost of anticipated environmental damage on the resource user in advance (Perrings 1989; Costanza and Perrings 1990).

6.24 Pay-by-the-bag System

The pay-by-the-bag system was designed to reduce the volume of solid waste (trash or garbage) being disposed of from domestic premises, and ease the pressure on increasingly overburdened and scarce landfill sites. Landfill is the most common form of disposal for household, commercial and industrial refuse, and it appears that 80–90% of the world's refuse will be disposed of by this method for several years to come. Clearly, a scheme to reduce the volume of material going to landfill is to be welcomed.

In many countries, the collection and disposal of garbage has been'free' in the sense that there is no specific charge at the time of removal; the overall cost has been met from local taxes or rates, or from property taxes. The fact that there is no specific charge means that it costs nothing to the householder to put out an extra bag or two (or fill an extra bin). Yet the cost of landfill operations rises from year to year due to the filling of sites, retirement of environmentally undesirable sites, and the increasing distance to new sites. The marginal cost of disposal increases annually, and there is no incentive to economize while communities are forced to spend too much on waste services.

The incentive problem has been corrected in some communities by charging households the full incremental cost of waste disposal, and in the USA hundreds of communities have introduced such schemes (Repetto 1993). It has been found that households that 'pay by the bag' respond positively to price signals: a community raising its collection fee from zero to $1.50, in line with marginal costs, may reduce waste generation by 20%. Fees combined with curbside recycling (of glass, paper, plastic and cans) may achieve a reduction of up to 30%. The revenue raised in this way eases the strains on local or property taxes, and reduces the scale of collection services.

The commonest system in US cities is to sell stickers or tags which householders attach to rubbish bags. The price of a sticker is set to equal the marginal cost of collecting and disposing of one bag. Only bags with these labels are picked up in the weekly collection. There are, however, costs in running such schemes, e.g. the costs of printing stickers and selling them. Further charging leads to a frantic dance called the 'Seattle stomp', in which householders cram more refuse into every bag and bin by stomping; so weight reduction will prove more modest than volume reduction. In addition, some people resort to illegal dumping, and this may prove to be substantial. However, it has been found that people prefer to recycle for free than pay to have garbage removed; hence recycling sharply improves. Incentive schemes can have some unforeseen consequences, both good and bad (*The Economist* 7 June 1997).

6.25 Debt-for-nature Swaps

Debt-for-nature swaps refers to the promotion of nature conservation projects out of the vast indebtedness incurred by many less-developed countries during the last twenty years, a period during which many lenders and investors found that they would never be repaid. It became possible to buy fairly worthless debt instruments from banks and institutions at prices much below face value. Lenders were glad to receive something, rather than lose all. Conservation organizations then found that it was possible to acquire some of this debt cheaply, with subscribed money. These bodies were able to negotiate a settlement of the debt with the debtor country on the understanding that the funds received to settle the debt would be invested in the country concerned on conservation projects. The indebted countries thus shed debt at a substantial discount, while the proceeds did not leave the country.

Conservation organizations have been able to achieve significant increases in the resources available for conservation in debtor nations. Debt-for-nature swaps have been endorsed by the World Bank, the Global Environment Facility, and the Group-of-Eight countries (the eight leading industrial nations: the USA, Japan, Germany, France, the UK, Italy, Canada and Russia). They are also known as debt-for-environment swaps. The procedure was applied initially in 1987 in Bolivia, and then in a number of countries including Poland. US$650 000 of Bolivia's debt was bought for US$100 000 (roughly 15 cents in the dollar) in the secondary financial market and retired. In exchange, the Bolivian government expanded protected areas around the Beni Biosphere Reserve by 1.5 million ha. It contributed US$100 000 to the protection programme which also attracted a US$150 000 grant from the US Agency for International Development (Zylicz 1992).

6.26 Road-pricing Schemes

The traditional response to traffic congestion has been the building of more and better roads; however, budgetary limitations and the ever-increasing demand for more roads as traffic increases have focused attention on demand management in general and road-pricing in particular. The introduction of toll charges has served both as a cost recovery technique and as a regulator of traffic, marginally

Box 6.8 Singapore's Road Pricing Scheme

- Singapore has an area of 646 km^2 and a population in 1996 of 3 045 000. The population density is 4713 persons per km^2. The population is likely to reach 4 000 000 by 2010.
- Singapore has over 365 000 passenger cars, and over 138 000 trucks and buses. It has 220 vehicles per road kilometre, which is one of the highest densities in the world.
- To curb growth, the Singapore government has imposed high duties on imported cars, coupled with high registration and road tax. Anyone buying a new car must have a permit, which is very expensive.
- Since 1977, to enter the central area, covering about 7 km^2, or to enter the expressways, drivers need to display a compatible licence; tolls must be paid at peak and off-peak times.
- In 1998, the licence system became automated with stored-value smart cards. With a built-in microprocessor chip, these can be preloaded to carry a credit. At toll plazas, the charging will be automatic by means of microwave communication between an electronic device in the car (with the inserted smart card) and an overhead gantry, with traffic moving at speed. Motorists are consequently charged quite precisely for their actual contribution to road congestion. Those who pass a gantry without a valid charge card are snapped by video camera and fined.
- The scheme will be supported by world-class public transport.

discouraging use. The main objection to tolls has been the need for drivers to stop and pay the toll either by cash or pre-paid ticket, detracting from some of the benefits of faster routes and recreating a measure of traffic congestion. In response to this problem, electronic automatic charging systems have been developed in Denver, Colorado; Cambridge, England; Singapore; Bergen, Oslo and Trondheim, Norway (see Box 6.8).

The Colorado Toll Highway E-470 was opened in 1991. The system here enables cars to drive through at speed, the toll booth automatically charging a toll to the driver's credit card by picking up electronic signals from the ID card located in the vehicle. The system has the capacity to vary the toll according to the level of congestion, charging higher tolls during rush hours. Drivers may respond to this by using alternative routes or travelling at alternative times. Congestion at critical times is much reduced, with pollution reduced because of higher speeds. In addition, the government or road-operator raises revenue for road maintenance and further construction works. The main objection has been that the individual's movements are monitored, although this objection can be offset by separate toll booths for those who wish to pay cash (Panayotou 1994).

Many transport planners believe that charging for road use is the best way of tackling urban traffic congestion, for which there are alternative approaches. Apart from the above, charges may be imposed on all private cars entering delineated central city areas. More cheaply perhaps, traffic congestion may be eased in the central business district by a combination of high parking charges and limited parking facilities, coupled with efficient public transport from the outer limits, achieving a balance of supply and demand through the medium of price. Traffic may also be excluded by a marked increase in pedestrian-only precincts. All such schemes need to be complemented by ample and inexpensive parking at the fringes of congested areas. In Britain, doubling the real price of petrol (gasoline) over the next decade has been recommended by the Royal Commission of Environmental Pollution, though this may not reduce urban congestion at critical times (*The Economist* 22 June 1996).

6.27 Peak-load Pricing

Peak-load pricing (or time-of-day pricing) endeavours to recover the capital costs of a facility at peak-load times, matching long-run marginal costs; at off-peak times prices recover only operating costs, or short-run marginal costs. The effect is to curb peak demand and boost off-peak demand, hopefully achieving a higher overall load factor, and reducing the need for additional capital investment to meet peak times. Peak-load pricing is well-established in electricity supply system tariffs, economizing in the use of generating equipment, and reducing pollution. The principle is also applied in public and freight transportation, by road, rail and air, and some buffer stocks.

6.28 Noise Fees

In some countries, charges are imposed at airports on levels of noise nuisance. The charges levied on aircraft may be on a per-aircraft or per-passenger basis, according to the type of aircraft and the noise level that is reached during take-off and landing.

The Netherlands has an aircraft noise charge or levy based on aircraft weight and noise characteristics. The revenues from this are used for noise insulation programmes.

In the USA, a tax on all airline tickets goes into a central fund, and a small proportion of this goes into environmental programmes around airports. In Japan, a special aircraft landing fee has been in operation since 1975 to finance noise abatement measures around airports. The fee is based upon aircraft weight and the sound levels on landing and take-off. In Switzerland also, an aircraft noise charge finances insulation programmes. Germany, Belgium, Hungary, Norway, Poland and Sweden also impose noise charges.

6.29 Transfer of Development Rights (TDRs)

Tradable permits issued by US local governments are increasingly being applied to land use in environmentally sensitive areas, redistributing development rights without much cost to government and without penalizing property owners. A transfer development right (TDR) involves the transfer of the rights to develop a given piece of land to a second piece of land, the first piece of land losing development rights, and the second gaining development rights it did not have before. Usually in such a transfer, the second property owner pays the first for development rights at a price which reflects the value of the rights had that land been developed. This enables the local jurisdiction to freeze large areas of environmentally sensitive land for public purposes, without the risk of giving ground for a 'taking issue' under the US Constitution that might require payment from the public purse of 'just compensation'. It also avoids much displeasure among farmers and other landowners when land is zoned in such a way as to involve substantial losses.

The transfer of purchased development rights to other areas requires local government approval, involving as it does development beyond the original environmental plan, but such transfers have a pragmatic value. The first owner makes nearly as much money by not developing as would have been made by developing, while the second landowner may undertake additional development which allows a recovery of the development costs. It avoids the need for government to compensate for the loss of development rights with public funds and ensures that zoning ordinances of a restrictive kind may be imposed without opposition. To government, such a transfer is essentially a trade-off, to achieve the most desirable public outcome. The customary procedure is for local government to create certificates of development rights which are made available to landowners in the affected district. Landowners who purchase these certificates become entitled to increased densities and floor areas in designated areas.

The concept was pioneered by New York City in the 1970s, when developers were allowed to build extra stories on one site, in exchange for preserving environmental amenities such as open space or heritage assets nearby. On Long Island, east of New York City, several local councils have created a system of TDRs to preserve the Pine Barrens, a 40 500 ha area that provides the region's water supply. Where new construction is forbidden, property owners have been issued with Pine Barrens Credits in compensation. The credits may be sold on an exchange or to a developer; they may be used to build more than would otherwise be allowed in approved nearby areas. Other states are developing TDR programmes, although not all schemes have been successful.

6.30 Transfer of Water Rights and Water Banking: The Colorado River

The Colorado River is a major river of North America, rising in the Rocky Mountains of Colorado and flowing for some 2330 km, into north-western Mexico. It has a drainage basin of some 632 000 km^2, embracing seven states: Wyoming, Colorado, Utah, New Mexico, Nevada, Arizona and California. The Colorado

River drains the most arid sector of the continent, through a number of extremely steep trenches, of which the Grand Canyon is the largest. The Colorado river system was the first drainage basin in which the concept of multiple use was put into practice: it is used for power, irrigation, flood control, recreation and navigation. In 1922, a Colorado River Compact was entered into by all the affected states for the fundamental uses of the water. The first major development was the Boulder (now Hoover) Dam, a major engineering feat of its time, followed by a range of hydroelectric projects. In all, more than 20 dams have been built on the Colorado and its tributaries. The river provides water to 25 million people (16 million in southern California alone).

Under a new plan approved by the seven state legislatures, the US Department of the Interior allows exchanges of river water between the lower-basin states of California, Arizona and Nevada. Water transfers may be sold. California is to restrict the wasteful use of water in farming areas such as the Imperial Valley, and try to meet the needs of San Diego.

The new policy is known as 'water banking'. Arizona is allowed to bank the river's water in aquifers for the Southern Nevada Water Authority. When Nevada needs water it takes its share of the river water plus some of Arizona's unused allotment. Nevada issues a credit, meeting all the costs. Arizona needs the money to help run the 540 km Central Arizona Project, a multi-billion dollar canal, the largest and costliest water transfer scheme ever constructed.

A parallel agreement exists between the Metropolitan Water District of Southern California (MWD) and the Imperial Irrigation District (IID), under which the IID transfers to the MWD a significant amount of its rights to Colorado River water, in return for which the MWD finances conservation measures for the IID. Obtaining water in this way appears to be the most cost-effective way for the MWD to make up for its declining share of Colorado River water when the federal government's Central Arizona Project began operation in 1985. The future transfer of water between IID and MWD depends in part on an assessment of the environmental effects.

6.31 Oil Pollution Compensation and Insurance

The International Oil Pollution Compensation Fund (IOPC Fund) is a worldwide intergovernmental entity created by the International Maritime Organization (IMO) under two conventions, the Civil Liability Convention and the Fund Convention, which came into force in 1975 and 1978 respectively. In 1995, some 91 nations were parties to the Civil Liability Convention and 66 nations were parties to the Fund Convention.

The Civil Liability Convention lays down the principle of strict liability for shipowners in respect of oil pollution damage and creates a system of compulsory liability insurance. The Fund Convention provides for compensating victims when compensation under the Civil Liability Fund proves inadequate. The IOPC Fund is administered by a Secretariat located in London, UK. The conventions apply only to spills of 'persistent' oil such as crude oil, fuel oil, heavy diesel oil and lubricating oil. Damage arising from non-persistent oil such as gasoline, light diesel oil and kerosine is excluded from the conventions. Non-persistent oils tend to evaporate quickly and do not require cleaning up. The conventions therefore cover incidents in which persistent oil has escaped or been discharged from a sea-going vessel actually carrying oil in bulk as cargo at the time of the spill, and within the territory or territorial sea of a nation which is a party to one or both conventions. Spills from tankers during ballast voyages, and spills from ships other than tankers are not covered.

Pollution damage includes the costs of measures taken after an oil spill has occurred to prevent or minimize pollution damage, and includes clean-up operations on shore and at sea. Reasonable costs for the disposal of collected material are allowed. Claims for pure economic loss by victims are also admissible. Initial and annual contributions are payable by parties to the conventions; the levels of contributions to the Fund vary from year to year.

In 1994, the USA introduced Certificates of Financial Responsibility (COFR) to be carried by all oil tankers entering US waters. Their purpose is to provide insurance against the possibilities of environmental damage. This requirement springs from the US Oil Pollution Act of 1990, legislation introduced following the *Exxon Valdez* disaster. The tanker *Exxon Valdez* ran aground in March 1989 in Prince William Sound, Alaska, on the Bligh Reef. Some 250 000 barrels of oil poured into the Sound. Over 2400 km of beach were fouled by the spill. A jury decided that the owners and the captain were reckless, allowing claims for damages.

6.32 Individual Transferable Quotas in Fisheries Management (ITQs) in New Zealand, Australia, Iceland and Canada

Without some form of restriction on the behaviour of individual fishing fleets, fish stocks tend to be exploited for immediate profit, with insufficient fish being left to maintain fishing stocks in the long term. In typical open-access unregulated situations, each operator decides on a level of fishing that will maximize short-term private profits, there being no incentive to have regard to the long-term effects of overfishing. Some regulation of fishing activities becomes essential to preserve the long-term profitability of the industry; in other words, to achieve and maintain sustainable fishing.

Two main forms of regulation have been practised: restraints on access to existing fisheries with limits on the size of boats; and *total allowable catch (TAC)* limits on the aggregate amount of fish that may be taken in a season. Both approaches have their limitations. As a consequence several countries have adopted economic instruments to discourage overfishing. These have taken the form of *individual transferable quotas (ITQs)*, which give individual fishing operators an entitlement to a particular level of catch over a given period. This entitlement eliminates the early-season race for fish, characteristic of the TAC policy, allowing efficient fishing over the season as a whole. The quotas issued restrict the total catch taken, while pressure is removed from individual operators. ITQ systems have been employed in New Zealand, and in a modified form in Australia, Iceland and Canada.

The ITQs are transferable property rights allocated to fishers in the form of a right to harvest any amount up to the level of the quota over a specified fishing period. The range of permissible trades is very wide.

The New Zealand ITQ system was established following the creation of the 200-mile EEZ in 1978, which brought under national control deep-sea fisheries previously exploited by foreigners. The ITQ system covers all the commercially significant fisheries, except for tuna fishing. A quota trading exchange was established to facilitate trades. Resource rentals are payable to the government for each quota, regardless of whether quotas are fully used or not.

Australia first introduced an ITQ system for blue fin tuna fishing in 1984, the system gradually expanding, although most of the fishing industry remains regulated through entry and other limitations. In Iceland, ITQs were introduced in 1984, and the system appears to have been of considerable benefit. In Canada, the ITQ system has been gradually introduced since the 1980s, but not all quotas are tradable and the system remains mixed with other forms of regulation.

6.33 Carbon Bonds: Certified Tradable Offsets (CTOs)

Presented as a model of north–south co-operation, certified tradable offsets were introduced in 1996 by the Costa Rica Government, as part of an effort to save what is left of the country's forests. The scheme is under the auspices of UNCTAD, the UN Conference on Trade and Development, and became part of the Kyoto Convention on Climate Change in 1997. In effect, industrial development is allowed provided the emission of carbon dioxide is offset by the creation of forest that will absorb an equivalent amount. The going rate is a ten dollar bond per tonne of carbon dioxide fixed by the forest. Many CTOs have been sold, about half having been bought by the Norwegian Government. Apart from reforestation schemes, tourism would also benefit greatly. However, forest management and the prevention of illegal logging remain a problem for Costa Rica. The scheme itself has been seriously criticized by national and international environmental groups.

6.34 Summary

This chapter has provided examples of a range of economic instruments used as ways and means of influencing human conduct in the areas of pollution control and environment protection.

Unless ear-marked for specific environmental purposes, as in the case of noise fees, charges and taxes will augment the coffers of government. In time, some charges will be avoided by changes in company practices, but a substantial flow of income will ensue.

This flow of income could be used by government to reduce other taxes falling on industry and polluters in general, such as labour charges (payroll taxes), company taxes, and other fiscal burdens. In respect of reducing payroll taxes, there is clearly a double dividend, for the burden of pollution is reduced while it becomes cheaper to employ labour at the margin. This probably slight fall in production costs will increase competitiveness. In this way, fiscal and environmental policies can be made mutually reinforcing. Most OECD countries have introduced various ecotaxes, yet few have implemented comprehensive green tax reforms (OECD 1997b).

A reservation expressed in this chapter concerns the ability to acquire credits for reducing emissions below the statutory limits imposed on a particular plant, and transferring those credits to plant more expensive to control and possibly in a less desirable location. Some plants have poor equipment and short stacks, resulting in higher ground level concentrations. To argue that the mass emission is all that matters is invalid, for it is the exposure of the public and the workforce that is crucial. To introduce pollution credits to a bad plant that needs urgent renovation or renewal, i.e. to aggravate hot-spots, cannot be considered good practice even though the mass emission target within a district or bubble may be met. With mass limitations, it is possible to maintain high ground-level concentrations at Plant A, without compensatory reductions at Plant B. The reduced costs of pollution control, often claimed, are in fact purchased through a reduced performance in environmental terms within the vicinity. The public suffers, while most economists appear to have missed the point.

Another reservation concerns the transfer of development rights, which can be purchased in one location, and transferred to another location. There development may be undertaken as of right without regard to environmental planning for the county or state, and in the face of opposition from residents. Once again the public suffer, while their rights to oppose are overridden by the imposition of the purchased rights to develop, without permission or consent. The market transcends the planning system and the welfare of the public.

Further Reading

Joskow, P. and Schmalensee, R. (1996) *The Political Economy of Market-based Environmental Policy: The US Acid Rain Program,* Centre for Energy and Environmental Policy Research, Massachusetts Institute of Technology, Cambridge, MA.

OECD (1997) *Evaluating Economic Instruments for Environmental Policy,* OECD, Paris.

Opschoor, J.B. and Turner, R.K. (eds) (1994) *Economic Incentives and Environmental Policies: Principles and Practice,* Kluwer, Dordrecht.

Panayotou, T. (1994) *Economic Instruments for Environmental Management and Sustainable Development*, UNEP Paper No. 16, UNEP, Nairobi.
Svendsen, G.T. (1998) *Public Choice and Environmental Regulation: Tradeable Permit Systems in the USA and Carbon Dioxide Taxation in Europe*, Aarhus School of Business, Denmark.
Tietenberg, T. (1990) Economic instruments for environmental regulation, *Oxford Review of Economic Policy*, vol. 16, Oxford University Press, Oxford.
Zylicz, T. (1992) *Debt-for-environment Swaps: The Institutional Dimension*, Beijer International Institute of Ecological Economics Discussion Paper No. 18, Stockholm.

7

Cost–Benefit Analysis and Valuation

7.1 Introduction

Cost–benefit analysis (CBA), or benefit–cost analysis (BCA) in North America, has attracted a great deal of attention in recent years. It may be viewed at three levels:

1. a financial assessment, by a potential investor, of the accountancy costs and returns of a venture, these findings indicating the expected returns on capital invested;
2. a broader assessment that takes account of external costs and benefits that may be generated by the venture locally and perhaps regionally, information essential to the decision-maker;
3. an assessment broader still of the implications for the economy as a whole and the national interest.

The narrowest assessment determines, among a range of options, the one likely to be pursued by the ambitious investor and developer; the second decides which project is most likely to meet with public approval within the planning system; while the third places the project within a national context and may influence the highest authorities to encourage or discourage such investments. Several boxes in this chapter draw on expert opinion on the value of CBA.

The chapter then dwells on the nature of discounting and the controversial issue of the choice of the discount rate. The views of Lawrence Summers of the World Bank are quoted. Cost-effectiveness analysis and the influence of the multiplier are discussed, followed by the all-important issue of CBA at the political level. CBA does not falter and fail at the minister's desk, but continues just as briskly in a different, political, mode.

The chapter then turns to the question of the valuation of environmental assets, including opportunity cost, preventive expenditure, surrogate market techniques, hedonic price techniques, wage differential approaches, travel cost approaches, survey-based approaches, contingent valuation and Delphi techniques. A comparison of techniques is attempted.

7.2 Cost–Benefit Analysis

Cost–benefit analysis (known in the USA as benefit–cost analysis) is a procedure for comparing alternative possible courses of investment or action by references to the net benefits that they are likely to produce. The term 'net social benefits' refers to the difference between social benefits and social costs, the one being subtracted from the other. The results may be positive or negative. As far as possible, the costs and benefits are measured in monetary terms; where costs and benefits cannot be readily assigned dollar figures (the 'intangibles') they are separately identified and described for assessment in a wider context by the decision-maker. In general, a programme having a high benefit–cost ratio will take priority over others with lower ratios, although political factors may intrude.

The procedure was first proposed in 1844 by the French engineer A.-J.-E.-J. Dupuit (1804–1866) in respect of public works, but it was not seriously applied until the 1936 US Flood Control Act, which required that the benefits of flood-control projects should exceed their costs, i.e. that projects be undertaken only 'if the benefits to whomsoever they may accrue are in excess of the estimated costs'. Efforts to implement this requirement led to the preparation of the *Green Book*; this report codified the general principles of economic analysis as they were to be applied in the evaluation of federal water resources projects. The authors drew on the then emerging subject of welfare economics for the theoretical background.

The year 1958 was noteworthy for the publication of three books on CBA (Eckstein 1958; Krutilla and Eckstein 1958; McKean 1958) These were followed in the early 1960s by a major work on water resource systems (Maass *et al.* 1962) and the first extensive work on the application of benefit–cost analysis in areas other than water resources (Dorfman 1965). The use of CBA spread rapidly to other countries, especially Britian. In 1965, the first comprehensive review of the subject appeared (Prest and Turvey 1965), followed by a major work, *Cost–Benefit Analysis* (Mishan 1976).

Hammond (1958) was one of the first analysts to apply the principles of CBA to pollution control; major advances then followed at Resources for the Future (Kneese 1962; Kneese and Bower 1968; Kneese *et al.* 1970).

Cost–benefit principles have since been applied to the design of public policies in various areas such as electricity power generation, irrigation, airports, road projects, rail services, shipping, urban development and new towns, health services, education and welfare, and to social issues such as equity and social justice, income distribution, and employment opportunities. It has also been extended to a whole range of environmental issues such as global warming, parks and open spaces, environmental planning, and to the framework of environmental impact assessment and sustainable development. In 1992, the *Journal of Economic Literature* published a comprehensive survey of environmental economics encompassing many topics and issues common to CBA (Cropper and Oates 1992).

In summary CBA may be viewed as an attempt to improve the quality of decision-making and social outcomes. It is an application of modern welfare economics, improving the economic efficiency of resource application, broadly considered.

However, CBA has not had a clear run. In fact, the Amendments to the US Clean Air Act in 1970 and to the US Clean Water Act in 1972 explicitly prohibited the weighing of benefits against costs in the setting of environmental standards. The former Act directed the US EPA to set maximum limitations on pollutant concentrations in the atmosphere 'to protect the public health'; while the latter Act set as an objective the 'elimination of the discharge of all pollutants into navigable waters by 1985'. Both Acts resulted in substantial reductions of pollutants to the environment.

Later, references to CBA began to appear in directives and US Acts. Under Executive Order 12291 of 1980, many proposed environmental measures became subject to CBA. Specifically, a Regulatory Impact Analysis (RIA) had to be carried out by federal agencies on all new major regulations to ensure that the potential benefits to society outweigh the potential costs of the proposed action. RIAs have become an indispensable tool in the analysis of the likely effects of government regulations, though they are still not the final determining factor as political considerations are not taken into account. Additionally, the US Toxic Substances Control Act 1976 and the US Insecticide, Fungicide, and Rodenticide Act of 1978 called for weighing benefits against costs in the setting of standards. In addition, economic incentives have begun to creep into US legislation. Amendments to the Clean Air Acts of 1977 and 1990 permitted the setting up of the US Emissions Trading Program.

CBA has been used by decision-makers (governments at all levels and their agencies) in the following ways:

- to accept or reject a single project, programme or policy;
- to choose one of a number of discrete alternative projects;
- to choose a smaller number from a larger number of discrete alternative projects;
- to accept or reject a number of projects;
- to choose one of a number of mutually exclusive projects;
- to help decide whether a proposed programme or policy should be undertaken, or an existing programme continued or discontinued;
- to help choose the appropriate scale and timing for a project, programme or policy.

The method is preferably and usually applied before it is undertaken (*ex ante*) but can be applied after it has been completed (*ex post*). As the future tends to be somewhat unpredictable, expectations and outcomes will often be different.

CBA is often used today in situations where the signals that market prices often provide are either absent or fail to reflect adequately the opportunity cost of the resources involved. Valuations may be made about such matters involving the following:

- savings in travel time resulting from alternative transport options;
- externalities or spillover effects, positive and negative, generated by the project;
- multiplier effects created in the community from a flow of investments;
- costs and benefits imposed on third parties;
- the effects on employment, and its character, in the wider community;
- the implications for infrastructure;
- the implications for the quality of life, intragenerational equity, intergenerational equity, and sustainable development.

All these considerations have implications for environmental impact assessment and regulatory impact statements which assess the likely costs and benefits of regulations created under a wide range of legislation.

The toolkit of CBA embraces the following concepts:

- marginal costs and benefits;
- diminishing marginal utility;
- marginal social and private net products;
- opportunity cost (the cost of the sacrifice of alternative activities, investments or earnings);
- contingent valuation (establishing a monetary valuation of a good or amenity by asking people how much they are prepared to pay for it);
- hedonic price techniques;
- travel costs;
- shadow prices (reflecting the true scarcity of environmental resources for society as a whole);
- discounting (in which all future costs and benefits are reduced to present values);
- optimization or making the most of resources;
- the Pareto optimum (when it is impossible to make anyone better off, without making someone worse off);
- the cost–benefit ratio;
- the role of time;
- the role of property rights;
- sustainable development and ecological evaluation.

Figure 7.1 sets out the key steps in the CBA process as outlined in the current handbook used by the Australian government, while Box 7.1 defines the nature of a cost–benefit (or benefit–cost) ratio. Box 7.2 provides examples of the costs incurred in a project, while Box 7.3 provides examples of the benefits provided by successful projects. Box 7.4 provides an interesting comment by Robert Dorfman

Box 7.1 Cost–Benefit Ratio (or Benefit–Cost Ratio)

The cost–benefit ratio is calculated as follows:

$$\frac{\text{Gross benefits (present value)}}{\text{Gross costs (present value)}}$$

The gross costs and benefits are discounted over the life of the project by a selected rate of interest. The difference between the two amounts is the present value of net benefits. The ratio of the two amounts is the gross cost–benefit (or benefit–cost) ratio. Both considerations are relevant in choosing between projects.

The implicit weighting of the present over the future (i.e. attaching a greater value to a dollar in the hand now, over the value of a dollar to be received at some future time, ignoring inflation) is known as 'discounting' and the rate at which the future is discounted is known as the 'discount rate'. With a discount rate of 10%, a dollar now is deemed equal to 1.10 dollars in one year's time. People tend to prefer the present to the future because of impatience, risk of death, uncertainty, and diminishing marginal utility of consumption. Discount rates may be individual or social.

Box 7.2 CBA: Examples of Costs

- capital expenditure
- costs of raising capital and interest payments
- operating and maintenance costs for the life of the project
- labour and management costs
- housing and social provision for the labour force
- costs of other inputs, e.g. raw materials, goods and services, transport and storage, energy
- costs of using public facilities including proponent contributions to roads, parks, fire services, and training
- research, design and development costs
- opportunity costs associated with use of land
- insurances
- pollution control costs in respect of air, water, land and noise; best available technology
- compensation for social dislocation; acquisition of properties
- protection of high-value conservation areas, threatened wildlife, and cultural heritage
- preparation of development applications and environmental impact statements
- buffer zones and other development consent conditions, including landscaping
- rehabilitation costs

on the subject. In Box 7.5 Munasinghe and McNeely carefully and usefully distinguish between economic returns and financial returns. In Box 7.6, Michael Jacobs indicates the limitations of the concept.

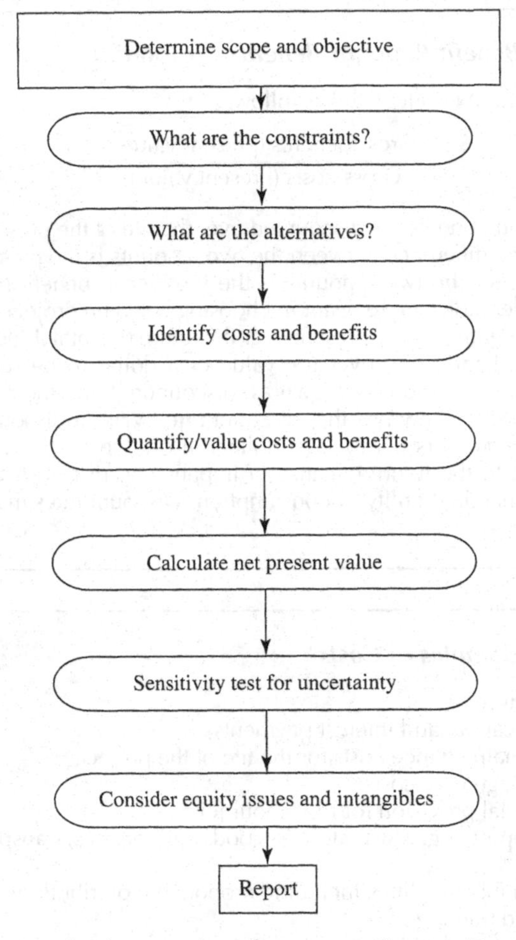

Figure 7.1 *Key steps in the cost–benefit process. Source: Department of Finance (1991) Handbook of Cost–Benefit Analysis, Australian Government Publishing Service, Canberra*

Box 7.3 CBA: Examples of Benefits

- the value of output as a contribution to economic growth
- the multiplier effect of local and regional expenditures
- provision of employment for a range of skills, full-time and part-time, male and female, with high real wages and salaries
- improved productivity with more efficient use of labour and skills

Continued on page 179

Continued from page 178

- the saving of costs that would have been incurred in a 'do nothing' situation
- environmental improvements and other social benefits
- the scrap value of the project's capital equipment and the final disposal of the site
- a possible contribution to sustainable development and the more efficient use of natural resources
- a possible boost to exports, or a reduction of imports
- a contribution to local rates and company taxes for the benefit of the community at large
- a contribution to social infrastructure, such as waterworks, sewage works, roads and highways, and telecommunications
- the return of rehabilitated land to the community, perhaps providing community facilities such as boating ponds, fishing, and other recreational facilities

Box 7.4 Dorfman on CBA

Benefit–cost analysis is closely analogous to the methods of investment project appraisal used by businesspeople. The only difference is that estimates of social value are used in place of estimates of sales value where appropriate ... the heart of the matter lies in deciding what benefits should be included and how they should be valued. The debate about benefit–cost analysis centers on the question of whether the social value of benefits can be estimated reliably enough to justify the trouble and effort involved in a benefit–cost computation.

Source: R. Dorfman (ed.) (1968) Measuring Benefits of Government Investments, The Brookings Institution, Washington, DC, pp. 7–8.

Box 7.5 Costs and Benefits

In purely economic terms, the production of a good is economically justified when the total benefits exceed the total costs; this must include the so-called external costs of dealing with pollution and environmental degradation. However the production of the good may well be profitable in a commercial sense, even if it is not economic. There are many situations where true *economic* returns are negative as nature is destroyed, while *financial* returns (that neglect external costs) are positive and private profits are made ...

Source: M. Munasinghe and J. McNeely (1994) Protected Area Economics and Policy, The World Bank, Washington, DC, p. 2.

Box 7.6 Jacobs on Cost–Benefit Analysis (CBA)

The hope that CBA provided a method by which all the conflicting values involved in environmental decision-making could be transformed into commensurable numbers and slotted into a universal calculus must be disappointed. It is now clear that, even though some environmental goods may be valued in money terms, many of the most important are not commensurable in this way. Effects of irreplaceable and large-scale environmental goods, like those on human life, must be judged directly ... there can be no mathematical short-cut to the "correct" solution ... ultimately such choices must be a matter of judgement, not computation.

Source: M. Jacobs (1991) *The Green Economy: Environment, Sustainable Development and the Politics of the Future*, Pluto Press, London and Boulder, Colorado, p. 219.

7.3 Discounting

The present value or worth of a sum to be received or paid at some future date is such an amount as will, with compound interest at a prescribed rate, equal the sum to be received or paid in the future. There are two main approaches to the choice of discount rate: the social rate of time preference, corresponding to society's preference for present as against future consumption; and the social opportunity cost of capital, corresponding to the rate of return on investment elsewhere in the economy. Estimates of the latter are often higher than estimates of the former. Box 7.7 indicates how present value or worth may be calculated, while Tables 7.1 and 7.2 show the effects of discounting at various rates of interest and periods of

Box 7.7 Present Value or Worth

The present value or worth is the discounted sum of a series of payments or receipts (costs and benefits), discounted annually at a selected rate. The present worth method offers a general and systematic approach to the problem of making comparisons between present proposals for investments, plant, programmes or policies, when future streams of costs and revenues are relevant to the comparison.

The present worth of a future sum may be calculated by multiplying the sum by a present worth factor. For a 12% discount rate, the present worth factor is calculated as follows:

$$\text{Present worth factor} = \frac{1}{1.12^m}$$

where m equals the number of years until the future sum is received or paid.

Some commentators have objected to discounting, certainly at high rates, on the grounds that it appears to discount the interests of future generations and is therefore incompatible with intergenerational equity.

Table 7.1 Discounted cash flow

Year	Percentage				
	5.0	7.5	10.0	12.5	15.0
5	0.784	0.697	0.621	0.555	0.497
10	0.614	0.485	0.386	0.308	0.247
15	0.481	0.338	0.239	0.171	0.123
20	0.377	0.235	0.149	0.095	0.061
50	0.087	0.027	0.009	0.003	0.001

Present worth factor $= (1 + r)^{-n}$ where $r =$ rate of discount and $n =$ number of years.

Table 7.2 Compound interest

Year	Percentage				
	5.0	7.5	10.0	12.5	15.0
5	1.28	1.44	1.61	1.80	2.01
10	1.63	2.06	2.59	3.25	4.05
15	2.08	2.96	4.18	5.85	8.14
20	2.65	4.25	6.73	10.55	16.37
50	11.47	37.19	117.39	361.10	1083.70

$S = (1 + r)^n$ where $S =$ sum, $r =$ rate of interest, and $n =$ number of years.

years, and conversely the effects of various rates of interest projected forward, the one being the reciprocal of the other.

The economic and social concept of discounting has resulted in considerable controversy over many years, particularly in situations in which the long-term effects on the environment and natural resources are uncertain or irreversible. The use of discount rates could lead to an almost total disregard of long-term environmental consequences after fifty years or so, even with typical rates adopted by governments. The application of rigid cost–benefit rules as the only criterion for decision-making could lead ultimately to significant and even catastrophic and irreversible damage to natural assets.

Low discount rates have been suggested so that the costs of possible long-term damage to the environment may be given greater weight in development decisions, in particular instances. However, a general use of low (or even zero) discount rates may result in higher rates of investment and create even greater damage in the future (Fisher and Krutilla 1975). Others have advocated special measures to protect irreplaceable environmental assets and critical resources (Ciriacy-Wantrup 1968; Krutilla and Fisher 1975; Bishop 1982).

7.4 Choice of Discount Rate

In making public investment decisions, future streams of costs and benefits are discounted to the present time so that they can be compared. The choice of discount rate often determines whether a project is worthwhile or not, and where it ranks compared with others.

There is no single rule to determine the choice of discount rate, but the schools of thought may be divided into five groups, favouring the following:

- the opportunity cost of capital, i.e. the discount rate should be set equal to the opportunity cost for government projects, viewed generally (Baumol 1968);
- the social time preference rate, i.e. the relative value which the community as a whole, through government decisions, assigns to present as opposed to future consumption at the margin;
- a variable discount rate, depending on the project, and the degree of social risk involved;
- the use of the discount rate to influence the level of investment in an economy, thus the discount rate is viewed as a policy tool. Both the UNIDO and OECD use this argument when calculating the social rate of discount;
- a discount rate that reflects the interests of future as well as present generations. The argument here is that any discount rate discounts the interests of those in the future. Sometimes a zero rate of time preference may be appropriate.

The actual rates used vary from country to country, industry to industry, sector to sector, and government to government. In many cases, it may be desirable to use several rates so as to determine how sensitive the project or programme is to the rates chosen (Hufschmidt *et al.* 1981). The general principles of CBA are widely applied in advanced economies in public investment decisions, natural resource allocation, and environmental planning and policy-making. In an extended form CBA stands at the heart of environmental economics.

The current standard rate of discount applied by the Australian Government is 8% with sensitivity testing at 6% and 10%.

7.4.1 Discounted Cash Flow

Discounted cash flow is a technique of comparing the profitability of alternative projects. It may be subdivided into (1) the yield method, and (2) the net present value method. Both of these techniques utilize as a measure of the *rating* of an investment the present cash value of a sum to be received at some future date, discounted at compound interest.

The present value or worth of a sum to be received at some future date is such an amount as will, with compound interest at a prescribed date, equal the sum to be received in the future. The yield method is based upon the assumption that the best investment is that from which the proceeds would yield the highest rate of compound interest in equating the present value of the investment with future proceeds.

In the net present value method an appropriate percentage rate is stipulated. The present value of the cash inflow is determined using this percentage, the original cost of the investment is subtracted therefrom, and the resulting surplus is the net present value of the investment.

Both the yield and the net present value methods give reliable guidance to the profitability of alternative schemes, on the assumptions that may be used. However, errors of a serious nature may occur in the assumptions made regarding capital and running costs or trading returns or benefits.

Box 7.8 summarizes the implications of positive discount rates, while Box 7.9 emphasizes the importance of the bird in the hand, and the justification of discounting. Box 7.10 summarizes the views of Summers of the World Bank.

Box 7.8 The Implications of Positive Discount Rates

- Fewer investments are undertaken which have large initial costs and long-term returns, benefits or pay-offs; these include hydroelectric and water irrigation projects, and some mining projects.
- Major future costs are not given their true importance, e.g. the long-term implications for salinity or the need to decommission large undertakings such as a nuclear or conventional power plant, or the cleaning up of contaminated land.
- The cost of storing and disposing of nuclear or toxic wastes is significantly minimized.
- High rates shift cost burdens forward to future generations.
- The effect of high rates may be to preserve some natural areas.
- On the other hand, the benefits of long-term projects such as hardwood plantations are grossly minimized in terms of present values. Hardwood trees may take 50 years to grow, as distinct from softwoods which may take only 10 years.
- The higher the rate of discount, the greater the discrimination against future generations; future generations may bear a disproportionate share of the future costs of the project.
- The higher the discount rate, the lower will be the level of capital investment and hence the lower the capital stock inherited by future generations.
- The lower the rate, the lower the discrimination against future generations; a zero-discount rate is the fairest of all.

Box 7.9 Discounting the Future: The Bird in the Hand

- Money available now can be put to productive use, at the very least earning interest; money in the hand offers scope, including *capital productivity* and the ability to purchase today.
- The bird in the hand is worth three in the bush, for the future is always uncertain with risks of death and illness, a return to high inflation rates, collapse of markets, bankruptcies, scams and fraud, etc.
- Money in the future may be worth less simply because people may be better off in the future; money may suffer *diminishing marginal utility*. The future dollar may be worth less, not simply as a result of inflation. To sacrifice now for the future may be pointless.

Continued on page 184

Continued from page 183

- People also prefer money in the present because of impatience and myopia.
- Some investments take a long time to mature.
- The general outcome is that people discount future benefits and returns, and to induce people to forgo present consumption requires positive rates of interest and other assurances.

Box 7.10 Lawrence H. Summers of the World Bank on Sustainable Growth

A summary of Summers' points:

- There is no intellectually legitimate case for abandoning accepted techniques of cost–benefit analysis in evaluating environmental investments, either by using abnormally low discount rates or by invoking special criteria regarding sustainability.
- The argument that a moral obligation to future generations demands special treatment of environmental investment is fatuous.
- We can help our descendants as much by improving infrastructure as by preserving rainforests, as much by educating children as leaving oil in the ground, as much by enlarging our scientific knowledge as by reducing carbon dioxide in the air.
- There is every reason to undertake investments that yield the highest returns.
- Once the costs and benefits are properly measured, it cannot be in posterity's interest to undertake investments that yield less than the best return, say between 10 and 15 per cent.
- What is the better course for rich countries: to put more aside for a posterity that will be far richer than we are, or to do more to help the world's poor now, the billion who survive on a dollar a day?
- Environmentalists who point to the damage done by dams, power plants and roads evaluated according the standard economic criteria have a point. The answer does not lie in blanket sustainability criteria, or in applying special discount rates, but in properly incorporating environmental costs into the appraisal of projects. The grim fact is that no careful analysis was done on many of the projects which environmentalists condemn.
- The world's problem is not too much cost–benefit analysis, but too little that is done well.
- Plenty of environmental improvements can pass rigorous cost–benefit tests. There is no need to cook the books.

Source: L.H. Summers (1993) The Economist Year Book 1993, pp. 255–6.

7.5 Multi-criteria Analysis

CBA is a procedure for comparing alternative possible courses or actions to the net social benefits that they are likely to produce. In general, a project or programme having a high cost–benefit ratio will take priority over others with lower ratios. As an application of modern welfare economics, CBA analysis may be regarded as an attempt to improve the quality of decision-making and social outcomes, improving the economic efficiency of resource use. However, in decision-making at a political level it is often necessary to take into account considerations other than those which may be incorporated in the extended CBA of development proposals: examples include the distribution of the associated costs and benefits among different groups in society, particularly where mass population movements are involved, as in the case of the China Three-Gorges hydroelectric scheme, or where major ecological issues are involved, affecting a wide community. Broader considerations are needed possibly involving a range of criteria more far reaching than even extended CBA.

Decision- and policy-makers often require a method of comparing options that assesses such options in terms of more than one criterion and illustrates why some options may be superior or inferior to others under prescribed assumptions or when subject to sensitivity analysis. Multi-criteria analysis (MCA) is such a technique (Resource Assessment Commission 1992).

MCA involves a set of procedures designed to identify and organize information relevant to various steps in a complex decision-making process. It outlines the options to be examined and the criteria that are deemed to be most relevant. The advantages and disadvantages of each option are then compared. Each effect is measured in a common unit across all options using quantitative or qualitative indicators. Weightings may be introduced indicating the relative importance of each effect. Such weights may be derived from surveys of community opinion or from policy statements by government.

MCA was used by the Australian Resource Assessment Commission to evaluate broad-scale options for forest use at a national level in its Forests and Timber Inquiry in 1992. The analysis underlined the extreme sensitivity of forest-use strategies to the weights that are attached to economic or ecological goals, and the difficulty of formulating national strategies that avoid trade-offs between the two goals. It also highlighted the crucial importance of the nature of the options identified for analysis in the first place. MCA was found to be most useful where there are a number of options in which there are gradations in the combination of the different criteria. However, MCA requires a level of data about resource uses and their impacts as well as weightings associated with objectives that may not often be available. Also subjectivity is often inevitably involved.

The final step in the CBA process is the writing of the report, with conclusions and recommendations to the decision-maker. The report should be short and concise. Background reports should be referred to in the main text, but only as supplements, available on request. Box 7.11 indicates the possible structure of a report.

Box 7.11 The Structure of a CBA Report

- a summary of the purposes and key findings of the report
- an introduction, providing backgrounds to and reasons for the CBA analysis
- the objectives of the analysis
- an outline of alternatives considered
- profiles of costs and benefits over time
- indications of intangible costs and benefits
- the sensitivity of profiles to variations in assumptions
- the valuation of the costs and benefits
- the identification of any constraints and uncertainties
- the distributional effects of costs and benefits say through different groups in the population or throughout a given sector of the population
- a conclusion summing up the findings
- the final conclusion and recommendation, with suggested conditions for the application of the project, programme or policy

NB It is suggested that a written report should be introduced orally to the decision-maker(s), to permit short explanations to pertinent questions and the briefest of introductions.

7.6 The Multiplier

The concept of the multiplier implies far-reaching benefits or disbenefits going well beyond the initial capital investment of a programme or project. Traditionally, the multiplier in economics is a ratio indicating the estimated effect on total employment or on total real income of a specified amount of capital investment or stream of expenditure.

Investment or expenditure becomes income for the factors of production (land or natural resources, labour and capital). After direct tax, some of that income will be saved and the rest spent on goods and services. This expenditure in turn will become the income of others, along an almost endless chain. If the factors were previously unemployed the benefits will be considerable. Thus employing one additional person may well create the equivalent of work for another. Whatever the factor, there is a chain reaction. The growth of income continues through the economy; the multiplier is a crucial concept in Keynesian economics.

Other benefits may flow from initial investments. For example, the treatment of water for human consumption protects the health of large numbers of consumers. The generation and transmission of electricity eliminates much pollution at the point of use. On the other hand, the production of cigarettes or alcohol contributes to a stream of disbenefits. Modern transport yields widespread savings in travel time. Refrigeration spreads widely the benefits of preserved foodstuffs. The car brings the benefits of mobility and the disbenefits of accidents. Vaccinations provide widespread benefits at small cost. Key surgical operations can add years to life.

7.7 Cost-effectiveness Analysis

In some situations, benefit–cost analysis may be impracticable or inappropriate, if not in theory then certainly in practice. Certain objectives may have corporate or political approval while the benefits, though obvious and real, cannot be readily measured in monetary terms or in some respects at all. The benefits may be beneficial in ecological terms (such as preserving the Great Barrier Reef, or establishing a system of national parks); social terms (such as setting standards for consumer protection or child care); defence terms (such as building warships or establishing bases); or educational or research terms (such as the establishment of university centres of excellence).

However, while formal cost–benefit analysis may falter, cost-effectiveness analysis (CEA) may remain highly relevant. Given the benefits, or alleged benefits, of a project, CEA is concerned with the least-cost approach to the objective; it involves an examination of the costs of alternative ways to meeting the objective(s) with a view to achieving the maximum value for the dollar invested. Of course, CEA cannot be used to compare projects with different objectives.

CEA is often used in the difficult area of social and educational infrastructure. For example, a decision to establish a new university to serve essentially a certain geographical area may have obvious benefits, in the medium and long term. The evaluation of those benefits in monetary terms over the next half-century would no doubt stretch cost–benefit analysis. CEA, however, remains highly relevant: whether to combine some existing facilities, to have one or several campuses, and the staging of construction are all essential considerations in making the project more affordable.

7.8 CBA at the Political Level

Few writers attempt to penetrate the veil of political decision-making, though most decisions about developments are made at a political level, by elected persons, either as members of local councils, county councils, state or provincial governments, or national governments. Some authorities or councils of a non-elected character, comprising officials and public servants, may enjoy some decision-making, subject to legislation enabling them to do so, and usually subject to rights of appeal to boards of a legal nature, or government itself. In countries less democratic in character, major decisions are largely a matter of presidential prerogative or left to trustworthy ministerial persons, colleagues or relatives.

The environmental assessor of a proposed project, programme or policy, may well be devoted to an objective analysis of the scheme, guided of course by any legal requirements as to matters for consideration and any government policies embodied in official planning policies and local environmental plans.

However, when the assessor or commissioner tenders advice at the ministerial level, handing over an official report, the minister will find not only a single recommendation to accept or refuse the proposal, but also, if the project is to proceed, a range of suggested conditions to be attached to any development consent. These conditions will affect several government departments and agencies. All these departments and agencies will need to be consulted.

The minister will need to reflect on a new kind of CBA, far removed from that of advisers who do not need to take into account the internal, political considerations of the government of the day. The minister must take account of the broad interests of the community, for sure, but also with a keen regard to political survival. To survive and thrive is the keynote above every minister's desk. Costs and benefits take on a different cast of meaning. Unfortunate is the assessor who finds that the minister reacts sharply to particular proposals, for reasons not accessible to the assessor. It is an uncanny experience to endure ministerial wrath for a crime you never knowingly committed. Box 7.12 attempts to embrace the costs and benefits of a political decision in respect of any development proposal considered to be controversial. Box 7.13 enlarges on the decision-making process, emphasizing that CBA is but one input to major decisions.

7.9 The Meaning of the Word 'Value' and Techniques for Capturing Environmental Values

Box 7.14 indicates the wide range of meaning that can be attached to the word 'value', within an economic, social and environmental context. Attempts to measure such values in physical or monetary terms now occupy a large literature, with some objective and many subjective outcomes. Since many development

Box 7.12 The Costs and Benefits of the Political Decision

- the likely costs and gains in terms of electoral popularity and the prospects of staying in power
- the views of cabinet, parliamentary party, backbench, party members, and other influential persons
- conformity with existing policies and earlier promises
- the likely effects on marginal constituencies and disadvantaged groups
- the likely effects on influential pressure groups and lobbyists
- the likely effects on regional and international interests
- the likely or certain effects on departmental advisers and consultants offering independent cost–benefit studies
- the likely effects on perks and facilities, the opening and closing of doors
- the effects on the opposition parties and concerned groups
- the kudos of the decision
- the rights and wrongs of the decision, including ethical considerations

Box 7.13 CBA and Decision-making

When cost–benefit analysis is used in a decision-making context, the decision-maker can expect to be presented with a quantitative comparison of options together with supporting information on the costs and benefits that have resisted quantification (intangibles) and on distributional or equity considerations. Thus the decision-maker may be presented with as many as three bundles of information – which can only be tied together with the use of individual or collective judgement.

This observation serves to reinforce the fundamental point that cost–benefit analysis is an *information aid* to decision-making, and never a substitute for it. There is invariably a large role for judgement, based on a wide range of considerations including social, ethical and political ones, as well as those relating to residual measurement uncertainties.

Nevertheless, the role of cost–benefit analysis as a decision-making aid remains an important and substantial one.

Source: Australian Department of Finance (1991) Handbook of Cost–Benefit Analysis, AGPS, Canberra, p. 101.

Box 7.14 Meanings of the Word 'Value'

- the market value, i.e. the exchange value or price of a commodity or service in the open market;
- the intrinsic value of personal possessions that may have little or no market value, but have use value;
- the value of assets and services not priced in the open market, such as environmental assets of many kinds;
- the value attached to the environment and animal life for their own sake, i.e. existence, non-use or bequest value;
- the value attached to relationships and to society;
- option value, i.e. a willingness to pay a certain sum today for the future use of an asset;
- the value today of a future asset, discounted to present worth or value;
- the value of intragenerational and intergenerational equity;
- moral, social and religious values;
- vicarious values, i.e. a willingness to pay to preserve the environment for the benefit of other people. Where other people include future generations, these values may also be known as bequest values;
- quasi-option value, i.e. the value of preserving options for future use assuming an expectation of the growth of knowledge about the functioning of the natural environment;
- the economic value of the environment measured as the sum of the actual use value, option value and existence value.

decisions are made at a political level (not necessarily at a party level), the values to be attributed to many aspects of proposals at a policy, programme and project level will derive from a whole range of considerations, many of them

Table 7.3 *Comparison of valuation techniques*

	Reliability of results	Data require-ments	Timing	Ease of appli-cation	Technical develop-ment	Accumulated expertise
Productivity approaches	High	Medium	Low	High	High	High
Opportunity cost	High	Medium	Low	High	High	High
Preventative expenditure replacement cost	High	Medium	Low	High	High	High
(*Special features: based on market transactions, assumes no distortions in market prices*)						
Hedonic pricing	High	High	Medium	Medium	High	Medium
(*Special features: assumes mobility and perfect information*)						
Wage differential	Medium	High	Medium	Medium	Medium	Medium
(*Special features: main techniques for valuing risks to life, assumes mobility and perfect information*)						
Travel cost	Medium	Medium	Low	High	High	High
(*Special features: use limited to recreation benefits*)						
Contingent valuation	High	Medium	High	Low	High	High
(*Special features: the only technique that covers existence values, can suffer from a lot of biases*)						
Delphi technique	Medium	Low	Low	Medium	Medium	Low
(*Special features: applicable to a wide range of impacts*)						

Source: NSW Environmental Protection Authority (1993) Valuation of Environmental Impacts, Chatswood, Sydney.

controversial. There is no simple approach to this subject, and compromise is often the key to progress. While most of the world favours development in the broadest sense, emphasis finally rests on the conditions attached to investments.

The outcome of much research, trial and error, has been a wide range of techniques and approaches to the problem of attaching values to environmental assets. These are summarized and elaborated on in the balance of the chapter.

Tables 7.3 and 7.4 attempt a comparative review of these techniques.

7.10 Opportunity Cost

The opportunity cost is the cost of satisfying an objective, measured by the value those resources would have had in another attractive alternative use. For example,

Table 7.4 *Applicability of valuation techniques to environmental effects*

	Health impacts		Aesthetic impacts	Ecosystem impacts	Recreation impacts	Production impacts/ material damage
	Illness	Mortality				
Productivity approaches	Yes	–	Yes	Yes	Yes	Yes
Opportunity cost	Yes	–	Yes	Yes	–	Yes
Preventative expenditure replacement cost	Yes	Yes	Yes	Yes	–	Yes
Hedonic pricing	–	Yes	Yes	–	Yes	–
Wage differential	Yes	Yes	Yes	–	–	Maybe
Travel cost	–	–	Yes	–	Yes	–
Contingent valuation	Yes	Yes	Yes	Yes	Yes	Yes
Delphi technique	Maybe	–	Maybe	Yes	Maybe	Maybe

Source: NSW Environmental Protection Authority (1993) Valuation of Environmental Impacts, Chatswood, Sydney.

if capital funds committed to a programme or productive use could have earned a rate of interest of 15% per annum in another use of similar risk, then that is the opportunity cost of those funds in their present or proposed use. There is an invisible though real cost for funds which could have been placed elsewhere, even though this cost is not revealed in any balance sheet. The opportunity cost of being self-employed is the salary that might be earned doing something else. In turn, the opportunity cost of leisure may be measured by the money that could have been earned instead. A government with a relatively fixed budget may choose to build better highways; such a decision will be made at the cost of other programmes such as schools, hospitals, low-income housing, recreational facilities and parks, railway modernization, and urban renovation. A consideration of opportunity costs assists in ensuring that resources are put to the best use and that all costs, implicit and explicit, are taken into account. In CBA analysis, the opportunity costs should be taken into account in the assessment. They provide some guidance to a suitable discount rate.

7.11 Preventive Expenditure/Replacement Cost

Some guidance to the value that people place on their environment can be estimated by observing how much is spent on preventing damage to it, and on its enhancement. The costs of sound insulation, air filters, water filters, sun blinds, decorations and garden maintenance, give some indication of values attached to the immediate environment, though of a minimal nature. Local rates also indicate how much people are prepared to pay for the removal of garbage and wastes, some for recycling programmes. The maintenance of properties otherwise dilapidated, through decoration, repair and progressive replacement, gives some indication of the contribution of these environmental considerations to the value of assets.

7.12 Surrogate Market Techniques

Surrogate market techniques use actual market prices to value a quality of the environment that is not separately marketed. These techniques aim to estimate the implicit price for environmental goods and services by observing the prices paid for other related goods which are marketed. Hedonic pricing, wage differentials, and travel cost approaches are examples of surrogate market techniques.

7.13 Hedonic Price Techniques

The hedonic price technique (or property value method) is a market approach, attempting to assess the value attributed by buyers to the environmental attributes of a dwelling. It is a generally accepted fact that, all other things being equal, a house located in a poor environment (broadly considered) will sell for a lower price than a similar house in a better environment. Thus the difference in house prices can be used as an estimate of buyers' willingness to pay for a better environment. The differential paid for the superior environment is known as the 'hedonic price'.

The proposition is simple, but complex in application. Houses are rarely exactly similar, varying in age, size, suitability, character, structure, number of rooms, quality, materials, and presentation; but what is true is that dwellings with similar costs of construction command different selling prices depending on location.

If the basic quality and character of a house with a similar cost of construction can be held constant, then the difference in market price, if significant, must be due to the quality of the environment. This is not, however, a simple entity confined to outlook, cleaner air, lower levels of noise, or remoteness to unsightly industrial plants. It also has much to do with access to better schools, shopping centres, recreational facilities and entertainment, having nicer neighbours and being in a district with a better reputation for law and order, with better standing. Such elements ensure that any given house will command a better price (and that the land chosen for it will also command a better price).

Ascertaining the size of the differential in the prices of similar houses is therefore a challenging matter; yet it reveals what financial value buyers are willing to place on environmental quality (and perhaps particular elements in environmental quality). Such a finding throws light on the weight in real proven terms given by members of the community to environmental matters impinging on immediate personal and family well-being; and perhaps throws some light on the values placed by the community on environmental elements somewhat further removed.

Research in this difficult area may be much simplified by talks with real estate agents: direct contact with the market enables them to form very sound opinions about the effects of different environments on property prices. Thus a hypothetical house can be placed in a variety of hypothetical locations with an experienced real estate agent (or valuer) offering very good guidance on the market value.

Hedonic pricing has been used for a variety of purposes; in the USA in respect of automobiles and in Australia on the effects of aircraft noise on property prices and the effects of a water-supply pipeline on the value of properties in the wheat-belt.

Relevant references include Hufschmidt *et al.* (1983), Hyman (1981), Pearce and Markandya (1989), Sinden and Thampapillai (1995) and Streeting (1990).

7.14 Wage Differential Approach

The wage differential approach is very similar to the hedonic price technique, save that it uses wage differentials in place of property prices as a guide to the values that people put on environmental qualities. In this approach, workers are assumed to be able to choose freely between jobs in different areas at wages that will maximize utility for the worker. However, these differentials in wages reflect a number of influences, one being the character and conditions of employment (and any hazards or special skills associated with it); others being the extent of urban amenities such as schools, access to shopping and recreational facilities, distances to hospitals. The value that people place on urban facilities should be reflected in the higher wages needed to attract workers to less desirable cities. Households seek comparable facilities in all towns and cities, or expect compensation for deficiencies or greater distances to facilities. More hazardous working conditions, in particular instances, should also attract additional compensation.

The study of wage differentials opens up an intriguing world of analysis to reveal the character of employment and social and environmental conditions. As seen by Roback (1982), the marginal value of an amenity to a resident is the sum of the partial derivatives of a hedonic wage function and a hedonic property value function.

7.15 Travel Cost Approach

The travel cost approach is another surrogate market technique, and is particularly useful for assessing the economic value of natural areas or recreational areas where no price is directly charged. In this case, the willingness to pay for an environmental amenity is assumed to be measured in the costs incurred by people when travelling to chosen locations.

Travel costs from each concentric zone around the site are used as a surrogate for price, the quantity being determined by actually counting the number of visitors from each zone. The relationship between cost and visitor rates becomes a demand curve for the recreational experience, which may involve a number of activities. The travel cost approach is well accepted, but probably underestimates real benefits since it does not endeavour to establish the maximum willingness to pay.

When the survey is taken of how frequently each person or group makes a visit to the area, it must be ascertained what travelling distance and cost was involved. The total population in each zone must be ascertained, and from this the rate of visits can be calculated. By fitting a regression equation to the data, a visitation rate as a function of travel costs can be determined. Sometimes an entrance fee may be simulated by including it in the regression equation; the process may be repeated for successively higher simulated charges until no visits are made, and an optimal charge can then be considered.

A travel cost study has been undertaken in respect of Lumpinee Public Park in Bangkok, details of which are reported in Dixon and Hufschmidt (1986). Other examples are provided in Sinden and Worrell (1979). The method is reviewed in Hyman (1981).

7.16 Survey-based Approaches

Other valuation methods have been developed in an attempt to value environmental benefits in the absence of data on market prices. These methods rely on asking consumers directly about their willingness to pay for a benefit, or to be compensated for a loss. The contingent valuation approach is the most well known, seeking personal valuations for increases or decreases in the quantity of some good or service, contingent upon a hypothetical market. Another approach, the Delphi technique, relies on the direct questioning of experts (or community leaders), and involves far fewer people. Other techniques such as trade-off games or lotteries, based on concepts of decision theory and utility theory have also been developed. In trade-off games, people are asked to choose between different bundles of goods. Several good surveys of techniques for valuing the environment have been published, among which are Cropper and Oates (1992), Hufschmidt *et al.* (1983), Pearce and Markandya (1989) and Smith and Desvousges (1986).

7.17 Contingent Valuation

Contingent valuation is a method of establishing a monetary value for a good or service by asking people what they are prepared to pay for it. The good or service in this context would be an environmental amenity, clearly of value but falling beyond the range of the market economy and therefore unpriced. The method seeks to determine a level of payment acceptable to most people; it can determine the willingness to pay for a better environment, or to accept compensation for a degraded environment.

The method takes several forms (as described by Freeman 1979; Sinden and Worrell 1979, Hyman 1981, Nichols and Hyman 1982, Hufschmidt *et al.* 1983, Sinden 1990, Imber *et al.* 1991, James 1993):

1. *Bidding games*: the simplest approach here is direct questioning, individuals being asked how much they are prepared to pay (or how much extra they are prepared to pay) to enjoy a particular environmental amenity, or bring about some environmental improvement. Bidding games can be applied also to the willingness to accept compensation for a loss of environmental amenity. Examples are to be found in Hufschmidt *et al.* (1983). Bidding games can embrace such concepts as (a) option value, which refers to the price that people are willing to pay to maintain the possibility of using or visiting an amenity or site; and (b) existence value, being the price that people are willing to pay just to know that certain things are being preserved but which the individual may never see such as whales, polar bears, tropical rain forest, coral reefs or Antarctica.
2. *Convergent direct questioning*: this is a development of (1). Each individual is given a high value likely to exceed any reasonable willingness to pay, and a low value that almost certainly would be paid. The higher value is reduced and the lower value increased until the two values converge at an equilibrium value. The answers can be analysed in various ways.
3. *Trade-off games*: each individual is asked to rank various combinations of two objects: a sum of money and some environmental attribute (such as water of a certain quality, or the preservation of a natural area). For any pair of combinations, the individual is asked to indicate either a preference of one over the other, or indifference. The marginal rate of trade-off of money for an environmental amenity is identified at the point of indifference.
4. *Moneyless-choice method*: instead of using money as one of the objects, only specified commodities are used in combination with environmental attributes. The trade-off is again identified at the point of indifference. The monetary valuation of an environmental attribute is then obtained by substituting the current market value for the commodities chosen.
5. *Priority-evaluation method*: in this method each individual is given a hypothetical sum of money to spend on conventional goods and environmental attributes at assumed prices. The game might comprise five objects (four goods and one an environmental attribute) in three different quantities (giving 15 possible choices). As the budget is limited, quite clear choices must be made. True preferences and marginal valuations can be derived.

All of these methods must be viewed with caution. A willingness to pay may be overstated to encourage reservation of an area (turning more into a public opinion poll); or may be understated to minimize the possibility of a significant user-charge or levy. The questions themselves may condition the responses, and bias may be introduced by the interviewer or into published briefings. The conduct of the survey itself may exaggerate the importance of the issue. Certainly, all questions and briefings should aim at being true, fair and reasonable.

Contingent valuation has been used in Australia for at least two decades by many kinds of agencies and organizations, to value a great variety of things (Sinden 1990). A contingent valuation study was undertaken in very recent times to estimate the value attached to a portion of land contiguous with the Kakadu National Park in the Northern Territory. This was part of a review by the Resource Assessment Commission to help the federal government to decide whether this slice of territory called the 'conservation zone' should be used at least in part for mining or incorporated in its entirety into the existing national park. Around 2500 Australians were asked to value the zone. It was a landmark event, as it was the first attempt to assess environmental values for the express purpose of being part of a report to the federal government (Imber *et al.* 1991).

The Kakadu contingent valuation was severely criticized by the Australian Bureau of Agricultural and Resource Economics and the Tasman Institute, and the Resource Assessment Commission itself had some reservations about it. However, it remained a significant contribution, although the final decision to incorporate the conservation zone in the national park was made out of concern for the views of the Aboriginal community.

7.18 The Delphi Technique

In seeking to identify environmental values, the Delphi technique relies on the informed opinions of a group of experts, being commonly used in many complex planning and decision-making situations (Hufschmidt *et al.* 1983). For economic evaluation, the technique depends heavily on the knowledge and background of the experts and on the skill with which the technique is used. Originally developed by Dalkey and Helmer (1963), the technique (and its variants) attempts, through the questioning of experts, to place values on particular environmental assets. Each member of the group responds independently, and the initial set of responses are summarized and presented in tabular or graphical form. This information is then fed back to the entire group, each member independently re-evaluating the figures and making new value judgements. Through successive rounds, the results may tend to converge.

All communication is indirect, so that personality problems are avoided. The quality of the outcomes depends on the quality of the panel, the panel's ability to reflect societal values, and the skill of the process. The view of the group may be compared with the results of more conventional surveys. It should be kept in mind that the outcome of such assessments can only be one input to the overall assessment of the project or programme. The decision-making body cannot be seen to be guided solely by 'experts', the current view of both decision-makers and experts being somewhat jaundiced.

The application of the Delphi technique in environmental assessment was initially discussed by Dalkey and Helmer (1963), with wider consideration in Linstone and Turoff (1975).

7.19 Comparison of Techniques

The wide range of techniques available for attempting to value environmental effects, positive and negative, raises the question as to whether some techniques are more advantageous than others. In comparing a number of valuation techniques, Pearce and Markandya (1989) suggested various criteria, including data problems, the relationship with economic theory, benefit functions, reliability of results, the number of studies carried out, and special features. A similar approach has been taken by the New South Wales, Australia, EPA and tables have been prepared to embrace such matters as the reliability of results, data requirements, timing, ease of application, technical developments and theoretical concerns, accumulated expertise, and the specific strengths and/or weaknesses of each technique. Table 7.3 provides a comparative assessment of the various valuation techniques with respect to these criteria, revealing that no single technique is the preferred one. Depending on the particular characteristics of the case at hand, some of the techniques might not be applicable, while others may be potentially useful. The technique that may eventually be selected depends on data availability, the overall timeframe, personnel and other resources available, the purpose of the study and the audience for which it is prepared.

Attempts to categorize valuation techniques in terms of their suitability for particular applications have been undertaken by Pearce and Markandya (1989). Table 7.4 shows that several alternative techniques may be used for the valuation of a given effect. However, the application of techniques to the valuation of human life remains a particularly contentious issue. Experience in the application of these valuation techniques has made much progress in recent years, embracing a wide range of issues such as improved water quality in rivers, lakes and wetlands; improved air quality and health and recreation; preferences displayed for reduced noise levels; and water-based recreational activities. The valuation of environmental costs and benefits is set to play an increasingly important part in development decision-making. However, the amount of resources devoted to valuation needs to be commensurate with the benefits likely to ensue.

An important impetus to valuation studies has been given by the rapid development of the regulatory impact statement, now common in the Western world. A good example has been the assessment of the US regulatory impact statement on lead in gasoline carried out by the US EPA in 1986. Using a market-based technique, the US EPA carried out a careful assessment of the proposed reduction of lead in gasoline. Using a linear programming model of refinery economics, the costs of complying with the proposed rule were first carried out. In assessing the benefits of the rule the study had regard to reduced medical costs associated with treatment to reduce blood levels, a reduction in the costs of compensatory education for children with learning difficulties due to exposure to lead, savings from reduced vehicle-maintenance costs, reduced crop damage and reduced costs of respiratory illness. The reduction in lead in the atmosphere in US cities has been dramatic.

7.20 Summary

Chapter 1 portrayed environmental economics as a grand array of costs and benefits presenting an allocational problem for decision-makers. This chapter deals with costs and benefits within a smaller framework: the costs and benefits of a business as seen by the investor; and the costs and benefits of that business seen through the eyes of society. The technique of cost–benefit analysis has a long and rewarding history, for it enables wise investment decisions to be made. It is a methodical approach to complex matters, involving the measurement of costs and benefits occurring over the potential economic life of a project, year by year, some of them intangible. Furthermore, all such costs and benefits are reduced to a present worth by the application of an appropriate discount rate. The choice of discount rate is itself highly controversial, for it is argued that high discount rates disadvantage late-maturing projects such as hardwood forests. They may also downgrade the future costs in the decommissioning of nuclear power stations and the ultimate disposal of radioactive wastes.

The valuation of assets outside the market economy also presents problems. A variety of techniques have emerged to assess values, some controversial in themselves. Many writers also omit the fact that the decision-maker, often political, also has a scheme of cost–benefit analysis. This approach translates everything into political values, either short or long term, with considerable emphasis on the issue of survival. Politicians prefer survival to martyrdom. Expediency is also often involved.

Further Reading

Markandya, A. and Richardson, J. (eds) (1992) *The Earthscan Reader in Environmental Economics*, Earthscan, London.

Mishan, E.J. (1982) *Cost–Benefit Analysis: An Informal Introduction* (3rd edn), Allen and Unwin, London.

Pearce, D.W. (1983) *Cost–Benefit Analysis* (2nd edn), Macmillan, London.

Pearce, D.W. (1993) *Economic Values and the Natural World*, Earthscan, London.

Pearce, D.W. and Markandya, A. (1989) *Environmental Policy Benefits: Monetary Valuation*, OECD, Paris.

Sinden, J.A. and Thampapillai, D.J. (1995) *Introduction to Benefit–Cost Analysis*, Longman, Melbourne.

Smith, V.K. (1996) *Estimating Economic Values for Nature: Methods for Non-market Valuation*, Edward Elgar, Cheltenham.

8

Air, Land and Water Issues

8.1 Introduction

Air pollution has a very long history, as illustrated by John Evelyn's thesis of 1661 on London smog. In Britain it intensified until the disastrous smog of 1952 when 4000 people prematurely died in London. The Clean Air Act 1956 marked the beginning of the environmental era in Britain, corresponding with measures in Pittsburgh and Los Angeles preceding much US federal legislation. Europe followed these leads. This chapter considers the current scene in respect of urban air pollution in developed and developing countries, and in respect of acid rain and the North American campaign to eliminate its more detrimental aspects. Global warming is addressed in Chapter 9.

This chapter then moves to the international question of biodiversity, its significance in respect of human welfare and future generations, respect for the intrinsic values of nature itself and the UN Convention on Biological Diversity. This commentary is reinforced by notes on forestry and the Amazon rain forest. Further text appears on deserts, both hot and cold.

The chapter then turns to water: the Baltic and Caspian Seas, the Murray-Darling Basin in Australia, immensely important but subject to intense salinity problems, and the Mekong River in Asia.

Two case studies follow relating to the OK Tedi mine operation in Papua New Guinea, and the Three-Gorges water conservation and hydroelectric project in China, the largest dam project in the world and possibly the most controversial.

The selection of examples in this chapter may be considered arbitrary and is certainly far from exhaustive, but the cases described do embody the key principles under consideration.

8.2 Urban Air Pollution

The Global Environmental Monitoring System (GEMS) was established in 1974 as one of the components of Earthwatch, the UN Action Plan for the Human Environment. GEMS comprises five closely linked major programmes, each containing various monitoring networks relating to pollution, climate, health, ecology and oceans. The WMO Background Air Pollution Monitoring Network and the WHO Program for Monitoring Air Quality in Urban Areas have been developed as part of GEMS in conjunction with UNEP. Initially, the GEMS urban air quality assessment concentrated on the five most common and serious air pollutants: sulphur dioxide, suspended particulate matter, nitrogen oxides, carbon monoxide, and lead. The GEMS assessment involved collecting data from 50 countries. Using data for the period 1973–84, initially trends in emissions and the quality of urban air were assessed, levels being compared with WHO guidelines. The cities included in the network were chosen to provide the broadest global coverage possible, representing different climatic conditions, levels of industrial and urban development, and pollution situations. Most of the cities have three monitoring stations located in industrial, commercial and residential areas respectively.

In 1988, the information that had been collected was compiled and analysed (UNEP 1991). Overall, only 20% of the world's urban dwellers live in cities where air quality is acceptable, these being, by and large, the richer OECD countries. Standards are poorer in Russia and its former satellites. In almost all developing countries, urban air pollution is worsening.

Rapidly growing cities (see Table 10.6), increasing traffic, reliance on outdated industrial processes, growing energy consumption, the lack of adequate zoning and statutory planning, and weak environmental regulations and enforcement, are all contributing to reduced urban air quality and deteriorating public health. The data reveal that much of the world's population live in cities where pollution levels exceed WHO guidelines. For instance, more than 1200 million people are exposed to excessive levels of sulphur dioxide, and 1400 million people to excessive particulate emission and smoke. The air in many cities contains excessive amounts of nitrogen oxides, carbon monoxide and lead. Bearing in mind that by the year 2005, half the world's population will live in cities, air pollution presents itself as a threat to future health.

However, in many developed countries the levels of pollution have been falling due to the cumulative effects of clean air legislation spread over some forty years or more. Indeed, there have some striking successes, such as smoke control areas in Britain, and the reduction of lead in the atmosphere of North American and European cities.

This demonstrates that economic growth, urbanization and industrialization may lead to better pollution outcomes than has often been predicted. However, in the developing world, without serious efforts to control pollution from all sources, the smogs of Victorian England may be spread world-wide.

Since the initial findings of GEMS, there have been even more disturbing findings. In 1996, WHO reported that a study had shown that smog claimed 350 lives a year in Paris and that pollution, mainly from vehicle exhausts, made it 10 to 100 times more dangerous to live in a city than inside a nuclear power plant. Air pollution, according to the study, not only causes cancer and lung diseases, but might be reducing male potency.

Concern has also grown over PM10, a category of airborne particles less than 10 μm (millionths of a metre) in size. In 1995, WHO reported that there was no safe level for exposure to PM10 and calculated that in a city of one million people, a three-day episode of PM10 at 50 μg m^{-3} would produce a thousand additional asthma attacks and four deaths. In Britain, the Expert Panel on Air Quality Standards found in 1995 that PM10 caused 2000–10 000 British deaths a year. The Committee on the Medical Effects of Air Pollutants also expressed warnings, announcing a new maximum limit of 50 μg of particulates per cubic metre of air averaged over 24 hours.

According to a 1996 report from the Quality of Urban Air Review Group entitled *Airborne Particulate Matter in the United Kingdom*, that level was exceeded in London on 139 days during 1992–94. In the same year, a report by the US Natural Resources Defence Counsel on 239 US cities estimated that each year about 64 000 deaths in US cities from cardiopulmonary causes could be attributed to particulate air pollution.

In respect of lead pollution, in 1996 the University of Michigan reported that most children in African cities had blood-lead levels high enough to cause neurological damage; in some cities more than 90% of the children suffered from lead poisoning. Africa accounted for 20% of the global emissions of atmospheric lead.

Box 8.1 summarizes some of the economic losses due to air pollution; Box 8.2 some of the environmental costs being experienced in the People's Republic of China; while Box 8.3 relates to the 1986 Chernobyl nuclear disaster in the independent republic of Ukraine. Box 2.6 refers to some early estimates by Sir Hugh Beaver and his committee on the economic costs of air pollution.

8.3 Acid Rain

The term 'acid rain' refers to the acidification of rain associated with the combustion of fossil fuels: coal, oil and natural gas. The constituents of flue gases which

Box 8.1 Economic Losses Due to Air Pollution

- direct medical losses
- lost income resulting from absenteeism from work
- decreased productivity
- increase in travel costs and time of travel due to reduced visibility
- increase in the costs of artificial illumination
- repair of damage to buildings and other structures
- increased costs of cleaning
- losses due to damage to crops and ornamental vegetation
- losses due to injury to animals of economic importance
- decrease of property values
- additional manufacturing costs due to pollution from outside sources
- losses due to the inefficient combustion of fossil fuels

Box 8.2 Environmental Costs in China

With a population of over 1.2 billion at a density of 330 persons per square kilometre, China is the world's third largest economy (after the USA and Japan) with plans for significant expansion. The predicted population for 2010 is 1.4 billion, with a doubling time of 67 years, notwithstanding that the birth rate is below the world's average. Life expectancy at birth is 69.1 years for males and 72.4 years for females (1992).

China's poor petroleum and natural gas reserves means that its industries must rely overwhelmingly on coal (with a coal production of 1.15 billion tonnes in 1993). Much of this coal is burned without emission controls, although power stations have been equipped with electrostatic precipitators. Fortunately, most of the coal is relatively low in sulphur (about 1%), but with low stacks. Much of it is burnt in inefficient boilers and household stoves, and often in an uncleaned state (run-of-mine quality). This results in high ground level concentrations of pollutants. Most waste water does not receive the simplest treatment: sewage disposal is by way of nightsoil removal and discharges to rivers; solid wastes are disposed of haphazardly.

Thus China experiences every conceivable environmental problem, with limited capacity to deal with those problems. It has extensive areas of badly damaged ecosystems. There has been an accelerated loss of farmland, increased soil erosion, and significant declines in the quality of cultivated soils. There has been progressive degradation of grasslands and increasing shortages of water, despite devastating floods.

Studies suggest that it was extremely unlikely that the economic cost of China's environmental pollution and ecosystem degradation was less than 5% of the nation's GDP in 1990. A range of 6–8% is the most likely conservative estimate. Values of 10% might reflect a more comprehensive, though not necessarily complete, assessment. The inclusion of a number of more elusive factors might raise the rate to about 15% of annual GDP, a share much higher than any estimate available for an affluent Western economy. Investment by China in environmental control during the 1980s was less than 1% of GDP, although it was announced by Qu Geping, Chairman of the National People's Congress Environmental and Resources Protection Committee in December 1995 that there was a plan to raise the share to 1.5% of GDP by the year 2000.

Source: Derived from S. Vaclav (1996) Environmental Problems in China: Estimates of Economic Costs, US East–West Center, Special Report No. 5, Honolulu, Hawaii.

contribute to the acidity of rain are the oxides of sulphur (mainly sulphur dioxide, SO_2) and the oxides of nitrogen (NO_x). Arising from the composition of the fuels themselves, they may react with water vapour to form acids. Some acids may adhere to particulates in the air to form acid soot, while most are absorbed by rain, snow or hail, and are carried far from the source of pollution. Figure 8.1 illustrates this process.

Significant areas affected by acid rain include north-eastern USA and Ontario, Canada, as well as parts of Scandinavia, notably Sweden and southern Norway.

Box 8.3 Chernobyl Nuclear Disaster

The Chernobyl nuclear disaster, which occurred in April 1986, was the worst accident in nuclear power station history. Located 104 km north of Kiev in the Ukraine (then part of the Soviet Union), the Chernobyl nuclear power station consisted of four reactors, each of 1000 MW capacity. The station was commissioned during 1977–83. During 25–26 April 1986, a test being run on the No. 4 reactor involved a breach of several safety precautions. These errors were compounded by others and a chain reaction in the reactor core went out of control. Several explosions and a large fireball followed, blowing the steel and concrete lid off the reactor. The result of these events and the subsequent fire in the graphite core released large amounts of radioactive material into the atmosphere, where it was carried great distances over Europe by air currents. The radioactive fallout reached as far as Norway and Scotland, but most people affected were in the Ukraine and the then Soviet Republic of Belarus, immediately to the north. About 5 million people are believed to have been exposed to the radioactivity.

On 27 April, the residents of Pripyat, the nearest town, began to be evacuated; more than half a million people were eventually displaced from towns in the Ukraine, Belarus and Russia. Thousands of square kilometres of land were heavily contaminated, as well as water supplies. Chernobyl caused the deaths of more than 30 people and many more were afflicted by radiation sickness. The incidence of thyroid cancer rose markedly in adults and children. The radioactivity was expected to cause several thousand extra cancer deaths over the coming years. The event strengthened opposition throughout the world to nuclear power plants.

In 1996, work began on the decommissioning of units 1, 2 and 3; waste management on site and in the exclusion zone; the storage of spent fuel and high-level waste; and the enclosure of unit 4. The aim was to close Chernobyl completely by 2000. At a meeting of the Group-of-Seven nations (the seven top industrial nations; now the Group of Eight) in 1996 in Moscow, just over US$3 billion worth of international assistance was offered to the Ukraine to close the station permanently by the end of the century and to complete reactors under construction at the Khmelnytsky and South Ukraine nuclear power station sites.

In 1998, official reports revealed that some 12 000 people had died as a result of the Chernobyl incident, including some three-quarters of the rescue workforce.

Other parts of Europe are also affected by acid rain, and it has possibly contributed to the dieback of the Black Forest in Germany.

The oxides of sulphur come mainly from the combustion of coal and heavy fuel oil in industrial plant and power stations (sulphur being a natural ingredient of these fuels), while the oxides of nitrogen arise both from stationary sources and transportation, most notably the automobile.

It has been estimated that over 80% of sulphur dioxide emissions in the US originates in the 31 states east of, or bordering, the Mississippi River, with a heavy concentration from states in or adjacent to the Ohio River Valley. These airborne emissions are transported by prevailing winds to the east across state and national borders.

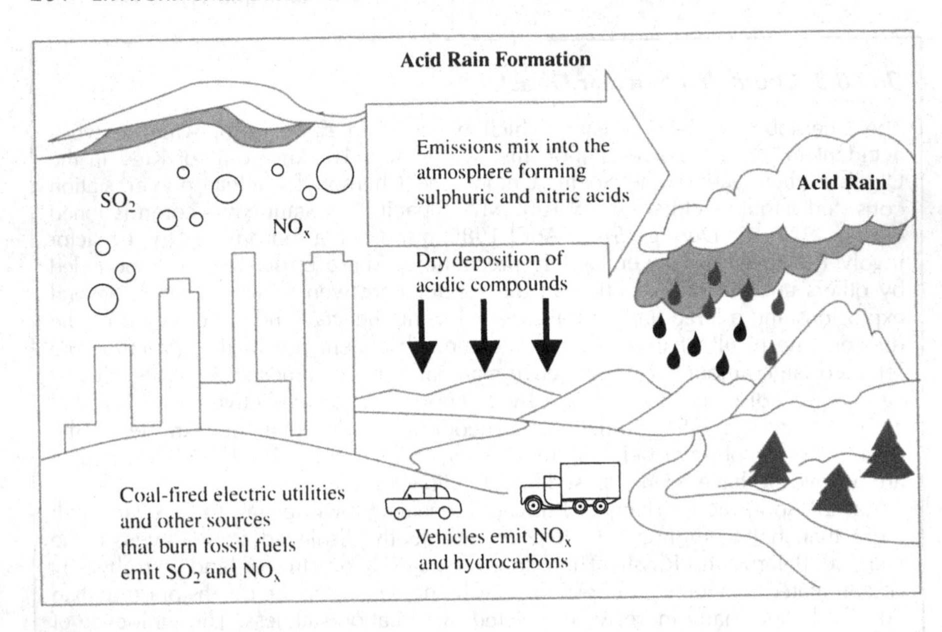

Figure 8.1 *Formation of acid rain*

In the west, sulphur deposition is relatively low. The US EPA's National Lake Survey, conducted as part of the National Acid Precipitation Assessment Program, found no lakes in the west to be acidic, with the exception of one hot spring. The survey also found no evidence of regional scale acidification.

It was during the late 1960s and early 1970s that a link was established between acid deposition and the acidification of lakes. Sweden blamed British power stations, both coal- and oil-fired, for part of their problem, and made much of this at the UN Conference on the Human Environment in Stockholm in 1972.

Enforcement of the US Clean Air Act from 1970 began a new era of progress towards reducing nationwide emissions of sulphur dioxide. All states were required to create State Implementation Plans for meeting national ambient air standards. By 1986, annual sulphur dioxide emissions in the US had declined by about 25%. At the same time, motor vehicle regulations slowed the growth of NO_x emissions.

The US Clean Air Act of 1990 aimed at cutting sulphur dioxide emissions virtually in half by the year 2000, establishing the US Acid Rain Program and the US Emission Trading Program (see Chapter 6). Phase 2 begins in 2000 with a further substantial reduction of sulphur emissions.

In 1998, Resources for the Future published the first integrated assessment of both the projected costs and benefits of Title IV of the 1990 Amendments to the US Clean Air Act, which mandates reductions in the sulphur dioxide and nitrogen oxide emissions that emanate from coal-fired electricity generation and contribute to acid rain. The dollar value of reduced mortality risk alone measured

Table 8.1 *RRF study of the projected costs and benefits of Title IV of the 1990 Amendments to the US Clean Air Act in the year 2010 (phase 2): acid rain reduction*

Benefits	Per capita US$ (1990 dollars)
Reduced human mortality risk	59.00
Improved visibility, residential and recreational	9.34
Reduced human morbidity	3.50
Improved aquatic benefits	0.62
Total benefits	72.46
Costs	<6.00

Source: Resources, Spring 1998, issue 131, Resources for the Future, Washington, DC.

Box 8.4 Acid Rain

In the early 1980s acid rain became the favourite cause of doom. Lurid reports appeared of widespread forest decline in Germany, where half the trees were said to be in trouble. By 1986, the United Nations reported that 23 per cent of all trees in Europe were moderately or severely damaged by acid rain. What happened? They recovered. The biomass stock of European trees actually increased during the 1980s. The damage all but disappeared. Forests did not decline: they thrived. A similar gap between perception and reality occurred in the United States. Greens fell over each other to declare the forests of North America acidified and dying. However, a ten-year official study costing $700m found no evidence of a general or unusual decline of forests in the United States or Canada due to acid rain.

Source: The Economist, 20 December 1997, p.20.

several times the expected costs of compliance with the acid rain programme. The total projected costs and benefits are summarized in Table 8.1. The views of *The Economist* are aired in Box 8.4.

8.4 The Costs and Benefits of Nuclear Power: The Swedish Dilemma

The Swedish nuclear programme was launched in 1966 and extended in 1975 by the Swedish Government with the approval of Parliament and (presumably) a broad majority of public opinion. The programme was completed in 1985 when the latest of 12 nuclear power units was connected to the electricity grid. Today, nuclear power supplies about half of the electricity used in Sweden.

Nuclear power first became a political issue during the election of 1976, when the Social Democratic Party lost its majority in Parliament while, at the same time, a minority faction of that party took up an antinuclear position. In 1979, following the Three Mile Island accident, the Swedish Government decided to organize a referendum on the question of nuclear energy in Sweden.

The choice offered to Swedish electors was limited to three nuclear options: (1) to decommission the six nuclear units already in operation over a period of 10 years; (2) and (3) to complete the nuclear programme approved by Parliament and decommission the 12 units at the end of their planned technical and economic lifetime (these two options varying only in some details, regarding timing). Almost 76% of the electorate participated in the referendum.

The first option received 38.7% support, while options 2 and 3 together obtained 58% of the votes (a few opposed all options). These results were interpreted by Parliament as endorsing the completion of the 12 units, but with all being phased out by 2010. The nuclear programme proceeded as originally planned.

However, the Chernobyl accident and its impact on Swedish public opinion led the Swedish Parliament in 1987 to decide to phase out two nuclear units by 1995. This was accompanied by a decision not to increase the consumption of oil and coal and to make available renewable sources of energy for electricity generation.

By the late 1980s, several official studies concluded that the phasing out of the two nuclear units could lead to a doubling of the price of electricity for heavy industry and the consequent loss of 100 000 jobs. Labour union leaders began to argue against the phasing out. Conflicts in policies began to emerge, for there had also been much opposition to the development of hydroelectric power on the major rivers.

In mid-1990, the Social Democratic Government decided to look for a more coherent and integrated energy policy and to seek a broad consensus on the energy issue. A three-party agreement was announced in January 1991; this cancelled the decision to phase out two units by 1995, but maintained the date of 2010 for the final phasing out of nuclear power in Sweden.

More recent public opinion polls have shown that a majority of the Swedish population favours the use of the 12 units, even after 2010, provided they remain competitive and safe. However, further deployment of nuclear energy in Sweden would require the endorsement of the public through a further referendum. By the time of the United Nations Conference on Environment and Development in 1992, the decision of the Swedish Parliament (Riksdag) remained that nuclear power in Sweden was to be phased out by no later than the year 2010.

In 1998, the Social Democratic Government then decided to shut a nuclear power station that is both efficient and safe, with another to be closed in 2001. It was felt that shutting the two stations would help boost renewable forms of energy. However, nuclear stations are expensive both to construct and to decommission, although cheap to run in between. Further renewable energy is too dear and too undeveloped to replace more than a fraction of the displaced power. Instead, more power would have to be generated from coal and natural gas, thus emitting more greenhouse gases. Nuclear power does not emit these gases. Further, Sweden has a good record of nuclear safety, and both the reactors the government intends to close have excellent records.

Another consequence of shutting Swedish plants would be to increase demand for power from the Nordic electricity pool, which is supplied with power from Russian and Baltic nuclear plants via Finland. Those plants do not meet Swedish safety standards.

The plant intended for immediate closure is the Barsback-1 600 MW nuclear station, located within the proximity of Copenhagen and Malmö. It has run efficiently since it was commissioned in 1975, and has years of useful and economical life left. The US Energy Information Administration has estimated that it will cost between US$1.2 billion and US$2.7 billion to close the place down. The Social Democrats, however, are trying to keep a campaign promise to close at least one of the 12 plants operating. Swedish industry and some trade unions are opposed to the closure. The decision to close the Barsback-1 nuclear station is also being opposed by the operators, through the courts. Most Swedes want all the nuclear stations closed down at least eventually.

The entire issue illustrates that the costs and benefits of nuclear power are complex, and never more so than when existing safe and economical stations are to be closed. Although Chernobyl has turned world opinion clearly against nuclear power, many dangerous plants are still operating, whilst Sweden intends to shut down two safe and efficient plants: Sweden's dilemma is certainly one of unusual proportions. The closing of nuclear power stations ahead of schedule could well aggravate the cost of power in Sweden and could damage the environment as well. The situation has been reviewed by William Nordhaus (1997).

8.5 Global Biodiversity

Biodiversity, or biological diversity, is an umbrella term to describe collectively the variety and variability of nature. It encompasses three basic levels of organization in living systems: the genetic, species and ecosystem levels. Plant and animal species are the most commonly recognized units of biological diversity, thus public concern has been mainly devoted to conserving species diversity. This has led to efforts and to legislation to conserve endangered species and to establish specifically protected areas. However, sustainable human economic activity depends upon understanding, protecting and maintaining the world's many interactive diverse ecosystems with their complex networks of species and their vast storehouses of genetic information. Most countries have now taken legal and physical measures to protect endangered species from extinction. Also the idea of protecting outstanding scenic and scientific resources, wildlife and vegetation has taken root in many countries and developed into national policies, embracing both terrestrial and marine parks. The World Heritage List, maintained by UNESCO, includes properties of great cultural significance and many geographical areas of outstanding universal value, augmenting the principle of biological diversity. Increasingly, wilderness areas are being identified and preserved, including some very substantial areas such as the Great Barrier Reef, Australia. However, major world resources such as the Amazon rain forest, covering about 40% of the land area of Brazil, is still only partially protected and subject to progressive logging and clearing.

The Convention on Biological Diversity signed at the UN Conference on Environment and Development in Rio de Janeiro in 1992, was the first international treaty on biodiversity and went further in addressing the issues involved. The convention began as a document drawn up by the World Conservation Union (IUCN). It was submitted to the UNEP Governing Council which recognized the need and value for an international biodiversity convention. UNEP redrafted it to broaden the proposed agreement. Formal negotiations involving some 75 countries began in November 1990. A final version was signed by 156 nations and the EU at the 1992 UN Conference. The Convention aims to save species of animals and plants from extinction, and their habitats from destruction. Other aims include the sustainable use of genetic resources and the fair and equitable sharing of the benefits. The parties are required to develop national strategies for the conservation and sustainable use of biological resources; to establish protected areas, restore degraded ecosystems, control alien species, and establish conservation facilities; to establish training and research programmes; to encourage technology and biotechnology transfer, particularly to developing countries; to provide financial resources to developing countries in support of the convention; to carry out environmental impact assessments prior to proposed projects that might reduce biodiversity; and to undertake reviews of national biological resources. The studies will allow an assessment of the costs and benefits of implementing appropriate measures, and the additional funding required from the Global Environment Facility. The possible extra financial support required by developing countries has been estimated at US$20 billion a year (UNEP 1993a).

Published in 1995, UNEP's *Global Diversity Assessment* was the world's first comprehensive review of biodiversity. It was presented to the second meeting of the parties to the Convention on Biological Diversity. It examined the current state of knowledge, ways to approach the problem, and possible solutions. Some 1500 scientists participated in the preparation of the report, which had been sponsored by the Global Environment Facility. Although the total number of species remains unknown, the report considered that a reasonable estimate is close to 14 million, of which about 1.7 million have been scientifically described. The working estimate for insects was about eight million, though the actual figure could be much higher. The report saw grave threats to many species, at a profound cost to society; it concluded that between 5% and 20% of some groups of animals and plants could be threatened with extinction in the foreseeable future.

Apart from its statistical assessment, the report reviewed strategies to protect biodiversity. The traditional approach had emphasized the separation of ecosystems, species and genetic resources from human activity through the creation of protected areas, prohibitions on the harvesting of endangered species, and the preservation of germ plasm in seed banks or cryogenic storage facilities (World Resources Institute 1996). Instead, it was now thought that preservation efforts must embrace a blend of strategies, including programmes to save species by creating controlled environments and developing policies to manage natural environments in ways that minimize adverse impacts. There is also a recognition of the need for more integrated approaches to conservation, including a consideration of entire ecosystems rather than just some protected areas within those ecosystems. Many existing international conventions relate in whole or part to the protection of biodiversity.

The 1996 Red List of Threatened Animals issued by the World Conservation Union identified 5205 species in danger of extinction. In tropical forests alone, it was estimated by biologists that three species were being eliminated every hour. Much of the decline, it was claimed, was caused by habitat destruction, especially by logging. Only 6% of the world's forests were formally protected, leaving 33.6 million km^2 vulnerable to exploitation.

In 1992, at the Seventh International Coral Reef Symposium, held in Guam, scientists estimated that some 10% of the world's coral reefs were effectively lost, and a further 30% were under immediate threat. To draw attention to this problem, 1997 was recognized as the International Year of the Reef.

Critical habitats continue to be progressively preserved, however. In 1997, the Bastak Nature Reserve was established in the Jewish Autonomous Republic of Russia. Other conservation areas included the Hawar Islands in Bahrain, the Masuala National Park in Madagascar, and the Northern Truong Son mountain range in Laos and Vietnam.

The threat to biodiversity is not solely a threat to the natural world, which people increasingly appreciate; it also poses a threat to economically valuable assets: wild and domestic plants, fish, animal products used for medicines, cosmetics, industrial products, building materials, and food. Species diversity also plays a role in the balance of nature and the functioning of ecosystems.

8.6 Forestry

Forestry is the business of managing forests for commercial wood production and other forest products. Forests also have important roles as water catchments and as a means of preserving natural ecosystems and biodiversity, and providing wildlife shelter. Forests serve as a genetic, scientific, educational and landscape resource. Forest areas can be managed to provide a balanced supply of many services and benefits; however, the point of balance is often a matter of intense controversy. Concern is often expressed that forestry means deforestation accompanied by erosion; the permanent loss of old growth forests; the loss of habitat and diversity; the destruction of mangroves and rain forests; and that reforestation often means replacing interesting and varied trees with uniform coniferous plantations.

In sum, forest products include a wide range of commodities and services valuable to affluent industrial and poor rural communities alike: timber for housing, construction and furniture; pulp for packaging and newspapers; timbers for mining and railways; fuelwood for villages; fodder, fruits, honey and pharmaceuticals; meat, fibres, resins, gums, dyes, skins, waxes and oils; beauty, amenity and recreational qualities; preservation of genetic diversity; and absorption of carbon dioxide from the atmosphere as food, helping to contain climatic change and contribute to a sustainable future.

With a growth of population from one billion in 1800 to over 5.5 billion today, human development has brought great pressures on forests, with a massive increase in cropland. The most dramatic changes are occurring in developing countries, with forest more or less stabilizing in most of the developed world. Between 1960 and 1990, it is thought that fully one-fifth of all natural tropical forest cover has been lost (World Resources Institute 1996).

The first complete assessment of 1990 forest cover was provided in the FAO report *Forest Resources Assessment: Global Synthesis* (Food and Agriculture Organization 1995). In 1990, forest and other wooded land covered 5.1 billion ha, i.e. about 40% of the Earth's land area. This estimate included 3.4 billion ha of forest, (defined by FAO as land with a minimum tree crown cover of 20% in developed countries and 10% in developing countries). Forests consist of natural forests, composed largely of native tree species, and plantation forests. The remaining 1.7 billion ha consisted of other woody vegetation such as open woodland, scrubland and brushland, together with areas under shifting vegetation.

FAO defines deforestation as the conversion of forest to other uses such as cropland and shifting cultivation. By using that definition, the area of the world's forests and other wooded land declined by 100 million ha (an area about the size of Egypt) between 1980 and 1990 (i.e. a decrease of about 2%). Almost all of the change occurred in tropical countries where forest and other wooded land declined by 3.6% (compared with a 0.04% decline in developed countries).

The loss of natural forest cover was offset by new plantation cover in tropical areas totalling 31.9 million ha and by 32.1 million ha in other wooded areas.

Tropical forests, including plantations, made up just over half of the world's forest cover, about 1.8 billion ha, in 1990. Rain forests were the more prevalent forest type in the tropics, covering about 714 million ha. According to the FAO data, the world lost 450 million ha of its tropical forest cover between 1960 and 1990. Asia lost about one-third of its tropical forest cover during the same period, while Africa and Latin America each lost about 18%. As a result of a massive tree-planting programme in China, afforestation actually exceeded deforestation in the 1980s in Asia and the Pacific.

The two major UN conventions, the Convention on Biological Diversity and the Framework Convention on Climate Change, both recognize the broad role that forests play in the maintenance of global ecosystems, as forests harbour the majority of the world's species and provide a major repository for carbon. In 1990, the International Tropical Timber Organization (ITTO) became the first intergovernmental body to produce guidelines for the sustainable management of tropical forests. The International Tropical Timber Agreement was renegotiated in 1993, establishing ITTO's Target 2000, the year by which all forest products should come from sustainably managed forests. Regional agreements followed, including the Tarapoto declaration of the Amazon Treaty Organization regarding the Amazonian forests and the need for sustainable management (see Box 8.5).

In 1993, the Forest Stewardship Council (FSC) was created. It is an assembly of NGOs, industry representatives, scientists and indigenous people established to promote the environmentally appropriate, socially beneficial, and economically viable management of the world's forests. It is governed by a nine-member board elected by the membership. In 1994, the FSC adopted a set of principles and criteria for the sustainable management of forests, as well as guidelines on how to conduct field inspections and the tracking of forest products from forest to store shelf. National certification procedures are being evolved.

An Intergovermental Panel on Forests was approved in 1995 during the third session of the UN Commission on Sustainable Development. Its purpose is to generate consensus and propose actions for the implementation of the UN Conference on Environment and Development's forest-related agreements at national and international levels. IPF's recommendations are likely to set the agenda for forest policy and international aid for the next several years.

Box 8.5 International Tropical Timber Agreement (ITTA)

In 1983, the International Tropical Trade Agreement established a system of consultation and co-operation between consuming and producing countries. The agreement has the support of some 47 countries, representing about 95% of the international trade in tropical timber. The ITTA recognizes in its charter the importance of conservation and the principles of 'sustainable yield'. In 1986, the ITTA created an International Tropical Timber Organization (ITTO). The ITTO has attempted to promote the sustainable management of tropical forests. In 1990, the ITTO adopted a set of guidelines for sustainable tropical forest management and set a target of the year 2000 by which the entire tropical timber trade should come from sustainably managed forests. However, by 1994, perhaps no more than 1% of tropical timber came from sustainably managed forests; further, the parties could not agree on the meaning of sustainable management. Clearly, incentives such as preferential entry to markets are needed to encourage good management. During a renegotiation of the ITTA in 1993, the producer countries again endorsed the putting of tropical forests under sustainable management by the year 2000, though with the added qualification that the target be also applicable to timber from all forests, tropical, temperate and boreal.

Box 8.6 Agenda 21 on Forests

The vast potential of forests and forest lands as a major resource for development is not yet fully realized. The improved management of forests can increase the production of goods and services and, in particular, the yield of wood and non-wood forest products, thus helping to generate additional employment and income, additional value through processing and trade of forest products, increased contribution to foreign exchange earnings, and increased return on investment. Forest resources, being renewable, can be sustainably managed in a manner that is compatible with environmental conservation. The implications of the harvesting of forest resources for the other values of the forest should be taken fully into consideration in the development of forest policies. It is also possible to increase the value of forests through non-damaging uses such as eco-tourism and the managed supply of genetic materials. Concerted action is needed in order to increase people's perception of the value of forests and of the benefits they provide. The survival of forests and their continued contribution to human welfare depends to a great extent on succeeding in this endeavour.

Source: Agenda 21 (1992) Combating Deforestation, UN Conference on Environment and Development, Chapter 11.

The World Commission on Forests and Sustainable Development was also created in 1995 by the UN Commission of Sustainable Development to generate consensus and resolve conflict on the dual role of forests in preserving natural habitats and promoting socioeconomic developments; and on the importance of co-operation between developed and developing countries in determining priorities on forest issues. A number of working panels have been created. The commission is based at Woods Hole Research Center, Woods Hole, Massachusetts.

The Convention on International Trade in Endangered Species of Wild Fauna and Flora (CITES) became effective in 1975 and was supported in 1998 by 128 countries. Box 8.6 quotes from Agenda 21 on forests, which was endorsed by the UN Conference on Environment and Development 1992. Box 8.7 quotes from Alexander Mather's assessment of global forest resources.

Box 8.7 Mather on the Fate of the Forests

While much damage has been done, most of the pre-agricultural forest area still remains. It is true that some forest types have been almost completely eliminated, but huge areas survive, and large areas survive almost intact. Much of the world's forest has proved to be resilient in the face of human use, and much of it lies in areas so inaccessible that exploitation is unlikely in the foreseeable future. Furthermore, while the current rate of deforestation is probably unprecedented in history, so also is the rate of reforestation and afforestation. The net trend is still one of contraction, but eventually it may give way to one of expansion, as afforestation continues in the developed world ... and becomes more rapid and more widespread in the developing world.

Source: A.S. Mather (1990) Global Forest Resources, Belhaven Press, London, and Timber Press, Portland, OR, p. 292.

In 1997, an agreement was reached between the World Bank and the World Wide Fund for Nature (WWF) to build a global network of protected forests. The alliance aims to conserve 10% of the world's forests by the year 2000, and to achieve 50 million ha of new protected forest areas by the year 2005. The alliance seeks broad application of the principles of sustainable logging. Brazil was already committed to protecting an additional 25 million ha. By 1998, 22 countries were committed to bringing 10% of their forests into protected areas by the year 2000.

8.7 The Amazon Rain Forest

The Amazon rain forest is a massive tropical rain forest, occupying the drainage basin of the Amazon River and its tributaries and covering an area of 7 000 000 km^2. It comprises 40% of Brazil's total land area, and is bounded by the Guiana Highlands in the north, the Andes Mountains in the west, the Brazilian central plateau to the south, and the Atlantic Ocean to the east. The rain forest stretches from the swampy mangroves in the east near the Atlantic Ocean to the tree line of the Andes.

The Amazon Valley is rather like an immense canyon opening into the Atlantic Ocean, with a mouth more than 400 km wide. With around 1000 tributaries, it is the largest basin area in the world. The Amazonian rain forest has a wide variety of trees including myrtle, acacia, rosewood, Brazil nut, rubber tree, mahogany, cedar and palm. Wildlife includes the jaguar, matee, tapir, red deer, capybara and several kinds of monkeys. The rich bird life of the forest includes parrots, toucans, haugnests, perdizes, cormorants and scarlet ibises. Fish include catfish, electric eels and piranhas.

The Amazon River is the greatest river in the world in terms of its volume and the area of its drainage basin. The Amazon flows some 6400 km across northern Brazil to its mouth in the Atlantic Ocean. It has been estimated that 20–25% of all the water that runs off the surface of the Earth is carried by the Amazon. The average annual discharge is roughly ten times that of the Mississippi River and about four times that of the Congo River. Its length is second only to that of the Nile.

The history of the exploration of the Amazon is briefly summarized in Box 8.8.

More than two-thirds of the Amazon Basin is covered by lush rain forest, which merges into dry forest and savanna on the higher northern and southern margins and into montane forest in the Andes to the west. The Amazon rain forest represents about one-half of the remaining rain forest of the Earth and the largest reserve of biological resources.

The vast Amazon forest has the appearance of being extremely lush, suggesting that the underlying soil must be extremely fertile. In fact, most of the nutrients are locked up in the vegetation which is continuously being recycled through leaf fall and decay. Soils are often sandy and of low natural fertility, lacking

Box 8.8 The Amazon: Exploration and Policy

1541	Francisco de Orellano (1490–1546), a Spanish soldier and the first European to have explored the Amazon, gave the river its name. Reported battles between tribes of female warriors
1799–1804	Visit of Alexander von Humboldt (1769–1859), mapping the connection between the Amazon and Orinoco systems, through the Casiquiare River. He and his colleague Aime Bonpland, a botanist, covered 9650 km on foot, on horseback and in canoes
1848–1859	The English naturalist H.W. Bates (1825–1892) spent 11 years exploring the Amazon, collecting nearly 15 000 specimens. In 1863, he published *The Naturalist on the River Amazon* in two volumes – a great classic
1854	Report submitted to the US Congress entitled *Exploration of the Valley of the Amazon*, following an official expedition by W.L. Herndon and L. Gibbon
1913–1914	The former US President, Theodore Roosevelt took part in an expedition that explored a tributary of the Madeira
1910–1924	Explorations by Harvard University's Institute of Geographical Exploration
Since World War II	Many expeditions to Amazonia by British, French, German, Japanese and North American groups
1979	Forest Policy Committee recommends reversal of policy on opening-up the Amazon with the establishment of national parks and ecological reserves
1997	President Cardoso undertakes to treble the area under national park-style control by the year 2000

phosphate, nitrogen and potash; they are also highly acidic. Some soils do not lack nutrients due to the deposit each year of fertile silt left as the waters recede, but their use for agricultural purposes is limited by periodic flooding.

Three-quarters of the population in Brazil are concentrated within 200 km of the coast, mainly in the south and the south-east in the four states of São Paula, Minas Gerais, Rio Grande do Sol, and Rio de Janeiro.

Since the Second World War, the opening of the Amazon has been high on political agendas; in the mid-1950s a decision was made to refocus the country towards its interior by constructing a new inland capital, Brasilia. This was accompanied by the initiation of a vast road-building programme aimed at integrating the northern states with the rest of Brazil, while providing relief for the crowded and drought-stricken North-east. A 1700 km highway linking Brasilia with Belem, the trade centre at the mouth of the Amazon, was completed in 1964. In all, a framework of nearly 36 000 km of highways was completed, including the Transamazonian Highway. The government planned to settle about 100 000 families along the Transamazonian Highway, but this target was not reached. Those who settled were faced with declining crop yields as a result of poor soils, weed invasions and plant diseases, problems that were aggravated by the distance to markets and lack of credit. Then cheap credit and tax breaks were offered to promote the creation of big cattle ranches in the Amazon.

Between 1966 and 1976, the official Amazon development agency Sudam approved the setting up of 354 ranches with an average size of 20 000 ha; tax rebates were granted to cattle companies. However, the African grasses planted on the cleared forest land grew less well after a few years, and large outlays on fertilizers and on measures against soil leaching and noxious weeds became necessary. Neither the cattle ranches nor the many small farms established have prospered. Between 1976 and 1979 only four additional cattle projects were approved.

In March 1979, President Figueiredo set up a forest policy committee to review the situation and analyse the causes of failure. The committee urged an almost complete reversal of the policy of opening up the 6 million km^2 of the Amazon forest. It recommended that no more forest land should be leased to companies, and that about 1.2 million km^2 should be designated as national parks and ecological reserves, with about 900 000 km^2 being retained as national forests. The failures, it was found, had been due in part to the belief that the tropical soil was prodigiously fertile; but the topsoil had been found to be very thin and ecologically fragile. The decaying vegetation did not create a fertile topsoil; further, half the rainfall was provided by evaporation from the forest itself. The committee reaffirmed that the Amazon forest was a valuable resource yielding drugs, fibres, fuel, crops and resins, as well as providing gene pools of rare and valuable species of plants and animals. In 1979, the committee's approach had strong support from the Brazilian President. In that year, three new national parks in the Amazon forest were designated.

Meanwhile, international concerns about the ecological consequences of continuing deforestation had been growing and were voiced at the UN Conference of Environment and Development held in 1992. International calls for conservation were based on the view that the Amazon forest was a global resource, one that serves as a control mechanism for the world's climate and as a genetic repository for the future. The extent and rate of continuing deforestation remains a subject of controversy.

The employment of radar to estimate deforestation more precisely led to the suggestion that by 1990 some 10% of the rain forest (selva) may have been cleared for pasture, crops, lumber and firewood. While the forest remains an efficient absorber of carbon dioxide, the volume of gas released when substantial areas are cleared and burned may offset this benefit to some extent.

The consequences of continuing deforestation have been much discussed in recent years. Undoubtedly, the unique gene pool of the Amazon Basin, containing perhaps two-thirds of the known organisms of the world, is threatened, together with unknown and unexploited pharmaceutics. Further, at stake is the survival of many indigenous peoples.

In 1992, Al Gore of the US asserted that 20% of the Amazon Basin had been deforested and that deforestation continued at a rate of 80 million ha a year. In 1997, *The Economist* (20 December, p. 19) asserted that the true figures were 9% (not 20%) and 21 million ha a year (not 80 million ha) at its peak in the 1980s, falling to about 10 million ha a year now.

In 1997, President Cardoso announced that about 25 million ha of the Amazon Basin would be added to the already protected areas by the year 2000, trebling the area under national-park style coverage. The proposal was truly remarkable in its size and content.

8.8 Deserts

Deserts are arid areas with sparse or absent vegetation and a low population density. Together with semi-arid regions, they constitute more than one-third of the Earth's land surface. However, only 5% of the Earth's land surface can be described as extremely arid. Such regions include the central Sahara and the Namib deserts of Africa, the coastal areas of Ethiopia and Yemen, the Rub'al Khali in Saudi Arabia, the Takla Makan desert in central Asia, the Atacama desert of Peru and Chile, parts of the south-west USA and northern Mexico, the Gobi desert of north-eastern China, and the great deserts of Australia. It has been alleged that the deserts are expanding, an issue that has received the attention of the UN in recent years.

The once-held belief that deserts have existed throughout geological time in their present locations has been disrupted by evidence that during the past 1.6 million years (the Quaternary period) some profound climatic changes have affected certain desert areas.

Artifacts and other evidence suggest that humans hunted large animals in areas that could not support such activity today; forests of oak and cedar grew in such highland areas as Tibesti, in the central Sahara; lakes and lake systems were once extensive in such regions as the Kalahari, the Iranian desert and the western parts of the USA. There is evidence of a former body of water, 180–210 m deep, in Death Valley, a structural depression in Inyo County, south-eastern California, which is now the hottest and driest part of North America. There is much evidence of former moist conditions in many of the world's desert areas.

There is much evidence in North Africa of human occupation in Roman times. The theatre at Leptis Magna, near present-day Al-Khums, Libya, was at the centre of a fertile area serving the interior of Africa. The theatre was designed to seat several thousand people; today Leptis Magna is a ghost town.

It has been concluded that the general cause of the arid regions of the Earth and their present locations is to be found largely in the dynamics of atmospheric circulation, i.e. incoming solar radiation combined with the effects of the Earth's rotation. However, the locations vary in the following characteristics: topography, hydrology, climatic variations, temperature variations, rainfall, subsurface water supplies, the variety of habitats, diversity of plant species amd soil microorganisms. Further, the natural productivity of the deserts is low compared with that of other ecosystems; to survive, human populations need to be sparse and nomadic. Subject to these conditions, Australian Aborigines have been able to survive in some localities, making use of a wide range of fruits, roots, insects, reptiles and mammals. Deserts have also been used for grazing domesticated livestock, moving the flocks to fresh pasture as the limited flora is grazed off.

The UN Conference on Desertification was held in Nairobi in 1977. It was the first world-wide effort to consider the global problem of advancing deserts. Some 95 countries participated, together with 50 UN agencies, 8 intergovernmental organizations, and 65 non-governmental organizations. The outcome of the conference was a World Wide Plan of Action to Combat Desertification, with 26 specific recommendations. In addition, the executive director of UNEP was asked to convene a Consultative Group for Desertification Control. The core members comprised representatives from 16 countries, with seven UN agencies including the World Bank. The purpose of the group was to marshall resources for the Plan of Action. In all these activities, particular emphasis was placed on the arid and semiarid regions of Africa and to a lesser degree on similar regions in south-west Asia and South America. Many of the plans involved better planning and management of soil resources.

One of the central problems emerging was that governments faced with desertification problems were confronted with conflicting demands on limited financial and human resources. Countries affected never managed to assign a high enough priority to anti-desertification measures.

China alone appeared to have made some progress. The State Council in the late 1970s decided to plant protective forests in the northern region, creating a Green Great Wall of benefit to present and future generations. According to the plan, some 5.3 million ha were to be covered by various kinds of protective forests, creating massive shelter belts in agricultural and pastoral districts menaced by sandstorms.

A renewed effort in relation to desertification emerged from the UN Conference on Environment and Development in 1992; a Convention to Combat Desertification was signed in Paris in 1994 and ratified later by individual countries. It involved the creation of national action plans to deal with the deterioration of farmland into desert.

In sum, more than 100 countries are suffering the consequences of desertification, or land degradation in dryland areas. Loss of productivity and other social, economic and environmental impacts are directly affecting the perhaps 900 million inhabitants of these nations (Hulme and Kelly 1993).

In a remarkably contrasting statement, *The Economist* stated on 20 December 1997 (p. 19): 'In 1984 the United Nations asserted that the desert was swallowing 21 million ha of land every year. That claim has been comprehensively demolished. There has been and is no net advance of the desert at all.'

Egypt illustrates some of the problems associated with arid areas. It is located in the north-eastern corner of Africa and covers almost 1 million km^2. It is bordered on the east by Israel, the Gulf of Aqaba, and the Red Sea; on the south by the Sudan; on the west by Libya; and on the north by the Mediterranean Sea. The River Nile traverses Egypt from north to south, fanning out into the densely populated lowlands north of the capital, Cairo. The Nile valley supports all of Egypt's agriculture and more than 99% of the population; in 1997 this was estimated to be over 62 million people. If the population continues to grow at the current rate of 2.1% a year, it may reach 85 million by 2010 – a doubling time of 29 years. The Egyptian birth and death rates are about average for North Africa, but the infant mortality rate has been high. The GDP per capita remains relatively low in comparison with other countries. Due to the amount of land devoted to cash crops, more than one-half of the country's food must be imported. The Aswan High Dam, completed in 1970, has permitted the annual Nile flood to be controlled. While Mediterranean coastal fishing has declined due to the High Dam cutting off nutrients, Lake Nasser, behind the dam, has been stocked with fish.

To ease congestion, the Egyptian Government is converting huge swathes of desert into farmland principally in the barren Western and Sinai deserts. The Al-Salam Canal, completed in 1997, carries Nile water across the northern Sinai Peninsula. To the south, the New Delta project will draw water from Lake Nasser behind the High Dam, and send it on a 500 km journey to link a series of desert oases that will form a parallel Nile Valley to the west. These schemes are intended to put more than 600 000 extra hectares of land under the plough, providing a living for hundreds of thousands of people. However, the New Delta scheme is controversial, taking a tenth of Egypt's available Nile water out of circulation. Water, it is argued, could be used more economically. Primitive irrigation methods, thirsty crops such as rice, leaky pipes, and the fact that water is free, have resulted in much wastage of water. Alternative schemes might well be cheaper, and a small charge for water could reap much economy.

8.8.1 Ice Deserts

Described by some as 'ice deserts', the Antarctic continent and the Arctic region should be mentioned, being fairly barren in terms of fauna and flora.

Antarctica is fifth in size among the world's continents, and lies concentrically around the South Pole. It is almost covered by a vast ice sheet, averaging about 2000 m thick. Antarctica comprises two subcontinents: East Antarctica consists mainly of a high ice-covered plateau, while West Antarctica consists largely of an archipelago of mountainous islands. The continent as a whole covers some 14 200 000 km^2. It has few economic prospects. Indeed, under the Convention for the Regulation of Antarctic Mineral Resource Activity, concluded initially in 1988 by the parties to the Antarctic Treaty and subject to a protocol in 1991, mining was banned for 50 years.

The cold-desert climate of Antarctica supports only a small community of plants, which are capable of surviving lengthy winter periods with near-total darkness. However, the coasts provide havens for immense seabird rookeries, with the emperor penguin, Antarctic petrel, and the South Polar skua breeding exclusively on the continent.

A great many metal ores are to be found, including large reserves of iron. Deposits of coal in the Transantarctic Mountains belong to one of the world's largest coalfields, with high quality anthracites.

The Antarctic Treaty, signed in 1959, allows freedom for scientific research and movement on and around the continent. Forty-two nations are parties to the Treaty. Important international conventions have been initiated by the Treaty nations promoting the conservation of flora and fauna, and other marine resources. The Treaty does not deny or support claims to territorial sovereignty in Antarctica (see Figure 8.2).

Research continues to indicate Antarctica's complex role in climate change. In 1996, the British Antarctic Survey scientists concluded that the large-scale retreat of Antarctic Peninsula ice shelves during the past 50 years may have been due to regional warming, rather than global warming.

Studies by William K. de la Mare of the Australian Antarctic Division, Environment Australia, have revealed that the mass of Antarctic sea ice remained constant from 1931 to the mid-1950s and then decreased by about 25% between then and the mid-1970s. Since then, sea ice levels have stabilized.

The Arctic region comprises portions of eight countries: Canada, the USA, Russia, Finland, Iceland, Sweden, Norway and Greenland (Denmark). The Arctic Ocean (over 14 000 000 km^2 in area) constitutes about two-thirds of the Arctic region, the remaining land area comprising a permanent ice-cap and extensive tundra. In 1996, the peoples of the circumpolar cultures numbered about 1 280 000. In that year, the eight countries created a new institutional agency, the Arctic Council.

The Arctic is characterized by distinctly polar conditions of plant life and climate; but unlike Antarctica which is surrounded by ocean, the Arctic is composed largely of an ocean almost completely surrounded by continents. Included in the Arctic region are the northern regions of Canada, Alaska, Russia, Norway, most of Iceland, Greenland and the Bering Sea. A useful line to mark the Arctic's southern boundary is the northernmost tree line of the continents.

Three shields comprise the main structures of the Arctic. The shields contain rich mineral reserves of iron, nickel, copper and zinc; while the flanking sedimentary rocks contain large reserves of coal and oil. In 1977, a pipeline from Prudhoe Bay to the southern port of Valdez, to carry oil from the North Slope oilfield, was opened. The period has also seen the establishment of several Alaskan national parks. Rich oil and gas reserves have also been tapped in northern Russia, notably in the republic of Komi.

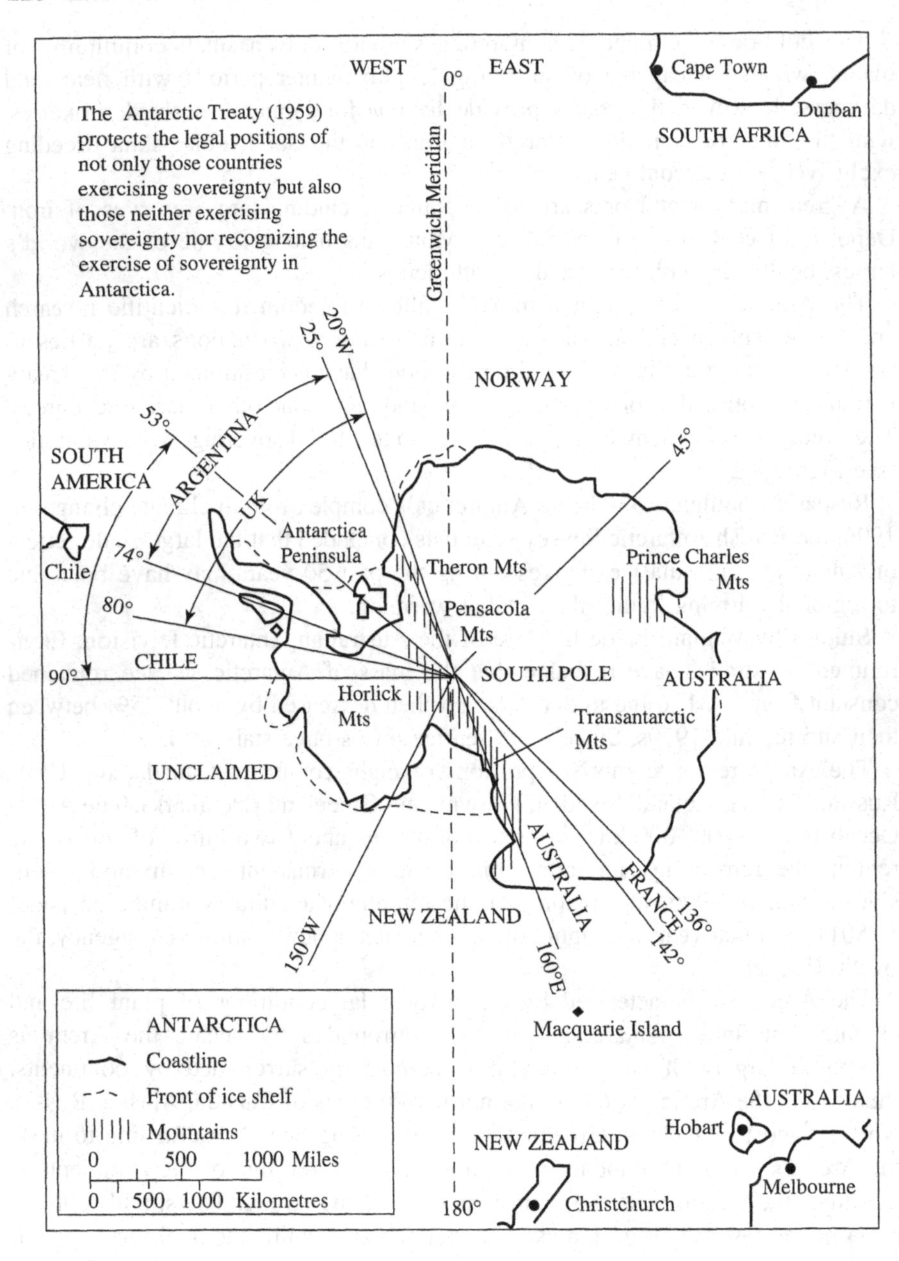

Figure 8.2 Antarctica

More than three-fifths of the Arctic is ice-free, with long, cold and relatively dark winters. Most of the zone is arid. An extensive area surrounding the glaciers in the Arctic is characterized by the presence of permafrost, i.e. permanently frozen earth. In North America, north of the tree line, there is a zone of continuous permafrost. In Russia, permafrost persists north of the tree line and east of the Ural mountains. Three concentric vegetation zones may be noted in the Arctic: low Arctic tundra, high Arctic tundra, and polar desert. Low Arctic tundra vegetation comprises dwarf shrubs, mosses, sedges, grasses and lichens. The high Arctic tundra vegetation consists of herbs and mosses. Polar desert vegetation consists mainly of scattered cushion plants.

The chief population groups include the Lapps, the Samoyeds, the Inuit or Eskimos, the Nordic Caucasians, the Greenlanders, and other groups of eastern Russia. The North Pole comprises floating and drifting pack ice over an ocean some 4000 m deep. It experiences six months of complete sunlight and six months of complete darkness each year. Many expeditions have reached the pole by dog sledge, airplane, dirigible and by submarine.

8.9 The Baltic Sea

The Baltic Sea is an arm of the North Atlantic Ocean extending northward from southern Denmark almost to the Arctic Circle, separating the Scandinavian peninsula from the rest of continental Europe. The Baltic Sea is encircled by Norway, Sweden, Finland, Estonia, Latvia, Lithuania, Russia, Poland, Germany and Denmark. Covering an area of some 420 000 km^2, the sea is served by the Vistula, Oder, Neva and Heman rivers. In all, one-fifth of the surface waters of Europe drain into the Baltic. The Baltic is the largest area of almost-fresh water in the world. Containing only about a quarter as much salt as the oceans, by volume, the waters freeze more readily; for long periods the Baltic is closed to navigation.

Today, Baltic Europe houses more than 80 million people conducting some 15% of the world's industrial production. Hence, the waters of the Baltic are becoming turbid due to increased nutrient flows from the land and the atmosphere. Bottom mud becomes loaded with phosphorus. Toxic industrial wastes from industry and transportation systems have greatly reduced populations of seals, otters and sea eagles. On the other hand, the mercury crisis of the 1960s was overcome, and DDT and PCB levels have been reduced remarkably. Sulphur emissions to the atmosphere have been reduced by 30%.

Meetings of the prime ministers of the Baltic countries have provided a means of handling environmental problems, while the Association of Baltic Sea Towns seeks to optimize socio-economic planning.

In April 1992, a Convention on the Protection of the Marine Environment of the Baltic Sea (the Helsinki Convention) was adopted by the riparain states Denmark, Estonia, Finland, Germany, Latvia, Lithuania, Poland, Russia and Sweden, to reduce pollution in areas around the Sea. An earlier convention in 1973 sought to introduce rational management of the marine resources of the area. In 1995, the Beijer International Institute of Ecological Economics completed a study of the Baltic with particular attention to the tendency to eutrophication. The study represents an important step towards achieving a sustainable Baltic.

The project involved some 20 scientists from Britain, Estonia, Germany, Lithuania, Norway, Poland and Sweden, and was partly financed by the European Commission. These scientists embraced ecology, marine biology, geography and economics. An important objective was to estimate the carrying capacity of various ecosystems in the Baltic drainage system, and to estimate the costs of decreasing nutrient loads, and the benefits from reduced eutrophication of the Baltic Sea in monetary terms.

A geographical information system (GIS) of the entire region has been developed, providing information on land use and population density; such information did not exist prior to this study. The 'ecological footprints' of the 30 largest cities have been delineated. The results indicate that the cities require substantial ecosystem support areas for food and other resources as well as for waste assimilation. A marine ecosystem six to seven times the size of the Baltic is required to sustain present seafood consumption in the region. The capacity of the region's wetlands to filter nutrients to the Baltic Sea has also been investigated.

One important objective in the management of the Baltic is to reduce the loads of nitrogen and phosphorus at minimum cost, i.e. cost-effective nutrient reductions. It is known already that the costs of reducing nitrogen loads can be ten times as high as the costs of reducing phosphorus. In the short run, the main cause of eutrophication is nitrogen; hence immediate improvements can only be made by means of normally expensive nitrogen reduction. However, it is known that wetlands serve as effective nitrogen sinks, and although conserving a wetland may preclude many types of commercial development, in many locations, it is argued (Gren and Tomasz 1993) that the conservation or restoration of a wetland may turn out to be a much cheaper means of nitrogen abatement than the alternative engineering solutions. Enhancing the natural absorptive capacity of ecosystems often offers an economically attractive option. Also, the conservation of wetlands offers other attractions, wetlands being particularly rich in biodiversity. Already a number of abatement projects are under way.

It seems likely that nutrient discharges need to be reduced to the levels of the 1930s, i.e. around 350 000 tonnes of nitrogen and 15 000 tonnes of phosphorus per year. This means a reduction of some 65% in nitrogen and 80% in phosphorus compared with current loadings. To approach such conditions a massive abatement programme is needed. To bring the price to an acceptable level with least-cost solutions being applied widely, soft options such as wetland restoration projects may turn out to be valuable contributions both in efficiency and environmental terms.

8.10 The Caspian Sea

The Caspian Sea lies to the east of the Caucasian Mountains of Russia. Its elongated shape measures some 1200 km from north to south, with an average width of about 320 km. It covers an area of 370 000 km^2, (i.e. a greater area than Japan) and has a maximum depth of some 1120 m. The Caspian is often described as the greatest salt lake in the world. The Volga River drains into the Caspian from the north; other tributaries include the Ural, Terek, Sulak, Samur, Kura and Araks rivers. The littoral states include Russia, Iran, Azerbaijan, Turkmenistan and Kazakhstan.

There are some 850 fauna and more than 500 plant species present in the Caspian. Fish life includes the sturgeon, herring, pike, perch and sprat, together with a variety of molluscs, crabs and clams. The Caspian seal is common. The Caspian was long famous for its sturgeon catch, though a decline in the sea level has reduced the catch in recent years, together with the output of caviare. However, the seal industry is expanding.

Following extensive geological surveys during the 1940s and later, oil and gas have now become the region's most important resources, second to those reserves lying under the Gulf. Oil is most readily tapped in the shallow waters of the new states, while Iran has the greatest difficulties with exploitation, lying on the steep side of the Caspian. The new states have signed major contracts with Western companies.

Azerbaijan, with a population approaching eight million, is already centred on the Baku oilfields, and has achieved a substantial output of oil and natural gas. Turkmenistan, with a population approaching five million, has also achieved significant outputs in these products. Kazakhstan, with a population approaching 17 million, has also achieved substantial outputs of oil and natural gas.

The Caspian also thrives as a major transportation route for the region, carrying petroleum, natural gas, wood, grain, cotton, rice and sodium sulphate. Plans also exist to augment the existing pipelines from Baku to the Black Sea, with possible pipelines connecting Baku with Armenia, Turkmenistan and Turkey.

The fragile Caspian ecosystem is buckling under the increasing tempo of exploitation. Parties to the Convention on International Trade in Endangered Species of Wild Fauna and Flora (CITES) were given warning in 1997 that over-fishing had put sturgeon in danger of extinction, with the possibility of losing 90% of the world's caviare.

8.11 The Murray–Darling Basin

The Murray River is the principal river of Australia, flowing some 2589 km across south-eastern Australia from the Snowy Mountains to the Great Australian Bight in South Australia. It has a total catchment area in excess of 1 million km^2 and a relatively small annual discharge. It forms most of the boundary between New South Wales and Victoria. Through most of its course through South Australia the river is bordered by a narrow flood plain, flowing between cliffs. The Murray's principal tributaries are the Darling River, Murrumbidgee River, Mitta Mitta, Ovens, Goulburn, Campaspe, and Loddon rivers.

The Murray Valley is of immense economic importance, lying across the great wheat–sheep belt, with extensive irrigation schemes. Its chief products are cattle, sheep, grains, fruit and wine. The Snowy Mountain hydroelectric scheme, completed in 1974, increased substantially the amount of water available for irrigation while generating large quantities of peak load electric power. However, irrigation has led to serious salinity problems. At times the city of Adelaide in South Australia, entirely dependent on the Murray for drinking water, receives water falling short of WHO standards, which is unfit for drinking.

The Darling River, the principal tributary, rises in the Great Dividing Range and flows generally south-west across New South Wales for 2739 km, until it joins the Murray at Wentworth on the Victorian border. The entire Darling system drains a 650 000 km^2 basin. The Menindee Lakes Storage Scheme, completed in 1960, created immense reservoirs offering water for irrigation and domestic use. In late 1991, a 1000 km algal bloom occurred in the Darling River, focusing world attention on Australia.

In 1915, the River Murray Commission was created, comprising representatives from the Commonwealth (federal) government and three state governments, to regulate the use of the river's waters. The River Murray Commission was replaced in 1988 by the Murray–Darling Basin Commission to combat salinity and environmental problems. The wider responsibilities of the new Commission included advice on the management of land and resources, and on environmental issues as well as water regulation. It was supported by a Murray–Darling Basin Community Advisory Committee, with 18 members. The Commission is under the guidance of a Murray–Darling Basin Ministerial Council comprising three ministers from each government, the Commonwealth and the New South Wales, Victoria and South Australia governments.

In 1985, the Murray–Darling Basin Commission estimated that 360 000 ha of irrigated land had shallow watertables, and 87 000 ha of land in the Victorian portion of the Basin were visibly salinized. However, the Murray–Darling Basin Commission, through its Natural Resources Management Strategy, is now providing an outstanding example of integrated catchment management.

Land use in the Murray–Darling Basin has involved steady clearing of native vegetation and the gradual intensification of agricultural activities. This has led to serious salinity and waterlogging on irrigated land, to substantial reduction losses on farms, and to substantial increases in the costs of treating water for downstream urban use. A variety of measures have been devised to reduce these problems, including better flow of water, better subsurface drainage, and improved interception of the salt. Fifteen major projects, each in a different location in the Basin, have been designed to implement these measures (Young 1988).

In 1998, the Australian Bureau of Agricultural and Resource Economics (ABARE) was building a profile of irrigated agriculture throughout the Murray–Darling Basin. Data collected have included physical and financial characteristics, farm performance, water use and irrigation management practices. ABARE will use the data to assess the preliminary impact of water reforms being introduced in relation to water use and sale and management practices. ABARE is also developing models for the horticultural and dairy industries. ABARE will be able to examine the impact of water trade on the demand for water within and across agricultural industries. Working out how to meet farmers' irrigation needs in years of high water prices will be key to the economically efficient use of water in the Basin. Water markets will have to operate efficiently if water availability is to be adjusted at the least cost and without greatly disrupting irrigation activities.

8.12 The Mekong River: Asia's Danube

The Mekong River is the longest river in South-east Asia, having a length of some 4350 km, and draining more than 795 000 km^2 of land. The headwaters rise at elevations of more than 4900 m in the Tibetan Highlands of Tsinghai Province of China. The upper Mekong, being about a quarter of its total length, descends in a southerly direction across the highlands of Yunnan Province, before forming part of the international frontier between Myanmar and Laos, as well as between Laos and Thailand. It then flows through Laos, Cambodia and Vietnam before entering the South China Sea in a wide delta south of Ho Chi Minh City in Vietnam (see Figure 8.3).

South of the Myanmar–Laos border, the lower Mekong basin receives the drainage from the Khorat Plateau of Thailand, from most of Cambodia, and from the western slopes of the Annamese Cordillera in Laos and Vietnam, before the river divides into two streams, the Mekong and the Bassac, in the delta section. Both Vientiane (Viangchan), the capital of Laos with a population exceeding 500 000, and Phnom Penh, the capital of Cambodia with a population exceeding 600 000, stand on the banks of the Mekong.

More than 45 000 000 people live in the lower Mekong Basin, with the heaviest concentrations of population being in the delta area and on the Khorat Plateau. Most are engaged in agriculture, producing rice. During the long dry period between November and May, however, rice requires irrigation. In 1957, the Mekong River Development Project was initiated by the UN Economic Commission for Asia and the Pacific (ECAP) (later the Economic and Social Commission for Asia and the Pacific (ESCAP)). The project required the co-operation of Laos, Thailand, Cambodia and Vietnam in completing facilities for the generation of hydroelectric power and for improvements in essential irrigation, flood control, drainage, navigation, water supply and management. During the 1980s co-operation continued in the development of new water resources in the basin.

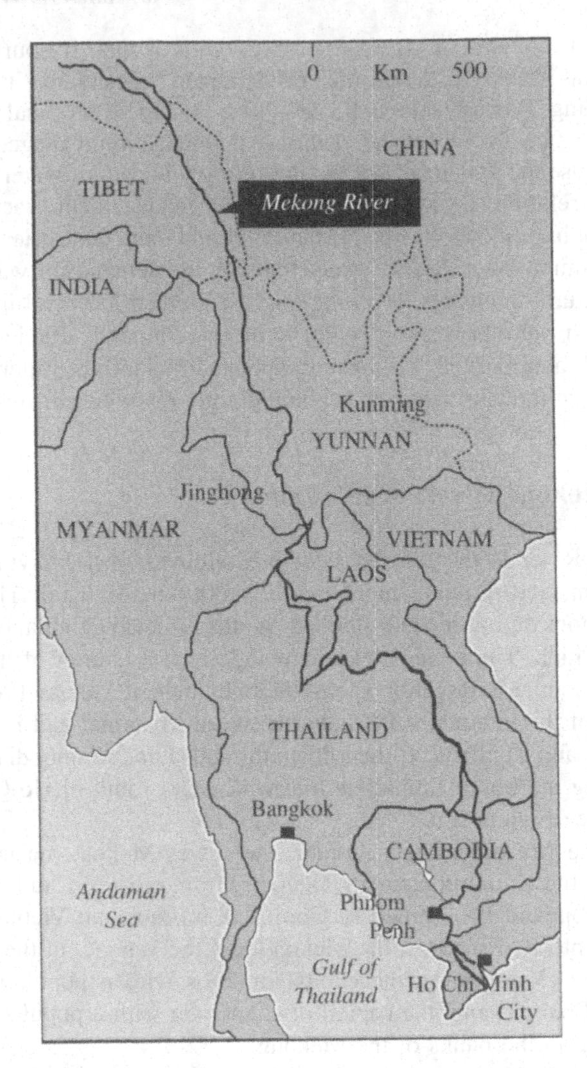

Figure 8.3 *The Mekong River*

Co-operation has been promoted through a ministerial council covering the six countries through which the Mekong flows. Known as the Greater Mekong Sub-region Forum (or GMS Forum), this collaboration has the support of the Asian Development Bank. The group hopes that the Mekong, home in all to some 230 million people, will become Asia's Danube, a channel of commerce and prosperity.

Schemes include dozens of hydroelectric plants on the Mekong. For China, the development of the Mekong offers a link with South-east Asia and the chance to develop Yunnan, one of its poorest provinces.

The Mekong River Commission, created in 1995, was the product of an Agreement on Co-operation for the Sustainable Development of the Mekong River Basin. The parties to the Agreement are Cambodia, Laos, Thailand and Vietnam (with China and Myanmar as observers). The agreement between the four governments covers the lower Mekong River Basin. The Commission itself comprises three permanent bodies: a council, a joint committee and a secretariat. The secretariat is based in Bangkok. The function of the commission is to implement the principles of sustainable development enunciated at the UN Conference on Environment and Development and the Rio Declaration on Environment and Development.

8.13 The Ok Tedi Mine Operation, Papua New Guinea

Ok Tedi is a large copper and gold mine located in the remote Western Province of Papua New Guinea. The mine has been in operation since 1984 and is run by Ok Tedi Mining Limited. Shareholders in Ok Tedi are the Broken Hill Propriety Company Limited (BHP) (52%), the Government of Papua New Guinea (30%) and Inmet Mining Corporation of Canada (18%). The development phase of Ok Tedi was long and costly, though it now makes a major contribution to the economy of the country. The biophysical environment is varied and subject to rapid changes; it is a difficult locality for people to live and work in. The area is subject to frequent volcanic activity, while the climate is equatorial.

The operation is based on a world-class copper–gold orebody that forms the core of Mount Fubiland in the central mountain range of Papua New Guinea. The development programme took over eight years and cost US$1.4 billion. Copper concentrate is sold under long-term contracts to smelters in Japan, Germany, Finland, South Korea and the Philippines. In 1991 an inaugural dividend was paid.

Most land in Papua New Guinea is owned in a complex and traditional system of private and clan ownership. Minerals and petroleum which lie underground are owned by the state, while forests and other resources above ground are owned by the clan. This leads to great legal complexities and disputes. Two groups of landowners have an interest in the Ok Tedi mine: the first group are the landowners in the mine area itself, and the second are the landowners living further downstream from the mine along the banks of the Ok Tedi and the Fly River (see Figure 8.4).

Around the mine, the people receive lease and royalty payments, at set rates. Compensation payments are also made for special purposes. The lower Ok Tedi–Fly River Development Trust was established in 1990, to spread benefits more widely.

Papua New Guinean government environmental regulations are based on comprehensive studies of the effects of mine sediment on the waters and fish of the Fly River. The primary environmental control is a maximum limit on

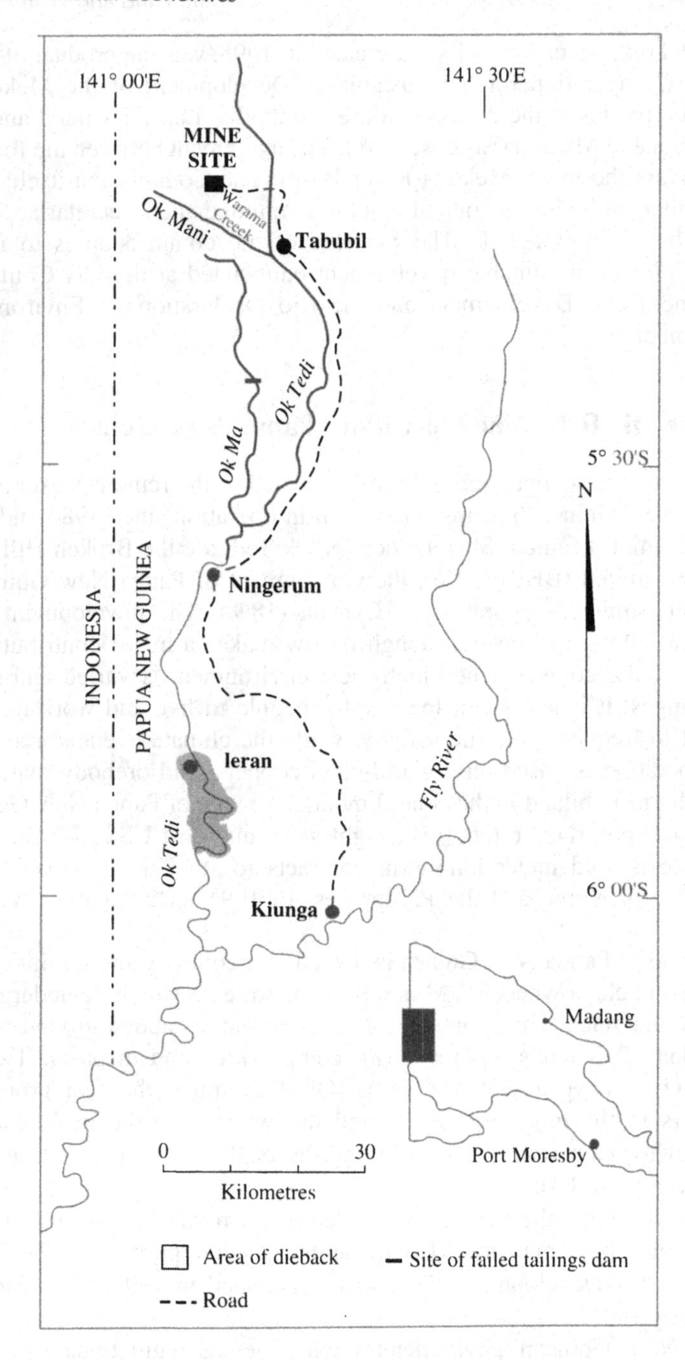

Figure 8.4 *Location of the mine site in the Ok Tedi area*

the mine's contribution to naturally high river sediment levels, known as the Acceptable Particulate Level (APL); here the APL is 940 mg l^{-1}. Other controls cover changes in fish biomass, dissolved and particulate copper concentrations and river bed sedimentation. Monitoring is designed to provide warning of potential impacts on water and fish in river channels, flood plains, off-river water bodies, and the Fly estuary.

An average of 60 million tonnes of ore residue and overburden is discharged into the Ok Tedi River each year; about 40 million tonnes of this reaches the Fly River. The Fly River carries 88 million tonnes of natural sediments each year.

The management and disposal of tailings materials has presented significant problems. In 1983–84, whilst work on a dam to contain the tailings from the mine was well under way, the dam foundations collapsed due to landslip activity. Since that time, some 120 studies have been undertaken on waste retention schemes. These studies led BHP to the conclusion that given the specific characteristics of the region, the stability and long-term safety of a tailings dam could not be assured. The search to limit impacts has continued, however.

The continued release of tailings has had some adverse effects. At the floor of the mountain range in which the mine is located, the river slows down and sediment deposits on its bed and along its banks. At times of heavy rain, this causes flooding so some sediment is left as mud over gardens and around riverside trees, causing dieback. The main effect of this deposition may be seen along a 20 km stretch extending over an area of some 30 km^2. Compensation has been paid to villagers.

Figure 8.4 indicates the location of the Ok Tedi Mine site, while Box 8.9 outlines the history of the operation, culminating in the successful legal action by villagers in the Supreme Court of Victoria, and the final announcement of agreement in 1996.

8.14 The Three-Gorges Water Conservation and Hydroelectric Project, China

The Changjiang (Yangtze) River is the longest river in both China and Asia, its length being over 6300 km. It rises in the Tibetan Highlands and traverses 12 provinces or regions, reaching the sea near Shanghai (see Figure 8.5). The Changjiang basin is the granary of China, contributing almost half the crops of the country. The potential resources for the generation of electrical energy from the river are great.

The three mountainous gorges of Tsyuytan, U and Silin are located in the middle course of the Changjiang River and are of high scenic value. A proposed major hydroelectric scheme could yield up to 40 000 MW, as well as help control destructive floods and develop new agricultural land. The development of the project is now in its earliest stages and remains highly controversial. It

Box 8.9 The History of Ok Tedi

1970 Geologists from the Kennecott Copper Corporation discovered a major mineral deposit at Mount Fubiland
1975 Kennecott withdrew from the project. Negotiations began between the Papua New Guinean Government and BHP
1976 Concession agreement signed
1981 Ok Tedi Mining Ltd (OTML) incorporated. Development of mine site infrastructure, and treatment facilities for copper ore; construction of a power station. Mining lease granted
1984 Production began as an open-cut operation; massive landslides destroyed tailings dam. Gold production commenced
1987 BHP appointed managing shareholder in OTML. Copper production commenced
1989 Papua New Guinean Government established maximum sediment level from mining to the Fly River
1990 Lower Ok Tedi/Fly River Development Trust formed
1991 Inaugural dividend paid
1993 Ownership restructure announced
1995 Between 1981 and 1995, 1548 employees received education and training. Ralph Nader campaigned against BHP. Villagers brought a $4 billion damages claim against BHP in the Victorian Supreme Court. The plaintiffs claimed that the mine had destroyed their traditional way of life. BHP offered compensation of $110 million to villagers, payment of legal costs, and renewed attempts at tailings-waste control, perhaps totalling $500 million, ending the two-year court battle
1996 On 11 June, BHP and the solicitors for the plaintiffs Slater and Gordon confirmed an agreement to resolve the litigation concerning the Ok Tedi Mine. The Papua New Guinean Government proposed an independent inquiry into a preferred option for a tailings disposal scheme.

Sources: BHP publications and press releases; Slater and Gordon press releases; PNG Government announcements; an article by Ralph Nader, 27 February 1996, in The Sydney Morning Herald; Annual Report (1995) of the Lower Ok Tedi/Fly River Development Trust.

commanded 10 papers at the 1993 Annual Meeting of the International Association for Impact Assessment held in Shanghai.

The project has been listed in the 10-year Programme for Economic and Social Development of China 1991–2000. The project involves the construction of a dam, a reservoir, a hydroelectric station, a navigation lock and other necessary auxiliary facilities. Apart from its advantages for flood control, power generation and navigation, the project is intended to bring benefits in irrigation, aquatic cultivation, tourism and economic development in general in the regions along the whole middle reaches of the Changjiang. The project was reaffirmed, despite reservations, at the Fifth Plenary Session of the Seventh National People's Congress of China in Beijing in April 1992.

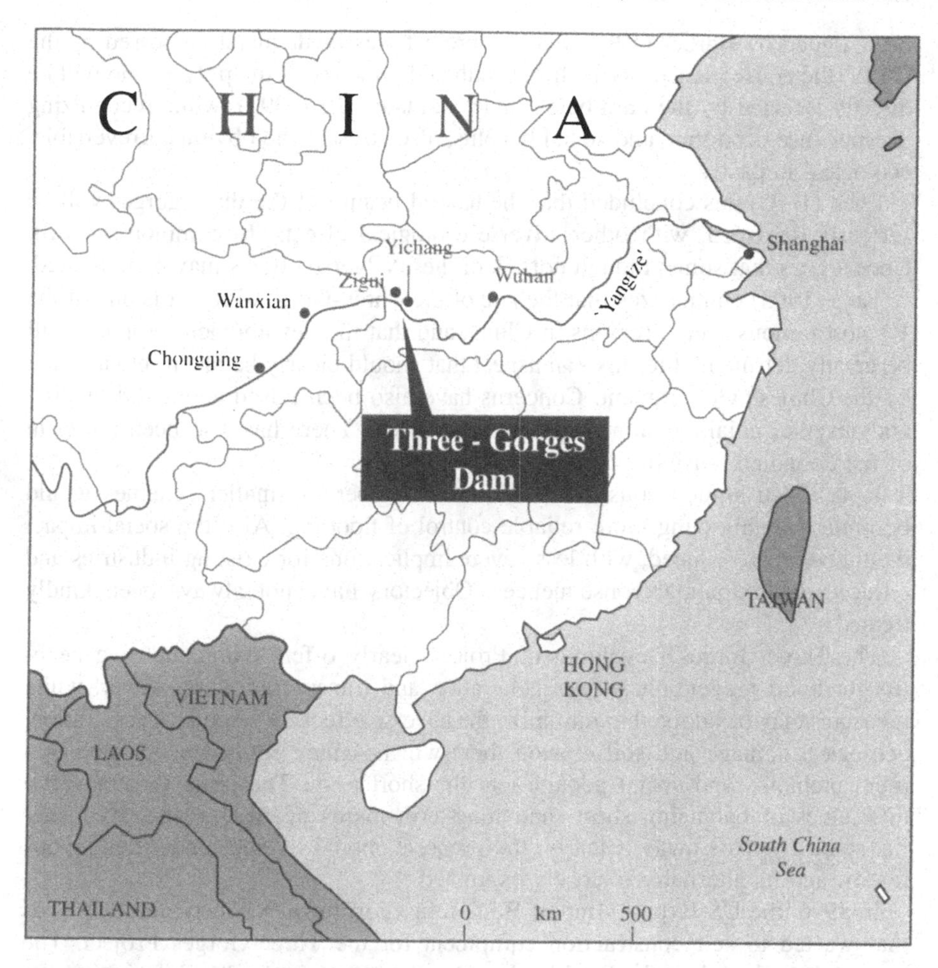

Figure 8.5 Location of the Three-Gorges Dam

The total length of the reservoir created by the dammed section of the river will be 600 km. Consequently some 725 000 people need to be resettled. More than 54% of these people are residents of towns and cities, the rest being rural residents. About 33 800 ha of farmland will be flooded, involving territory in 19 counties and cities in the provinces of Sichuan and Hubei. The major construction programme is expected to take 15 years, including the hydroelectric station with its 26 hydraulic generators. Over half the cost of the total project will be absorbed in the resettlement programme. Some 70 years have now passed since the Three-Gorges Project was first proposed.

The technical and economic feasibility study for the project was financed initially by the Canadian International Development Agency and completed by a Canadian consortium called the Canadian Yangtze Joint Venture (CYJV). It has been argued by Barber (1993) that the project was inadequately assessed by the CYJV, the consequences to the life, health and property of the people who will be directly affected by the dam being underestimated. Bi (1993), while recognizing the immense economic and social benefits, also stresses the obvious, irreversible, ecological impacts.

Chen (1993) has concluded that the natural beauty of the three-gorges will be seriously damaged, with other adverse ecological effects. In common with Bi, Chen argues that some, though not all, of these adverse effects may be mitigated.

Yang (1993) emphasizes that the site of the Three-Gorges Project is one of the 10 most famous scenic regions in China, and that the development proposal will be greatly detrimental to this – an aspect that should clearly be taken into account by the Chinese Government. Concerns have also been raised about soil quality, and adverse social, economic and cultural issues. There has also been criticism of the Canadian feasibility studies.

It has been argued constructively that a number of smaller schemes would be preferable, allowing more reliable control of flooding. Also the social impact would be much reduced, with less severe implications for existing industries and reduced environmental consequences. Objectors have not always been kindly treated.

The Three-Gorges Development Project clearly offers comprehensive benefits for flood prevention, power generation and transportation. However, while measures may be adopted to minimize the adverse effects of large-scale migration, ecological damage and soil erosion there will be many environmental management problems and social penalties in the short term. The issue illustrates the difficulties of balancing short- and long-term losses against prospective benefits some 15 years away. Clearly, the proposal should be kept under continuous review and its alternatives closely examined.

In 1996, the US Export–Import Bank refused to finance American companies that wanted to sell construction equipment for the Three-Gorges Project. The decision was based on the Bank's view that the environmental and social aspects of the proposed dam had not been adequately addressed. The Board's decision would be reconsidered in the light of new evidence.

Box 8.10 summarizes the history of the project since 1919, while Box 8.11 gives the key features of the Three-Gorges Project. Box 8.12 summarizes the criticisms of the proposal over time, many of these being expanded in Dai Qing's book, *The River Dragon has Come!*

8.15 Summary

This chapter embraces a diverse range of problems falling into the contexts of air, land and water. Sections 8.2 and 8.3 deal with the interrelated subjects of urban air pollution and acid rain, illustrating that the developed OECD countries

Box 8.10 The Three-Gorges Project, China

1919	Damming of the Three-Gorges was first proposed by Dr Sun Yat-sen. Five times this century Yangtze floods ravaged the middle and lower valley, killing more than 300 000 people, with millions made homeless
1973	China held its first national conference on environment protection
1979	The Law for Environmental Protection introduced
1986	First feasibility study conducted by the State Planning Commission
1989	Second feasibility study conducted by a Canadian consortium, financed by the Canadian International Development Agency, and supervised by the World Bank: it was recommended that the dam be built. Dai Qing went to prison for a book on anti-dam interviews
1990	Revised Environment Protection Law enabling the National Environment Protection Agency to promote environmental management through the country, including EIA
1991	Three-Gorges Project listed in the 10-year programme for the Economic and Social Development of China
1992	Project reaffirmed at the Seventh National People's Congress of China in Beijing. The International Water Tribunal in Amsterdam (established to review cases of alleged water mismanagement) heard a case against the Three-Gorges proposal: the jury ruled that the very high ecological and socio-economic risks of the dam had not been adequately assessed by the feasibility studies
1993	Ten papers on the Three-Gorges Project were presented at the 13th Annual Meeting of the International Association for Impact Assessment, Shanghai, 12–15 June 1993
1994	Preparatory construction work began
1997	Blocking of the Yangtse was completed, sending the river through a huge diversion channel while works proceed and the first generators are installed. Premier Li Peng presided over ceremony
2009	Work scheduled for completion, at which time the water will reach its projected height. Relocation of people will be completed

Box 8.11 The Yangtze Joint Venture Three-Gorges Water Conservation and Hydroelectric Project, China

Primary need	Flood control
Location	Sandouping Village Xiling Gorge (Three Gorges)
Cost	US$ 11.5 billion
Construction time	18 years
Reservoir length	600 km
Urban centres to be submerged	Fuling (population 80 000) Wanxian (population 140 000)
Number of towns to be submerged	114

Continued on page 234

Continued from page 233

Number of factories to be submerged	1599
Resettlement over 10 years	Probably more than one million people
Cultivated land to be submerged	14 500 ha
Historical sites to be lost	1208 (back to the Stone Age)
Dam height	185 m
Dam length	2150 m
Spillways	53
Normal pool level (maximum height of reservoir in dry season)	160 m
Installed generating capacity hydro	16 750 MW
Flood control storage	31×10^9 m^3
Average annual output	68.8 terawatt-hours
Navigation locks	Twin five-stage flight locks; 20 m lift per stage

Box 8.12 Criticisms of the Three-Gorges Dam

- The consequences to the life, health and property of the people directly affected have been significantly underestimated.
- There will be obvious, irreversible, ecological impacts.
- The natural beauty of the Three-Gorges area will be seriously damaged, it being one of the 10 most famous scenic regions in China.
- There will be adverse social, economic and cultural effects.
- The feasibility study was inadequate, as confirmed by the International Water Tribunal in Amsterdam.
- Over one million people will need to be relocated.
- The dam will not solve flooding problems; some people around the reservoir will suffer an increased incidence of flooding.
- The dam could trigger landslides and earthquakes.
- The concentration of hydropower creates a prime target for military attack.
- A series of smaller dams on the Yangtze tributaries would be more efficient.
- The dam, in impeding the flow of water, will shed silt to the bottom of the reservoir, gradually turning the water into mud.
- The retention of sediment will reduce the deposition of nutrients further downstream, reducing fertility and promoting erosion, as in the case of the Egyptian Aswan High Dam.
- In the event of dam failure, some 75 million people would be immediately at risk downstream.
- Cumulative effects have not been considered.
- Rare wildlife species such as the Chinese sturgeon, the Chinese river dolphin and the Chinese alligator and the finless porpoise are threatened.
- The only replacement land for the population is too poor, too steep and too elevated, for farming.
- The most fertile land will be lost to the reservoir.

Continued on page 235

Continued from page 234

- People who cannot be resettled in the area may be transported to distant areas; forced relocation to minority areas in the deserts of western China remains a probable outcome.
- A large number of historical and archaeological sites will be lost.
- The affected residents were rarely consulted; criticism of the dam has been crushed; no benefits are available to unregistered residents (around a fifth of the population).
- In 1992, 32% of the delegates to the National People's Congress failed to support the project.
- The World Bank and the US Government have shunned the dam on environmental grounds.
- The whole of the proceedings have been shrouded in secrecy; there has been no public inquiry or adequate public consultation.
- In 1998, Dai Qing (1998) published *The River Dragon has Come!* which is a collection of essays on the problems that can be expected if the dam is constructed as planned. It examines the cost of the human resettlement plan, the loss of artistic treasures, and the previous history of dam collapses in China. Dai Qing has been an outspoken leader in opposition to the dam as proposed.

have made much progress in this area since the 1950s, while the problems of the developing world progressively worsen. It seems that the developing nations must relive the experiences of those who industrialized earlier.

The issue of nuclear power generation receives attention, within the context of Sweden, a nation committed to the phasing out of its 12 nuclear power stations. Chernobyl and Three-Mile Island gave nuclear power a bad name, yet nuclear power remains a major source of power that does not emit greenhouse gases. Alternatives have also proved unpopular, hydroelectric power being an example. Nuclear power supplies over 20% of the world's electricity, some countries such as France, Japan and Taiwan being very dependent on it. Much plant has operated safely, and will continue to do so. The Swedish commitment to shut down nuclear plants is understandable, yet fraught with problems. It seems that shutting down any plant ahead of time may worsen the overall environmental situation, and not improve it.

We then address the complex subject of global biodiversity, forestry and the Amazon rain forest, all turning on the issues of fauna and flora and the interests of future generations.

The chapter then turns to the issue of deserts and desertification, with particular attention to Egypt, and finally the ice deserts, particularly those of the Arctic and the Antarctic.

The chapter then looks at the current state of the Baltic Sea and the Caspian Sea. These are followed by the Murray–Darling Basin in Australia, and the Mekong River, known as Asia's Danube.

The chapter concludes with two case studies of diverse character: the controversial Ok Tedi mining operation in Papua New Guinea, and the greatest dam project in the world, the Three-Gorges water conservation and hydroelectric project in the People's Republic of China.

Further Reading

Barber, M.S. (1993) Inadequate IA threatens life, health and property: the Three-Gorges water control project feasibility study in *Development and the Environment,* Proceedings of the 13th Annual Meeting of the International Association for Impact Assessment, 12–15 June, Shanghai, China, IAIA.

Chen, G. (1993) Impacts of the Three-Gorges project on the eco-environment and countermeasures in *Development and the Environment,* Proceedings of the 13th Annual Meeting of the International Association for Impact Assessment, 12–15 June, Shanghai, China, IAIA.

Fearnside, P.M. (1993) The Canadian feasibility study for the Three-Gorges dam proposal for China's Yangtze River: a grave embarrassment to the impact assessment profession in *Impact Assessment,* **12** (1), Spring 1994. A paper presented at the 13th Annual Meeting of the International Association for Impact Assessment, 12–15 June, Shanghai, China.

Food and Agriculture Organization (1995) *Forest Resources Assessment: Global Synthesis,* FAO, Rome.

Gren, I.-M. and Tomasz, Z. (1993) *Cost-effectiveness of the Baltic Sea Clean-up: Will the Wetlands Reconcile Efficiency with Biodiversity?* Beijer International Institute of Ecological Economics, Beijer Discussion Paper Series 24, Stockholm.

Mahar, D. (1989) *Government Policy and Deforestation in Brazil's Amazon Basin,* World Bank in co-operation with WWF and the Conservation Foundation, Washington, DC.

Mather, A.S. (1990) *Global Forest Resources,* Belhaven Press, London; Timber Press, Portland, OR.

Pearce, F. (1995) The biggest dam in the world, *New Scientist, January*: 25–29.

Qing, Dai (1998) *The River Dragon has Come! The Three-Gorges Dam and the Fate of China's Yangtze River and Its People,* Sharpe, New York.

The Economist (1998) The deep green sea: a survey, *The Economist,* London, 23 May 1998.

Yang, H. (1993) Analysis and assessment of the Three-Gorges project on the natural scenery of the Three-Gorges in *Development and the Environment,* Proceedings of the 13th Annual Meeting of the International Association for Impact Assessment, 12–15 June, Shanghai, China, IAIA.

9
Global Warming

9.1 Introduction

The so-called greenhouse effect is produced by the trapping of heat in the lower atmosphere. Next to water vapour, carbon dioxide is the most important gas in this process, followed by methane. This greenhouse effect is essential to life as it keeps the Earth tolerably warm. However, in recent years it has been the enhanced greenhouse effect, due to increasing levels of carbon dioxide in the atmosphere, that has given rise to international concern and investigation. Fears have been expressed that global mean surface temperatures could rise by up to 5 °C by the end of the 21st century, with many implications for life and the environment. These are discussed here. Many nations are now committed to a reduction in greenhouse gas emissions during the next few years, though the need for this remains controversial, particularly with grave economic consequences flowing from such policies involving a dramatic reduction in the consumption of fossil fuels, for which the world is far from ready. Furthermore, such major countries as India and China have been exempted from the process. The US Senate rightly disagrees with this. In any event, research has revealed no more than a discernible human influence on global climate. Other factors may be at work in this complex area.

Further, measures to curb global warming may be introduced at huge expense, at a time when evidence suggests we are entering a new glacial stage.

9.2 Changing Climates

Geologists and astronomers appear to agree that the Earth is about 4600 million years old, evolving and revolving within a much older universe of perhaps 15 billion years. Our knowledge of the climate of the Earth throughout that long period cannot, therefore, be much more than fragmentary, relying heavily on proxy evidence, recently reinforced by dating techniques of various kinds and the study of oxygen-18 isotopes in remains of planktonic microorganisms or foraminifera (Bolin 1989). Clearly, the Earth moved from an initial phase in which the atmosphere would be essentially volcanic, to later stages when oxygen in sufficient concentrations and primitive life forms emerged. It appears that the climate on Earth has always been subject to variations in its elliptical path round the Sun, the consequent variations in solar radiation, and the drift of the continents. Changes in the albedo or reflectivity of the Earth would also play a role. Vegetation and ocean water have a low albedo, as they absorb a large fraction of incoming energy, whereas ice and cloud surfaces have a high albedo, nearer to unity, as most of the incoming solar energy is reflected or scattered. Climatic variations over millions of years are essentially complex, as is the question of global warming today. However, some of the major climatological changes are known.

At one time, ice-sheets covered vast areas of the Earth, the earliest known ice age taking place during the Precambrian era, dating back some 530 million years. The most recent glacial periods occurred during the Pleistocene epoch, from 1.6 million years ago to as recently as 10 000 years ago (see Box 9.1).

Earlier warm periods have also occurred. At the beginning of the Cenozoic era, about 66 million years ago, the Earth's climate began to warm. For a time the climate became relatively mild and moist. Evidence indicates that deserts existed and that the area round the poles was ice-free. These mild conditions then appeared to end, when rapid cooling may have contributed to the extinction of the dinosaurs and other forms of life. Cooling continued into the Pleistocene, when great continental glaciers reappeared.

The Nebraskan Glacial Stage is regarded as the oldest Pleistocene episode of widespread glaciation in North America. The Nebraskan was named after deposits in Nebraska, although such deposits are also found in Missouri, Iowa and Kansas. Such deposits are rich in clays. The Nebraskan Glacial Stage preceded the Aftonian Interglacial Stage.

The Aftonian Interglacial Stage was a time of relatively moderate climatic conditions, preceding the Kansan Glacial Stage. The Aftonian was named after deposits studied in the region of Afton, Iowa. In some places these deposits consist of ancient soils that may also include peat and windblown deposits; elsewhere the deposits consist of sand and gravel deposited in streams.

The Kansan Glacial Stage of the Pleistocene followed the Aftonian Interglacial Stage and preceded the Yarmouth Interglacial Stage, both these interglacial periods being characterized by moderate climatic conditions. The Kansan Glacial Stage was named after deposits studied in north-eastern Kansas and near Afton. The Kansan glaciations appear to have been very extensive. However, away from the ice margins in Kansas and Indiana well-developed forests and mammalian faunas existed. Areas that are now generally arid or semi-arid were clearly moist, with diverse and abundant fauna. Characteristic fossil forms of the Kansan included the mammoth, musk-ox, moose and giant beaver. In the Midwest is evidence of a widespread volcanic ash bed, probably produced by a volcanic eruption in the western Cordilleras.

The Yarmouth Interglacial Stage followed the Kansan and preceded the Illinoian stages of widespread continental glaciation. The Yarmouth Interglacial was named after deposits that were studied in the region of Yarmouth, Iowa, being equivalent to the Mindel-Ris Interglacial Stage of Alpine Europe. The Yarmouth

Box 9.1 Geological Time-scale

Era	Period		Epoch	Date BP (M of Yrs)
CENOZOIC	Quaternary		Holocene	0–2
			Pleistocene	
	Tertiary	Neogene	Pliocene	2–5
			Miocene	5–24
		Palaeogene	Oligocene	24–37
			Eocene	37–58
			Palaeocene	58–66
MESOZOIC	Cretaceous			66–144
	Jurassic			144–208
	Triassic			208–245
PALAEOZOIC	Permian			245–286
	Carboniferous	Pennsylvanian		286–320
		Mississippian		320–360
	Devonian			360–408
	Silurian			408–438
	Ordovician			438–505
	Cambrian			505–545
PRECAMBRIAN				545–2500
				2500–3800
				3800–4600

Interglacial deposits are characterized by the remains of ancient soil horizons developed on Kansan glacial deposits and by deposits of peat. Fossil vertebrates have been found in some districts, indicating that Yarmouth climates were at least as warm as contemporary climates. Some Yarmouth climates may have been semi-arid. The Illinoian stages of glaciation followed.

The Sangamon Interglacial Stage followed the Illinoian Glacial Stage and preceded the Wisconsin Glacial Stage, both these glacial stages being periods of widespread continental glaciation and severe climatic conditions, in contrast to the moderate conditions of the Sangamon. The Sangamon was named after deposits studied in Sangamon County, Illinois. The ancient soils of the Sangamon are widespread throughout much of central USA. The Illinoian Glacial Stage ended with a cool, moist period that gradually became drier and then warmer; a warm dry climate marked the height of the Sangamon period, during which oak–hickory forests were dominant. Towards the latter part of the Sangamon the climate again became cooler, then wetter, finally passing into the next glacial period, the Wisconsin. Sangamon vertebrates in the Great Plains region included the dire wolf (now extinct), the short-faced bear, the giant bison, mammoths, sabre-toothed cats, giant ground sloths, camels, jaguars, horses and others. Other creatures, now extinct, included numerous small animals such as rodents, insectivores and lizards.

The Wisconsin Glacial Stage marks the most recent major division of the Pleistocene period. It is named after rock deposits studied in Wisconsin. The Wisconsin Stage corresponds to the Würm Glacial Stage of Europe. The Wisconsin Stage followed the Sangamon Interglacial Stage, and represents the last time that major continental ice-sheets advanced across North America. Its end coincides with the close of the Pleistocene Epoch in North America, about 10 000 years ago, although glacial ice did not recede evenly in all places at the same time. There is no conclusive evidence that the ice will not return. The Wisconsin Glacial Stage is the best known glacial episode, and is divided into early, middle and late episodes. In the Great Lakes Region, where Wisconsin deposits are well represented, five substages, where glaciers advanced and retreated, have been recognized. The Great Lakes are in fact remnants of glacial lakes that bordered the vast continental ice-sheets. The rich soils of the Great Plains are largely derived from the silts deposited by streams of glacial meltwater. Sequences of pollen grains have revealed much about the Wisconsin period. Studies have revealed the presence of tundra on the fringes of the ice masses. These were occupied by the woolly mammoth, caribou and musk-ox. The Great Plains were occupied by bison, horses, giant ground sloths, peccaries, sabre-toothed cats, mastodons and camels. The first clear traces of human presence became evident during the Late Wisconsin period.

A feature of the end of the Wisconsin Glacial and of the Pleistocene Epoch in general was the extinction of numerous species of large mammals in North America, including the horse, mammoth, mastodon, camel, giant armadillo and sabre-toothed cats. A major factor may have been overhunting by humans, but climatic and environmental factors may have been involved. The Wisconsin/Würm glaciation reached its summit about 18 000 years ago.

Another warming period followed. By about 9000 years ago, during the Holocene, summer temperatures and precipitation were above current values in much of the Northern Hemisphere, especially in the Middle East and Asia, encouraging pastoral and agricultural activities.

From about AD 600 to 1400, there is general agreement that European and North Atlantic climates became warmer, temperatures being somewhat higher than today. Viking settlers were able to establish colonies in Iceland, Greenland and Newfoundland, the Crimea and eastern North America. The first immigration of the Maori peoples into New Zealand probably occurred during this mild period.

An interval beginning in the 16th century, and advancing and receding over three centuries, has been known as 'The Little Ice Age'. Its coldest phase was reached around 1750, but temperatures showed large fluctuations until 1850. Glacial advances were recorded in the Alps, Alaska and the Sierra Nevada. We are now in a new warming age, with human assistance.

Meteorological measuring instruments came into existence as early as the 17th century, but it was not until 1850 that enough reliable instruments were available for adequate analyses. In 1951, the World Meteorological Organization (WMO) was created as an agency of the UN, evolving from its predecessor, the International Meteorological Organization established initially in 1873. The WMO's primary role is to maintain a world-wide meteorological system and to assist individual nations with the creation of national networks. It has a distinct role in grappling with environmental problems of a global nature. Its contribution has found expression in the Global Atmospheric Research Program and in an operational monitoring system called World Weather Watch. In 1979, WMO undertook a World Climate Program in co-operation with UNEP. In addition, the WMO Background Air Pollution Monitoring Network has been developed as part of the Global Environment Monitoring System.

In recent years much investigation and research has been undertaken by the Intergovernmental Panel on Climate Change (IPPC) which was created in 1992 to address the implications of global warming.

Atmospheric monitoring has, in recent years, been augmented by such programmes as the World Ocean Circulation Experiment, a project supported by 25 nations, that has been designed to measure ocean temperatures and trends; and the Greenland Ice-core Project (GRIP), a programme supported by eight European partners, which aims to obtain insights into the history of the atmosphere through ice-core drilling. The records in the ice extend back for several hundred thousand years. Evidence has been found to show that the whole of the last glacial period was interrupted by a series of warm periods. This has much relevance to the current debate on global warming and climate change. See Box 9.2 for the comments of Pearman on climatic variations over time.

9.3 The Enhanced Greenhouse Effect

In the general atmosphere surrounding the Earth, a beneficial warming effect is due to the selective absorption of solar energy by certain gases such as carbon dioxide, methane, nitrous oxide, tropospheric ozone, chlorofluorocarbons (CFCs) and water vapour. These greenhouse gases prove transparent to incoming

Box 9.2 Climatic Variations

The climate of the Earth has varied in the past and will continue to do so in the future because of a number of factors. Some of these factors can influence climate globally or regionally; they are effective over time periods from a few years to thousands or even millions of years. Significantly, the only systematic and highly probable climate change over the coming decades and even centuries will be that driven by the accumulation of greenhouse gases in the atmosphere, the enhanced greenhouse effect. While there are still many uncertainties, the IPCC has found that the balance of evidence suggests a discernible human influence on global climate.

Source: G. Pearman (1998) Chief, Atmospheric Research CSIRO, to a conference Greenhouse beyond Kyoto: Issues, Opportunities and Challenges, Bureau of Resource Sciences, Canberra.

short-wave solar radiation but relatively opaque to long-wave radiation reflected back from the Earth, the result being a warming effect reminiscent of a greenhouse. Without this greenhouse effect, the Earth, certainly throughout the history of humanity, would have been very much colder; life has depended therefore on the beneficial effects of these greenhouse gases. They are defined in greater detail in Box 9.3.

Concern has arisen in recent decades because of evidence that the levels of greenhouse gases, most notably carbon dioxide, in the atmosphere have been

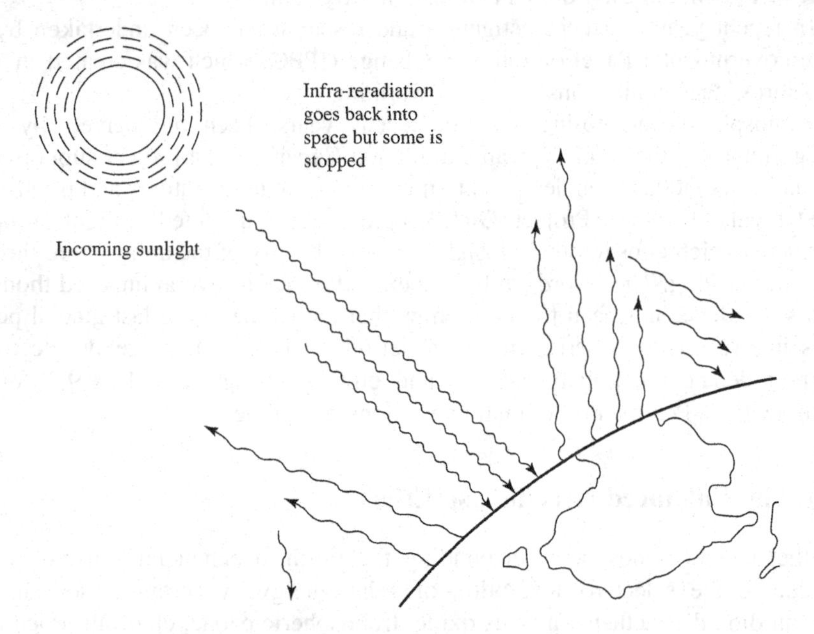

Figure 9.1 *The greenhouse effect*

Box 9.3 The Principal Greenhouse Gases

Carbon dioxide (CO_2)

Carbon dioxide is a product of combustion and respiration (breathing). It is a colourless gas, quite innocuous at ordinary levels. Indeed, plants synthesize their component tissues in part from atmospheric carbon dioxide. Carbon dioxide is absorbed from the atmosphere through the stomata of leaves and is converted into complex compounds such as sugar, starch and other carbohydrates, proteins and fats. However, at higher levels carbon dioxide is the most significant of the greenhouse gases contributing to global warming. Carbon dioxide concentrations have increased from about 280 ppmv in the late 18th century to 358 ppmv in 1994 – an increase of almost 30%. The increase is primarily due to combustion of fossil fuel in steam generators and industrial furnaces, in cement production, and in changes in tropical land use. However, about half the carbon dioxide produced does not linger in the atmosphere but is absorbed by natural sinks, notably vegetation and the oceans. Measurements from four Antarctic ice-cores combined with those from Mauna Loa, Hawaii, confirm that the concentration of carbon dioxide in the atmosphere did not vary much over a period of 1000 years up to the industrial era, but grew rapidly from 1800.

Methane (CH_4)

Methane is a non-toxic gas, though it does contribute to the greenhouse effect. It is a principal constituent of natural gas, as well as being produced during the process of digestion in sludge digestion tanks, occurring in coal mines as firedamp, and being generated at the bottom of lakes and marshes as marsh gas. More generally, it emanates from rice paddy fields, livestock, landfill waste dumps, the inefficient burning of biomass and coal, and agricultural activity generally. The increase of methane in the atmosphere is closely linked to the world's growing population and its growing demand for food, as expressed, for example, in the increasing numbers of ruminant animals (cattle and sheep), and increasing hectares of rice paddies. An average cow emits 280 litres of methane each day. When burnt, methane yields carbon dioxide and water vapour. Atmospheric methane has been increasing at the rate of about 1% each year. The atmospheric concentration of methane has increased from about 700 ppbv in pre-industrial times to 1721 ppbv in 1994.

Nitrous Oxide (N_2O)

Nitrous oxide is another of the gases contributing to the greenhouse effect. Sources are fossil-fuel combustion, biomass burning, and the use of fertilizers. Other compounds of nitrogen formed by the oxidation of atmospheric nitrogen at high temperatures include nitric oxide, nitrogen dioxide, nitrogen pentoxide and nitric acid. They are discharged into the atmosphere from motor vehicle exhausts, furnace stacks, incinerators, and other sources. Nitrogen oxides are all potentially harmful, catalysing in bright sunlight to form photochemical smog. Atmospheric concentrations of N_2O have increased from about 275 ppbv in pre-industrial times to 311 ppbv in 1992. Natural sources are probably twice as large as anthropogenic ones.

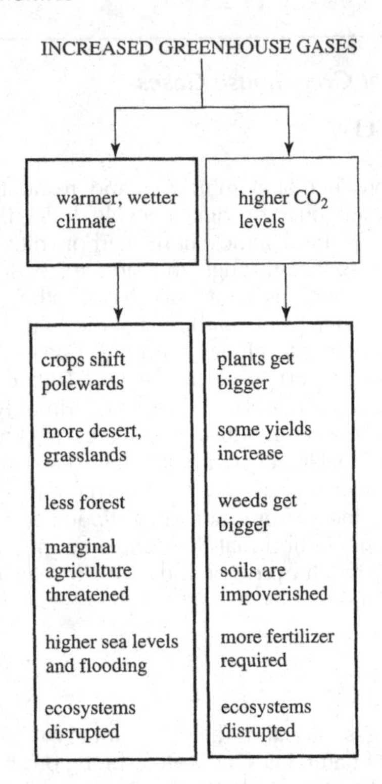

Figure 9.2 *The social and environmental effects of increased greenhouse gases*

progressively increasing due, it is thought, to the activities of humanity, both industrial and agricultural. The precise contribution of humanity is a continuing topic of debate, and the ability of humanity to control such increases is a matter of dispute.

Carbon dioxide enters the atmosphere largely as a waste product from the burning of fossil fuels, from naturally occurring fires, through the decomposition of plant and animal tissues, and in exhalation from the lungs; it is removed by absorption through the leaves of plants and trees, while the oceans act as a sink. The concentrations of carbon dioxide in the atmosphere appear to have increased since 1890 from around 280 ppmv to over 350 ppmv. If this trend continues, climatic changes may occur which are of benefit to some regions but detrimental to others. The sea levels may rise due to the thermal expansion of the oceans. The Arctic and Antarctic polar ice-caps might eventually melt. At several international conferences, measures to curb the increase in carbon dioxide have been urged through greater efficiency in the use of energy and in the conservation of forests.

Figures 9.1–9.3 illustrate the greenhouse effect, the social and environmental effects of increased greenhouse gases, and the global climate system.

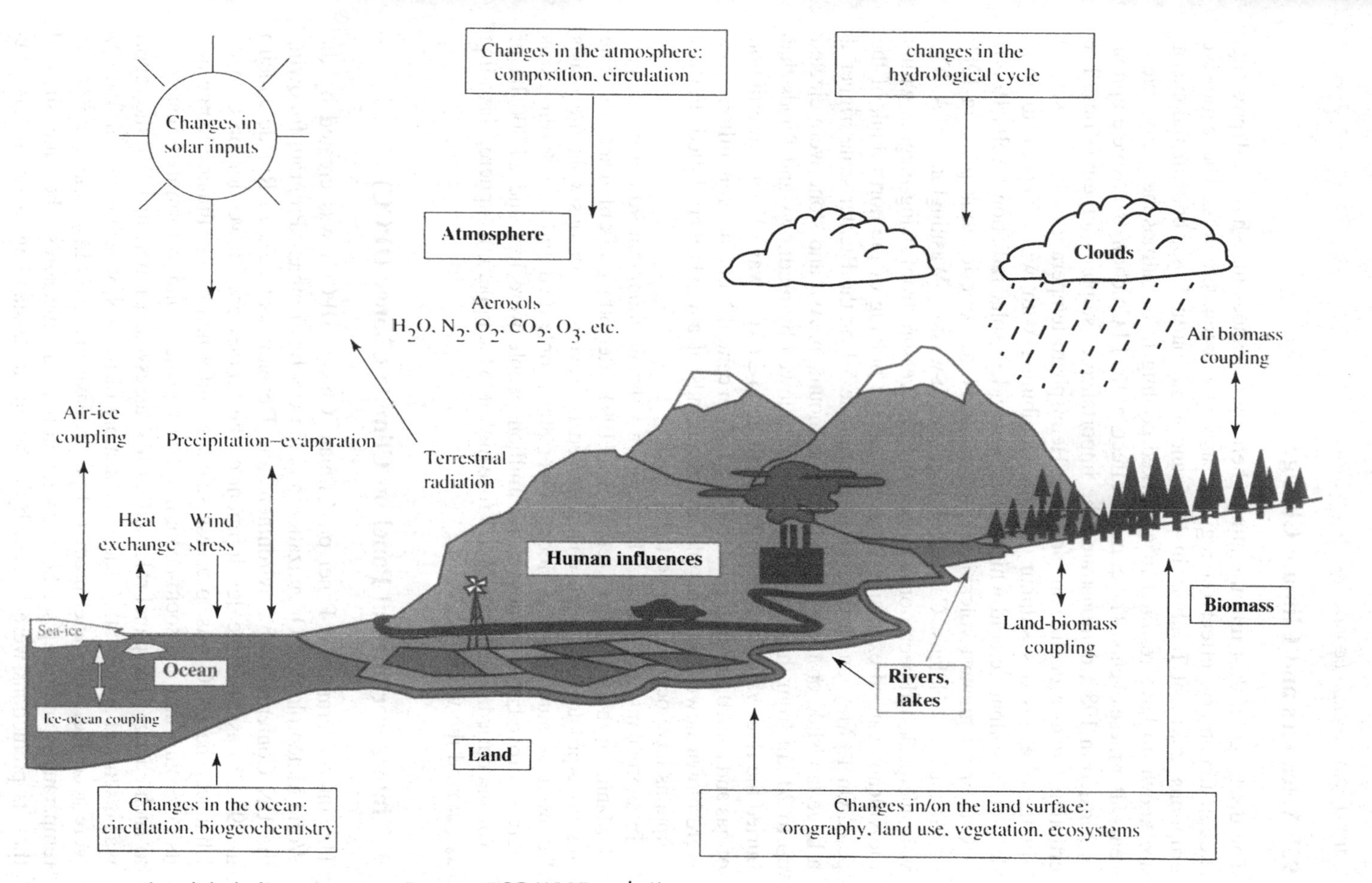

Figure 9.3 *The global climate system. Source: IPCC (1995, vol. 1)*

9.4 Volcanoes and Climate Change

The discharge of dust into the atmosphere by volcanoes may have a depressing, though temporary, effect on regional and global temperatures. The explosive eruptions at Mount Tambora in 1815 and Krakatoa in 1883, both in Indonesia, are known to have resulted in a marked cooling in the atmosphere for several years after these events. The climatic effects of the El Chichon explosive eruption, in Mexico in 1982, demonstrated the importance of sulphur-rich emissions. The eruption led to a decline in Northern Hemisphere temperatures lasting several months, the gaseous sulphur dioxide combining with water vapour to form droplets of sulphuric acid which reflected incident solar radiation (Wells 1989).

One of the greatest volcanic explosions in North America took place in 1980 at Mount Saint Helens, Cascade Range, south-western Washington, USA. The volcanic peak had been dormant since 1857. Pressure from rising magma within the volcano caused extensive fissuring and the bulging of the north flank of the peak. On 18 May, an earthquake of magnitude 5.1 on the Richter scale triggered a huge landslide on the north face of the mountain. Ash and stone were ejected up to 20 km from the summit. This was followed by mudflows and floods that buried the river valleys with mud and silt up to 27 km away. A vertical eruption of gas and ash 20 km high produced ashfalls in central Montana. Ten million trees were blown down. Sixty people and thousands of animals were killed. Further eruptions have occurred since then.

In another instance, Mount Pinatubo, a volcanic mountain 90 km north-west of Manila in the Philippines, erupted in 1991 after 600 years of inactivity. The months April to June were characterized by several enormous volcanic blasts followed by mudflows and floods. Over 800 people died and two dozen towns were destroyed; the homes of 1.2 million people were lost and 81 000 ha of farmland buried. The plume of ash rose high into the upper atmosphere, spreading ash around the world.

9.5 Intergovernmental Panel on Climate Change (IPCC)

The Intergovernmental Panel on Climate Change (IPCC) was created by the World Meteorological Organization and the UN Environment Program following the UN Conference of Environment and Development held in Rio de Janeiro in 1992, to assist in the development of the Framework Convention on Climate Change. The IPCC was to assess scientific information on climate change and its environmental and socio-economic consequences, and to formulate responses to such change. The panel released its first assessment report in 1990; the report was discussed at the 45th General Assembly of the UN, where the main findings were accepted. The panel predicted progressive increases in global atmospheric temperatures and rising sea levels, without remedial measures. The panel stressed that its predictions were limited by a lack of understanding of greenhouse gas sources and sinks; of clouds and oceans and how they influence climate change; and of polar ice-sheets and their impact on sea-level rises.

The second major assessment by the IPCC was released in 1995. The report was written and reviewed by some 2000 scientists and technical experts from about 130 countries over a two-and-a-half year period. The complete assessment is about 1500 pages long, with some 10 000 references. There are three volumes to the report, each volume being the report of one of the three working groups. Working Group 1 evaluated the scientific understanding of climate change; Working Group 2 considered climate change impacts and options for adaptation and mitigation; while Working Group 3 reviewed the economic and social dimensions of climate change. Early in 1996, Cambridge University Press published the full second assessment report in three volumes.

Key findings of the IPCC Working Group 1 were as follows:

- Increases in greenhouse gas concentrations since pre-industrial times (since about 1750) have led to a positive radiative forcing of climate, tending to warm the surface and to produce other climate changes. The upward trends can be attributed largely to human activities, mostly fossil-fuel use, land-use change and agriculture.
- Global mean temperature has increased by between 0.3 °C and 0.6 °C since the late 19th century, recent research having confirmed this range.
- Recent years have been among the warmest since 1860 (the period of instrumental record), despite the cooling effect of the 1991 Mount Pinatubo volcanic eruption.
- Regional changes are also evident. The recent warming has been greatest over the mid-latitude continents in winter and spring, with a few areas of cooling such as the North Atlantic Ocean.
- Global sea level has risen by between 10 and 25 cm over the past 100 years and much of this rise may be related to the increase in global mean temperature.
- The 1990 to mid-1995 persistent warm phase of the El Niño-Southern Oscillation (which causes floods and droughts in many areas) was unusual in the context of the previous 20 years.
- Night-time temperatures over land have generally increased more than daytime temperatures.
- If carbon dioxide emissions are maintained at current levels, they will lead to an increase in atmospheric levels reaching 500 ppmv by the end of the 21st century (i.e. approaching twice the pre-industrial concentration of 280 ppmv).
- Since the 1990 IPCC report, considerable progress has been made in attempts to distinguish between natural and anthropogenic influences on climate. This progress has been achieved by including the effects of sulphate aerosols in addition to greenhouse gases. The observed trend is unlikely to be entirely natural in origin.

- Our ability to quantify the human influence on global climate is currently limited, because of uncertainties in key factors and in the difficulty of separating out the noise of natural variability. Nevertheless, the balance of evidence suggests that 'there is a discernible human influence on global climate'.
- The limited evidence available from proxy climate indicators suggests that the 20th century global mean temperature is at least as warm as any other century since at least AD 1400; data prior to AD 1400 are too sparse to allow reliable estimations of global mean temperature.
- The models predict an increase in global mean surface temperature relative to 1990 of about 2 °C by 2100. This estimate is about one-third lower than the best estimate in 1990. The range is between 1 °C and 3.5 °C, allowing for considerable natural variability. Regional temperature changes could differ substantially from the global mean value.
- Average sea level is expected to rise as a result of thermal expansion of the oceans and the melting of glaciers and ice sheets. Sea levels might rise by 50 cm from the present to 2100. This estimate is about 25% lower than the best estimate in 1990, due to lower temperature projections and improvements in the climate and ice-melt models. The range suggested is from 15 cm to a maximum of 95 cm between now and 2100.
- 'Climate change' in the IPCC Working Group 1 report refers to any change in climate over time, whether due to natural variability or to variability as a result of human activity. This differs from the usage in the Framework Convention on Climate Change where 'climate change' refers to a change of climate that is attributed directly or indirectly to human activity which alters the composition of the global atmosphere and which is in addition to natural climate variability observed over comparable time periods.

Key findings of IPCC Working Group 2 were as follows:

- Human-induced climate change represents an important additional stress, particularly to the many ecological and socio-economic systems already adversely affected by many factors, such as pollution and non-sustainable management practices.
- The technological potential to achieve significant gains in energy efficiency is large: 10–30% energy efficiency gains could be achieved at little or no net cost, through conservation measures and improved management practices. Gains of 50–60% are achievable in many countries using best practice technologies, but strong policy direction would be needed.
- Deep emission reductions could be achieved within the coming century with more efficient conversion of fossil fuels and switching to low-carbon fossil fuels, e.g. from coal to natural gas, and to renewables.
- Global carbon dioxide emissions could be reduced by about two-thirds by the year 2100, through a strong investment and research programme.

- A number of measures could sequester substantial amounts of carbon over the next 50 years (in the forestry sector alone).
- Governments do have attractive options for mitigating greenhouse gas emissions with the help both of existing and emerging technologies.

Key findings of IPCC Working Group 3 were as follows:

- The literature indicates that significant no-regrets opportunities are available in most countries, and that the risk of aggregate net damage due to climate change, consideration of risk aversion and the precautionary principle, provide rationales for action beyond no-regrets.
- The costs of damage, adaptation and mitigation may be borne inequitably between nations.
- The literature on the social and economic damage that would arise from increased greenhouse emissions is controversial and conflicting. Monetary valuation should not obscure the human consequences of anthropogenic climate change damage.
- The assessed literature quantifying the total global damages of a 2–3 °C warming tends to give an estimate of a few per cent of world GDP (though higher shares in developing countries). These aggregate estimates are subject to considerable uncertainty.
- Published estimates of the marginal damage of carbon dioxide emissions range from US$5 to US$125 per tonne of carbon.
- There is agreement that energy efficiency gains of perhaps 10–30% below baseline trends over the next two or three decades could be realized at negative to zero net cost.
- Analyses for OECD countries suggest that the cost of substantial reductions of carbon dioxide below 1990 levels could be as high as several per cent of GDP. However, studies also show that appropriate timing of abatement measures and the availability of low-cost alternatives may substantially reduce the size of the total bill.
- A number of studies reveal that global emissions reductions of 4–18% together with increases in real incomes are possible from phasing out fuel subsidies.

The IPCC has set in motion a work programme that will culminate in a third assessment report in 2000–2001. The report will review the state of the science of climate change, the environmental and socio-economic impacts, the options for adaptation and mitigation, and the economics of climate change.

Some scientists have speculated that global warming may be due less to human influence than to changes in the output of the sun, i.e. to variations in the average number of sunspots over the course of a typical cycle of 10–12 years. Sunspots are transient dark patches on the face of the sun. These dark patches are accompanied by brighter patches; a spottier sun is actually a brighter sun. Several similar theories are being developed, without firm conclusions, and may not, in any case, account for the full temperature change.

The economist, William Stanley Jevons (1835–1882) associated sunspots with variations in harvests, establishing a link between harvests and the business cycle. He established a clear link between the concentration of sunspots over a typical cycle and climatic conditions on Earth.

Other doubts raised about the findings of the IPCC have been expressed, most notably that the world's oceans and land-based ecosystems will be less able to absorb carbon dioxide than has been thought. Particular reservations have been expressed about the capacity of oceans to act as a sink. Doubts have been expressed by teams at Stanford University and the University of Sheffield.

9.6 Kyoto 1997

Many international meetings have been held since the inaugural Villach Conference on Climatic Change in 1985 in an attempt to create a world approach to the issue of growing concerns about progressive global warming and its ultimate consequences. In 1992, the UN Conference on Environment and Development created a Framework Convention on Climate Change to protect the world's climate system most notably against the effects of greenhouse gases and their warming influence. Initially, the developed nations (or Annex I nations committed to emission abatement) were required to reduce their emissions of carbon dioxide and other greenhouse gases to 1990 levels or below. The non-Annex I nations (the most heavily populated but least developed) were exempted from this process. In 1995, a climate conference held in Berlin reviewed progress towards achieving the objectives of the Framework Convention on Climate Change. Representatives from 150 countries attended this conference, but there was little to report. However, the conference received and endorsed an initial report from the Intergovernmental Panel on Climate Change (subsequently revised in 1995 with some significant re-assessments). At the Geneva talks in 1996 progress was then reviewed again, but few countries could agree on possible targets.

However, at the Kyoto Climate Change Convention held in Kyoto, Japan, in December 1997, Annex I nations at last undertook specific commitments. The collective global target was to cut greenhouse gas emissions by a little more than 5% of the 1990 baseline, by the year 2012 – a modest but attainable target (see Boxes 9.4 and 9.5). However, Australia, Norway and Iceland were allowed to increase emissions over the same period, Australia by 8%. Australia had initially wanted much more. The agreed emission targets are set out in Table 9.3.

In the meanwhile, the Australian Bureau of Agricultural and Resource Economics (Brown *et al.* 1997) published a report entitled *The Economic Impact of International Climate Change Policy*. The purpose of the report was to contribute analytical input to the international climate change policy development process by providing an assessment of the economic impacts of policies to reduce carbon dioxide emissions from fossil fuel combustion in Annex I countries over

Box 9.4 Approaches to the Kyoto Protocol

- *Villach Conference on Climatic Change*: convened in October 1985, a conference of scientists from 29 countries held in Villach, Austria, at which scientific recognition was given to the greenhouse effect.
- *World Conference on Climate and Development*: an international conference organized by WMO, held in 1988 in Hamburg, Germany. The result was the Hamburg Manifesto on Climatic Change, urging a 30% reduction in the emission of carbon dioxide by the year 2000, and a 50% reduction by the year 2015, largely through improved efficiency in the use of energy. The Second World Conference on Climate and Development was held in Geneva, Switzerland, in 1990 to review progress.
- *Toronto Conference on the Changing Atmosphere*: convened in June 1988, a conference of scientists and policy-makers from 48 countries held in Toronto, Canada, which gave much emphasis to the greenhouse effect.
- *London Conference on Climatic Change*: a conference convened in 1989 and attended by the representatives of 188 nations, to discuss the protection of the ozone layer and the control of the greenhouse effect. The conference recognized the need for a major reduction in the emission of carbon dioxide from the burning of fossil fuels.
- *Hague Conference on the Environment*: a summit of world leaders convened by the prime ministers of France, Norway and the Netherlands, to consider ways to combat global climate change arising from the greenhouse effect. In its Declaration of the Hague Conference on the Environment 1989, the conference urged the creation of a new institutional authority to devise means of combating global climatic change.
- *Intergovernmental Panel on Climate Change*: a panel created by the World Meteorological Organization and the UN Environment Program to assess scientific information on climate change and its environmental and socio-economic consequences, with the formulation of responses to such change. The panel released its first assessment report in 1990; the report was discussed at the 45th General Assembly of the UN, when the main findings were accepted. The panel predicted a rate of increase of global mean temperature during the next century of 0.3 °C per decade, without remedial measures, with sea-level rises of between 3 and 10 cm per decade. By 2030, the global mean sea level might rise by 20 cm and might reach 65 cm by the end of 2100. The panel said its predictions were limited by a lack of understanding of greenhouse gas sources and sinks, of clouds and oceans and how they influence climate change, and of polar ice-sheets and their impact on sea-level rises. Supplementary reports were published in 1992 and 1994. *Climate Change 1995* was the first full sequel to the original assessment. The new report concluded that many aspects of climate change were effectively irreversible and suggested a 'discernible human influence on global climate'. Predictions of future temperature and sea-level rises were, however, modified (reduced a third in respect of global warming, and by a quarter in respect of sea-level rises).
- *UN Conference on Environment and Development 1992*: an international conference with representatives from some 167 countries held in Rio de Janeiro, Brazil, in 1992. The primary objective was to review progress since earlier conferences

Continued on page 252

Continued from page 251

in 1972 and 1982 in safeguarding the human environment and promoting human welfare. One of the products of the conference was a Framework Convention on Climate Change. The aim was to protect the world's climate system most notably against the effects of greenhouse gases (principally carbon dioxide) and their warming influence. Initially, developed nations were expected to reduce their emissions of carbon dioxide and other greenhouse gases to 1990 levels.

- *International Conference on the Economics of Climate Change*: a conference convened in June 1993 by the OECD and the International Energy Agency to help ensure that the UN Framework Convention on Climate Change was based on sound economic reasoning, and to help OECD governments to consider practical options and priorities for action. The conference was attended by about 250 economists and policy-makers.
- *Berlin Climate Conference*: a conference of the world's environment ministers held in Berlin in March 1995 to review progress towards achieving the objects of the Framework Convention on Climate Change. The aim became to reduce emissions of carbon dioxide to 1990 levels by the year 2000. Representatives from 150 countries attended, but there was little to report in the way of solid achievement in respect of carbon taxes, promotion of renewable energy sources, or the curtailing of actual carbon dioxide emissions from transport or industry. The conference received an initial report from the Intergovernmental Panel on Climate Change.
- *The Geneva Talks, 1996*: international discussions to review progress since the Berlin Climate Conference, and to examine new evidence. Few countries could agree on possible targets.
- *Kyoto Climate Change Convention*: held in Kyoto, Japan, in December 1997, a conference of 160 nations which produced the Kyoto Convention, the first international treaty relating to global warming, binding key nations to reductions in carbon dioxide emissions to the atmosphere. The collective global target was to cut greenhouse gas emissions by a little more than 5% of the 1990 baseline, by the year 2012. This was a modest but attainable target. However, Australia, Norway and Iceland were allowed to increase emissions over the same period, by 8% in the case of Australia. The talks began with the European Union demanding cuts of 15% on 1990 levels, by the year 2012; although the US wanted stabilization at 1990 levels by the year 2012. Australia wanted an increase of 18% over 1990 levels, by 2010. The US introduced emissions trading into the protocol.
- *Buenos Aires Conference of the Parties*: a follow-up meeting on the Kyoto Convention, held in Argentina in November 1998, to review progress and make further decisions.

The conference received a fresh report from Britain's Hadley Centre for Climate Change, based in Berkshire, making the following predictions:

- parts of the Amazon rain forest will turn into desert by 2050, threatening the world with an unstoppable greenhouse effect;
- land temperatures will increase by up to 6 °C by the end of 2100;
- the number of people on the coast subject to flooding each year will increase from 5 million in 1998 to 100 million by 2050 and 200 million by 2080;

Continued on page 253

Continued from page 252

- another 30 million people will be hungry by 2050 because it will be too dry to grow crops in large parts of Africa;
- an additional 170 million people will live in countries with extreme water shortages;
- malaria will threaten much larger areas of the globe, including Europe, by the year 2050;
- the US prairies will be detrimentally affected;
- global sea level will rise by 21 cm by 2050.

Other authorities such as the World Wide Fund for Nature warned that global warming could trigger an explosion of life-threatening infections in areas where they are not now present. These threats include not only malaria, but cholera, dengue fever, yellow fever and encephalitis.

It was confirmed that it will be the developing world that will produce the bulk of greenhouse gases by the year 2050, yet China and India refused to even discuss the matter. However, within an atmosphere of doubt and uncertainty, the private sector emerged as an influential contributor to appropriate technical measures, including such companies as DuPont, United Technologies, Enron and Royal Dutch Shell. The Conference proved another step forward in reducing greenhouse gases.

Box 9.5 Key Elements of the Kyoto Protocol

- Developed (Annex I) countries have collectively agreed to reduce greenhouse gas emissions by at least 5% below 1990 levels by 2008–2012.
- To achieve this collective target, individual nations were allocated differentiated targets, ranging from an 8% reduction for the European Union to a 10% increase for Iceland from 1990 levels by 2008 to 2012. The USA and Japan have agreed to reduce greenhouse gases by 7% and 6% respectively, over the same period. The target for Australia is an 8% increase.
- Six greenhouse gases are covered by the protocol: carbon dioxide, methane, nitrous oxide, hydrofluorocarbons, perfluorocarbons and sulphur hexafluoride.
- Reductions in greenhouse emissions from sources and removal of carbon from sinks can be used to meet target commitments.
- Joint implementation, emissions trading and emissions banking can be used by the parties to meet their targets. Joint implementation projects may involve Annex I and non-Annex I countries.
- Some basic rules for emissions trading were established; further principles and guidelines are yet to be agreed.
- The Protocol recognizes that nations will need to implement policies and measures in accordance with their national circumstances if they are to meet their target commitments.
- The Protocol provides in principle for the establishment of bubble arrangements between any group of Parties which choose to fulfil their commitments jointly.

Continued on page 254

Continued from page 253

- No agreement was reached on how to deal with non-compliance.
- The entry into force of the Protocol requires ratification by at least 55 parties to the Framework Convention on Climate Change; and such parties must represent at least 55% of total Annex I carbon dioxide emissions in 1990.
- No agreement was reached on the future involvement of developing countries.
- The Kyoto Protocol was adopted by the international community on 10 December 1997, the result of two-and-a-half years of negotiations under the Berlin Mandate.

Table 9.3 gives the emission targets for each Annex I country, expressed as a percentage change from a 1990 base of 100.

the period to 2020. The assessment was based on the MEGABARE model of the world economy in a context of analysing alternative emission abatement policy scenarios. The report looks at business-as-usual scenarios (i.e. without any carbon dioxide restraints); less stringent scenarios in which Annex I countries reduce their carbon dioxide emissions to 1990 levels by 2010 and further reduce emissions to 10% below 1990 levels by 2020; and more stringent scenarios where Annex I countries stabilize their carbon dioxide emissions from fossil-fuel combustion at 15% below 1990 levels by 2010 and hold emissions at those levels by 2020.

Under the business-as-usual scenario, global carbon dioxide emissions from fossil fuel combustion are projected to double over the period 1990 to 2020. This growth derives in large measure from non-Annex I countries. Indeed, non-Annex I emissions will overtake Annex I emissions by 2016. The rapid growth in non-Annex I emissions is attributed to their relatively high projected rates of economic growth, increasing levels of industrialization, and continuing population growth. In view of recent developments in Asia, these figures may be somewhat moderated (see Figures 9.4 and 9.5; also Table 9.1).

Figure 9.5 also examines the effects of a less stringent scenario and a more stringent scenario. The results show that the more stringent emission reductions lead to moderately greater reductions in global emissions relative to business-as-usual, than the less stringent emission reductions over the medium term. However, over the longer term the difference between the impacts of the more or less stringent policies becomes increasingly small. This is partly due to the fact that Annex I emissions as a share of global emissions are expected to decline, and that emission abatement action in Annex I countries is projected to encourage fossil-fuel use and emissions in non-Annex I countries as fossil-fuel intensive industries relocate to non-Annex I regions, where emission abatement targets do not currently apply. This process is known as 'carbon leakage'.

Owing to significant differences in economic structures and trading patterns, uniform emission abatement targets do not lead to uniform economic costs

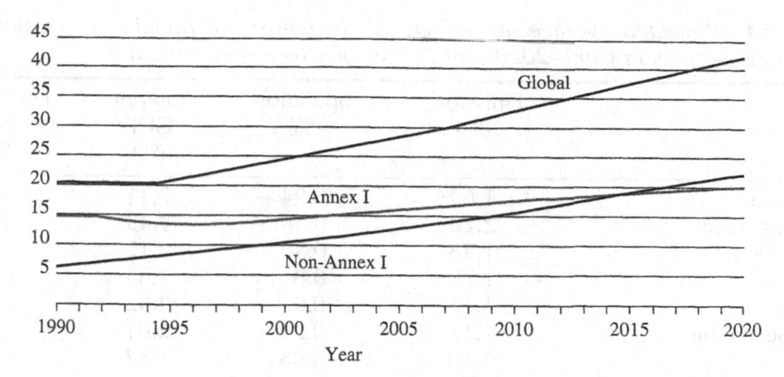

Figure 9.4 *Carbon dioxide emission from fossil fuel combustion: business-as-usual*

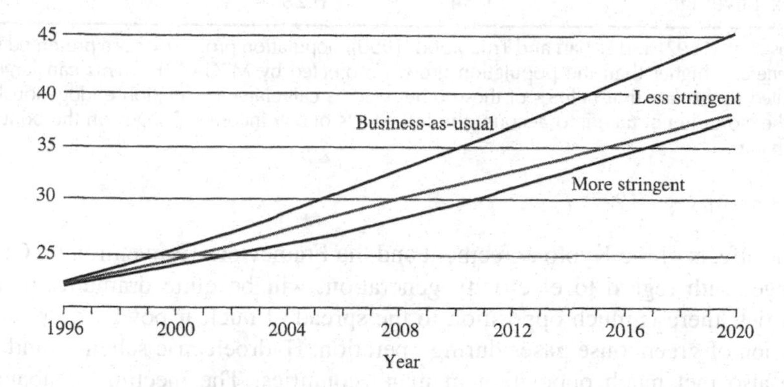

Figure 9.5 *Global carbon dioxide emissions from fossil fuel combustion under uniform emission reduction scenarios*

between Annex I countries. For example, the Australian Bureau of Agricultural and Resource Economics (ABARE) points out that the projected economic costs for Australia, Norway, New Zealand and Japan are many times higher than those projected for other Annex I regions. Japan will experience high costs because Japanese industries have already taken major steps to improve energy efficiency and reduce fossil-fuel use; further steps would involve large costs.

In the very long term, ABARE argues, the UN Framework Convention on Climate Change will be judged to have been effective if a balance has been achieved between the net damages from climate change itself and the economic costs imposed as a result of emission abatement and adaptation. One of the necessary conditions for such a balance is that major emitters are part of an agreement to reduce greenhouse gases. This type of participation will be encouraged only if emission abatement actions undertaken by the signatories to the Kyoto agreement are equitable and least cost. In the outcome, the concept of uniform abatement targets was abandoned at Kyoto.

Table 9.1 *Projected average annual growth in emissions, population, output and emissions per person 1990–2020: Annex I regions, business-as-usual*

	Emissions (%)	Population[a] (%)	Output (GDP) (%)	Emissions per person (%)
Australia	1.63	0.94	2.31	0.68
New Zealand	2.20	0.34	2.43	1.86
United States	1.38	0.36	2.13	1.03
Canada	1.28	0.85	1.83	0.43
Japan	1.16	−0.07	2.52	1.22
European Union	1.12	−0.24	2.01	1.36
EFTA	1.10	−0.08	1.47	1.18
Former Soviet Union and Eastern Europe	0.85	0.22	1.34	0.63
Annex I average	1.34	0.29	2.01	1.05

[a]In Bos *et al.* (1992) and Urban and Trueblood (1990), population projections are presented which are generally higher than the population growth projected by MEGABARE. This can largely be attributed to the fact that neither of these other studies calculates population endogenously and they therefore do not take into account the full effects of any income changes on the population growth rate.

Source: Brown et al. (1997).

The effects of the Kyoto agreement and the Framework Convention on Climate Change, with regard to electricity generation, will be quite dramatic. In many countries, there is much opposition to the spread of nuclear power, despite zero-emission of greenhouse gases during operation. Hydroelectric schemes and dams have also met much opposition in many countries. The meeting of abatement targets might well require halving coal-fired electricity production in countries as diverse as the United States and Australia. Some countries, already shifting to a reduced dependence on coal and an increased reliance on natural gas, are fortunate in finding economics in the narrower sense in harmony with abatement policies. Even oil-based electricity generation has a lower emission intensity, carbon-wise, relative to coal. In other words, renewable technologies relying on solar, wind, geothermal and hot-rock sources will take on a significantly enlarged role over the coming decades. Table 9.2 shows three data sets produced by ABARE: the initial 1992 shares of various electricity generation technologies in Annex I countries; the projected technology mix for electricity generation in 2020 under the business-as-usual scenario; and the projected technology mix for electricity generation in 2020 under the less stringent uniform emission reduction scenario.

9.7 Mitigation Cost Studies

The contribution of Working Group III to the Second Assessment Report of the Intergovernmental Panel on Climate Change (IPCC 1995, vol. 3, chapters 8 and

Table 9.2 *Percentage of total electricity generated by various technologies*

	Australia (%)	New Zealand (%)	United States (%)	Canada (%)	Japan (%)	European Union (%)	EFTA (%)	FSU/EE (%)
Percentage of total electricity generated by each technology, 1992								
Coal	79.2	1.3	53.2	17.5	17.3	34.8	0.1	26.1
Oil	2.3	0.4	3.3	2.9	25.4	10.2	0.4	8.8
Gas	8.8	25.1	13.0	2.6	21.6	6.6	0.2	37.4
Nuclear	0.0	0.0	20.1	15.5	25.1	34.4	13.1	13.7
Hydro	9.2	73.1	8.3	60.7	9.5	13.4	85.3	13.8
Renewables	0.4	0.1	2.1	0.8	1.1	0.6	1.0	0.2
Total	100.0	100.0	100.0	100.0	100.0	100.0	100.0	100.0
Percentage of total electricity generated by each technology under business-as-usual assumptions, 2020								
Coal	79.0	0.9	56.4	42.3	22.2	46.1	0.1	36.4
Oil	1.4	0.2	2.3	1.8	18.3	7.9	0.3	5.7
Gas	13.5	51.5	19.2	6.0	32.0	8.0	13.8	28.8
Nuclear	0.0	0.0	13.9	9.6	18.1	26.7	11.4	16.1
Hydro	5.7	47.4	5.7	37.5	6.8	10.4	74.4	12.9
Renewables	0.4	0.0	2.5	2.9	2.6	0.9	0.0	0.1
Total	100.0	100.0	100.0	100.0	100.0	100.0	100.0	100.0
Percentage of electricity generated by each technology under the less stringent emission reduction scenario, 2020								
Coal	21.9	0.9	8.2	0.0	0.0	1.6	0.1	2.8
Oil	3.0	0.2	1.9	0.2	0.1	28.3	0.1	1.1
Gas	9.1	3.0	8.9	11.7	43.6	18.7	8.3	13.3
Nuclear	0.0	0.0	14.5	9.2	18.9	27.7	11.5	45.4
Hydro	6.2	50.4	6.0	36.0	7.2	10.8	75.1	37.1
Renewables	59.7	45.4	60.5	43.0	30.1	12.9	4.9	0.4
Total	100.0	100.0	100.0	100.0	100.0	100.0	100.0	100.0

Source: Brown et al. (1997); MEGABARE database and projections.

9) addresses the cost of mitigating greenhouse gases and reviews mitigation cost studies to 1995. Key findings were as follows:

- Mitigation costs will be affected by a wide range of factors, including population growth, consumption patterns, resource and technology availability, geographical distribution of activity, land use, transportation patterns, and trade.
- Substantial disagreement still exists over the existence and size of a significant no-regrets potential for greenhouse gas emission reduction. Some studies show significant no-regrets potential while other studies show little no-regrets potential. Whether a no-regrets potential for emission reduction exists depends on whether the economy under consideration is on or below its theoretical production frontier.

Table 9.3 *The Kyoto Agreement: greenhouse emission targets between 1990 and 2012, Annex 1 countries*

Country	Emission target (% change from 1990)
Australia	108
Austria	92
Belgium	92
Bulgaria	92
Canada	94
Croatia	95
Czech Republic	92
Denmark	92
Estonia	92
European Community	92
Finland	92
France	92
Germany	92
Greece	92
Hungary	94
Iceland	110
Ireland	92
Italy	92
Japan	94
Latvia	92
Liechtenstein	92
Lithuania	92
Luxembourg	92
Monaco	92
Netherlands	92
New Zealand	100
Norway	101
Poland	94
Portugal	92
Romania	92
Russian Federation	100
Slovakia	92
Slovenia	92
Spain	92
Sweden	92
Switzerland	92
Ukraine	100
United Kingdom of Great Britain and Northern Ireland	92
United States of America	93

Source: Kyoto Agreement

- Despite significant differences in views, there is agreement that some energy efficient improvements (perhaps 10–30% of current consumption) can be realized at negative to slightly positive costs.
- Energy-related emissions can be reduced through both demand-side (energy efficiency) and supply-side (alternative sources of supply) options. In the short term, demand-side options are cheapest in most countries.
- It is necessary to distinguish four types of cost concepts used in costing analysis: the direct engineering and financial costs of specific technical measures (e.g. the cost of switching from coal to gas in electricity production, of improving the thermal efficiency of homes, or of planting trees in reforestation projects, reported in present-value terms); the economic costs for a given sector, in which sectoral models compare the relative costs of different scenarios; macroeconomic costs which measure the impact of a given strategy on the level of the gross domestic product (GDP) and its components such as household consumption and investment; and welfare costs, those costs which may not be reflected in GDP variations, for example environmental degradation reduces welfare but does not result in a corresponding reduction in GDP.
- There has been much more analysis to date of emission reduction potentials and costs for industrialized countries than for other parts of the world.
- The overall cost of abatement programmes will be determined ultimately by the rate of capital replacement, the discount rate, and the effect of research and development.
- It is important to identify those paths that minimize the costs of achieving a particular concentration target.
- The overall impact of a carbon tax will depend not only on the size of the tax, but also on the uses to which the revenues are put.
- Given differences in marginal emission reduction costs among countries, international co-operation can significantly reduce the global price tag for emission reduction; emission reduction could be carried out where it is cheapest to do so.
- Estimates of the costs of stabilizing carbon dioxide emissions at 1990 levels in OECD countries vary widely. Many bottom-up studies suggest that the costs of achieving this target over the next few decades may be negligible. However, given rising baseline emissions over the longer term, many top-down studies suggest that the annual costs of stabilizing emisions may ultimately exceed 1–2% of GDP.
- Analyses suggest that there may be substantial low-cost emission reduction opportunities for developing countries. However, these are unlikely to be sufficient to offset increasing emission baselines associated with economic growth. The likely magnitude of emission reduction will be particularly costly for developing countries.
- In recent years, there have been numerous studies of the costs of reducing carbon dioxide emissions. Unfortunately, estimates have spanned such a wide

range that they have been of limited value to policy-making. Some studies place losses at several per cent of GDP, while others question whether there will be any losses at all.
- The establishment of international trade in carbon emission rights requires a decision on the allocation of emission rights among nations, which in turn has major implications for the international distribution of wealth. This is a political rather than an economic issue.
- Although much of the work on greenhouse gas emission reductions has focused on energy-related emissions, a growing literature is emerging on non-energy emissions.
- The atmosphere is an international public good, in that all countries benefit from each country's reduction in greenhouse emissions; greenhouse gases are an international externality, in that emissions by one country affect all other countries to some extent.

Box 9.6, quoting from the report of IPCC Working Group III, sums up rather well the complexities of the subject in its application to policy.

Box 9.7 outlines the tradable permit system as proposed in the Kyoto Protocol, at least in principle.

To round out the chapter, Box 9.8 summarizes the measures that may be adopted to restrict the emission of carbon dioxide to the atmosphere. Box 9.9 is a telling comment on global warming conferences while Box 9.10 is a comment on the economic effects of subsidies. Box 9.11 quotes a commentary by *The Economist* on the Kyoto compromise. The inclusion of these quotations is not meant to suggest that *The Economist* is always right; but simply that it is rarely off the mark.

Box 9.6 Climate Change: Economic and Social Dimensions of Climate Change

Climate change presents the decision-maker with a set of formidable complications: large uncertainties, the potential for irreversible damage or costs, a very long planning horizon, long time lags between emissions and effects, a global scope, wide regional variation, and multiple greenhouse gases of concern. Irrespective of the possible consequences of climate change, policies that mitigate or assist adaptation to climate change and have zero or negative net costs ("no-regrets" policies) are clearly justified. If the evidence suggests that damage can be expected from climate change, then the expectation of damages provides a rationale for going beyond no-regrets to those that incur positive net costs.

Source: From the contribution of Working Group III to the Second Assessment Report of the Intergovernnmental Panel on Climate Change (IPCC 1995, vol. 1, chapter 1).

Box 9.7 Controlling Carbon Dioxide Emissions: The Tradable Permit System as Proposed in the Kyoto Protocol

- The UNCTAD secretariat have indicated that tradable permits are both an efficient means of controlling carbon dioxide (CO_2) emissions at minimum cost, and an effective mechanism for transferring resources to developing countries.
- A viable strategy should begin with a simple pilot scheme and evolve gradually into a more complete global system; some of the major emitters would need to act as pioneers.
- The system would be under the control of a new UN body; it would be responsible for creating and allocating new permits, while market mechanisms would be in the private sector.
- The UN agency would arrange self-monitoring by the individual countries, but with independent auditing.
- The system involves defining an overall target for emissions for all participating countries, being significantly lower than current usage.
- Individual countries would then be issued with permits to emit carbon to a defined level. The sum of these permits would equal the overall target; participants would have to hold permits equal to the value of their emissions.
- If the number of permits proved insufficient, permits would have to be bought or leased from other participants who had permits in excess of their requirements.
- The aim is to sell at a profit through commodity exchanges.
- Initially, the developed countries would have a deficiency of permits, while developing countries would have a surplus. In the ensuing trading, resources would be transferred to the poorer countries.
- Improvements in control techniques lead to reduced emissions and surplus, saleable permits.
- The Intergovernmental Panel on Climate Change considered that, of all instruments, tradable emission rights offered flexibility, efficiency in pollution abatement, direct control of total emission levels, and incentives for research.

Source: UNCTAD (1995) Controlling Carbon Dioxide Emissions: The Tradeable Permit System, United Nations, Geneva.

9.8 Alternative Energy Sources

Solar energy, the energy of the Sun, reaches the surface of the Earth in the form of visible light, short-wave radiation, and near ultraviolet light. After penetrating the atmosphere, some of the energy heats the surface of the Earth while part of it is re-radiated into the atmosphere in the form of long-wave radiation and absorbed by carbon dioxide and water vapour in the atmosphere. The utilization of solar energy for the generation of electricity using photovoltaic cells has been developed in recent years, providing power for houses and satellites. In another application, many solar energy heating and hot-water units have been installed in domestic homes in countries with appropriate climates. Solar steam generators have been built using movable mirrors to concentrate large amounts

Box 9.8 Measures to Restrict the Emission of Carbon Dioxide (CO_2)

- The progressive introduction of more energy-efficient technology, reducing energy consumption per unit of production, e.g. the introduction of steam generators with higher thermal efficiencies, or furnaces with heat recovery, or more efficient household appliances and vehicle engines.
- The substitution of fuels which yield less carbon dioxide per unit of output, e.g. the substitution of natural gas for coal.
- The increasing use of hydropower.
- The increasing use of solar power, wind power, wave power, tidal energy, geothermal energy, or hot dry rock energy, in appropriate circumstances.
- The use of waste methane to generate electricity.
- The substitution of ethane for naphtha as feedstock in chemical processes.
- The reduction of gas flaring.
- The introduction of low-energy cells in aluminium production.
- Cogeneration plants to recover heat from gas turbines, converting it into useful high-pressure steam; they offer much higher thermal efficiencies.
- The removal of energy subsidies world-wide. Subsidies tend to lower the price of fossil fuels, leading to profligate use.
- The imposition of carbon taxes on emissions, also leading to the re-optimization of resource use.
- Peak-load and off-peak pricing policies to enhance the utilization of plant, perhaps obviating the need for new plant.
- The more rapid retirement of old and inefficient plant.
- The acceleration of research and development into alternative sources of energy.
- The utilization of nuclear power in remote locations, emitting no carbon dioxide.
- The introduction of energy standards.
- Targets for the introduction of renewable energy into power supplies by the year 2012.
- Loans and grants to support important renewable energy initiatives.
- The promotion of plantations and plantation-based industries.
- Revegetation projects to further develop carbon sinks.
- The reduction of methane emissions from livestock through vaccines.
- The reduction of emissions from land clearing and burning-off.
- Improvements in industrial and chemical processes and plant.
- The introduction of superconducting cables.
- Improvements in the design of buildings and dwellings to achieve energy conservation.
- Improvements in public transport in city areas, including light rail systems and monorails.
- Improvements in urban developments and city planning; energy-efficient homes.
- Measures to enhance carbon dioxide sinks, such as oceans.
- The promotion of viable recycling.
- Better batteries.
- The effective use of waste biomass and coal wastes.
- The development of fuel cells.

Box 9.9 Too Much Hot Air

There is something about environmental conferences that brings out the hypocrisy in politicians. Five years ago, at the earth summit in Rio de Janeiro, world leaders agonized about a host of planetary problems, from the destruction of forests to the loss of biodiversity. The centrepiece of their efforts was the global warming treaty, a promise to reduce the output of greenhouse gases from rich countries to 1990 levels by the decade's end. Then they went home and did virtually nothing about it: almost all rich countries are expected to overshoot their 2000 target, some by as much as 40 per cent.

Source: The Economist, 15 March 1997, p. 17.

Box 9.10 Subsidies to Users of Fossil Fuels

Greenpeace, one of the world's most vocal environmental groups, recently released a report commissioned from the Institute of Environmental Studies of the Free University, Amsterdam, which found that governments in Europe fork out around $10 billion a year in direct subsidies to users of fossil fuel. According to the World Bank, the global figure for such subsidies is around $230 billion. And recent work by the OECD suggests that removing support for the coal industry in particular could cut carbon dioxide emissions from rich countries by hundreds of millions of tonnes a year.

Source: The Economist, 14 June 1997, p. 99.

Box 9.11 The Economist on the Kyoto Compromise

Put diplomats from 159 countries, plus lobbyists with at least as many agendas – hidden or open – into one place and the result is bound to be hypocrisy, gamesmanship and sheer nonsense. Such was the case during the negotiations in Kyoto to reduce greenhouse gas emissions .... It was not until well past the formal deadline, amid gestures and language that were scarcely diplomatic, that delegates were able to cut a deal. Too bad it will never fly.

Source: The Economist, 13 December 1997, p. 14.

of solar radiation to convert water to steam. Biological systems use sunlit algae to convert carbon dioxide and water into oxygen and protein-rich carbohydrates. Solar energy is also used to produce salt from seawater by evaporation. Solar energy is non-polluting and constantly renewable.

Geothermal energy is power obtained by using heat from the Earth's interior, usually in regions of active volcanoes. Hot springs, geysers, pools of boiling mud, and fumaroles (vents of volcanic gases and heated groundwater) are often sources of such energy. The Romans used hot springs to heat baths and homes. The greatest potential use for geothermal energy, however, lies in the generation of electricity. By the late 20th century, geothermal power plants were in operation in Iceland, Italy, Japan, Mexico, New Zealand, Russia and the USA. Many more are now in operation. Water and steam hotter than 180 °C are the most easily utilized for electricity generation, and are used in most existing geothermal power stations. Geothermal power generation is non-polluting, competitive, and offers an alternative in certain locations to polluting fossil-fuel and nuclear power stations. Much depends on local geology and sources are limited to a depth of about 5 km for practical reasons.

Hot dry rock (HDR) systems are geothermal reservoirs in which granite or other suitable rock is artificially fractured at an accessible depth of say 3–5 km. Water is injected into the artificial reservoir and returns to the surface through a production well as steam to generate electricity. HDR systems differ from natural geothermal energy systems in that there is no natural reservoir of hot water and steam, as occurs at Rotorua in the North Island of New Zealand.

Partially successful HDR projects have been completed in the USA, Britain and France. However, difficulties still remain in creating large artificial heat exchangers with horizontal fractures at depths of several kilometres, having low hydraulic resistance. Research work is being undertaken by the Australian National University in Canberra, with test bores in the Hunter Valley, New South Wales.

HDR systems offer a route for converting the limitless energy of the Earth's mantle for human and environmental purposes. If established as a competitive source of energy, HDR systems would displace the use of fossil fuels for generating electricity, offering a very large reduction in the emission of greenhouse gases, and retarding global warming. However, the problems of harnessing this source of energy are not only technical, but economic.

Wind energy entails the conversion of the kinetic energy of the wind into electricity, by means of wind turbines or windmills. Wind is generated by atmospheric pressure differences in the atmosphere, driven by solar power. Studies reveal that the wind could supply an amount of electrical energy equal to the world's electricity demand. Turbines have been supplied for the Californian windfarm market and in relation to programmes being developed in the European Union, and in China, India, Russia and Israel.

Tidal energy involves the use of the rise and fall of the tides to generate electrical energy. A significant commercial plant is a 240 MW power plant in the La Rance Estuary, France, which also functions as a dam and a bridge. Studies have indicated that changes in tidal height, periodicity, salinity, pollutant dispersion and other factors could cause major alterations to the general ecology of an estuary or coastal habitats, with consequent effects on sea fisheries and wildlife in general.

Wave energy is the utilization of the energy of waves, as distinct from tides, for the generation of electricity. The energy of waves is very diffuse, and the harnessing of this energy requires large and robust structures to achieve even low conversion rates. Most proposals for using wave energy comprise a system of floats allowing the movement of waves to compress or elevate a fluid which drives turbo-generators. Several devices have undergone sea trials, and prototype devices have been built in Norway and Britain. Wave energy could replace fossil fuels in time and have a positive influence on reducing the harmful greenhouse effect, while supporting sustainable development.

Hydrogen (H) is a colourless, odourless, tasteless, highly flammable, gaseous substance; it is mostly found combined in water, hydrocarbons and other organic and biological compounds. It is the lightest of all gases. The *hydrogen economy* is a concept in which hydrogen plays a central role in energy production and use. Hydrogen would be generated by electrolysis, be stored or transported and then used to produce energy in processes. It is a concept of the future in which hydrogen would be generated on a vast scale and be used instead of petrol (gasoline) in internal combustion engines, as a source of energy in fuel cells, in the home for domestic purposes, and as a feedstock for hydrogen-based industrial processes. It envisages a cheap production process for hydrogen, not yet realized.

It seems that the fuel cell (see Box 10.23) will be the most likely channel for the future use of hydrogen, and there appears a case for public investment coupled with private capital in the mass production of fuel cells. There could well be a double-dividend in a low-emission (or zero-emission) vehicle just at a stage when the supply of petroleum diminishes and its price rises. For 100 years, the source of power was coal, then for the next 100 years it has been oil; the third age of fuel could be hydrogen. It could be the key to future energy sources, with a much reduced contribution to global warming.

Natural gas is a hydrocarbon gas from underground sources, often found in association with crude oil. It is generally rich in methane, with varying amounts of ethane, carbon dioxide, helium and nitrogen. Commercially, natural gas is transported by pipeline or in a refrigerated liquefied form. There are three main types of reservoir: (1) reservoirs in which only gas can be produced economically; (2) condensate reservoirs, in which the gas is associated with light liquid hydrocarbons; and (3) reservoirs in which gas is found dissolved in crude oil (solution gas) and in some cases also in contact with underlying gas-saturated crude (cap gas). Due to its economic and environmental advantages, natural gas is currently experiencing faster growth in consumption than any other energy source. Natural gas now supplies about one-quarter of global commercial energy, with estimates of global natural gas reserves increasing rapidly. Russia holds the largest natural gas reserves, having more than ten times the reserves of the USA. Ample reserves also exist in the Middle East. There has been an increasing use of gas for power generation, as in the case of British power stations. Qatar has pressed ahead with the development of its huge natural gas reserves, completing the first phase of the Qatargas North Field Project.

9.9 Summary

The chapter opens with a note on the history of changing climates, for during the last two million years or so the Earth's climate has swung dramatically between glacial and interglacial periods. During interglacial periods, deserts existed, the poles were ice-free, vegetation thrived, fauna flourished, volcanoes erupted, and camels roamed the Great Plains. We are currently in a new warming period, possibly with human assistance. Temperatures are influenced by the path of the Earth around the Sun, the incidence of sunspots, and the greenhouse effect. The essential character of the greenhouse effect for human survival is noted, together with the actual or potential effects of the enhanced greenhouse effect.

The findings of the Intergovernmental Panel on Climate Change (IPCC) are summarized, and some of the reservations expressed about these findings. It should be noted that the findings of the IPCC have varied between its first assessment report in 1990 and its second assessment report in 1995 quite significantly. The estimates of future global warming have been reduced by about one-third, and the estimates of increases in sea level have been reduced by about one-quarter. The second report argued that there were many opportunities for economies in the use of fossil fuels, which could significantly reduce the emission of greenhouse gases. Further, whatever the ultimate outcome of global warming predictions, these measures could be undertaken as justified in themselves, i.e. undertaken on a 'no-regrets' basis. Fuel economy tends to be ignored largely because energy is often not a major item in corporate budgets, and comes too low in the agenda to receive much, if any, attention. The IPCC's third assessment will appear by 2001.

The chapter concludes with a summary of the undertakings of nations at the 1997 Kyoto Summit. For most nations, meeting these targets will involve a major transition to alternative pollution-free energy sources such as solar energy and hot rock. Clearly some important economies will be achieved readily, but later the costs will probably rise steeply. The time span of about a decade appears much too short to achieve these ambitious targets. Those nations most likely to succeed, such as Britain, will do so simply because coal has been substantially replaced by natural gas, with a lower carbon dioxide emission. Here a fundamental shift for economic reasons coincided with the meeting of emission targets, as occurred with the smoke control programmes of earlier years. Most countries will probably fail, and dramatically so. Political will is likely to falter in the face of rising costs and impracticable time-tables.

Further, it is not at all clear what the human contribution is to global warming. It should be kept in mind that the balance of evidence as judged by the IPCC suggested no more than a 'discernible human influence on global climate'. A 'discernible influence' means no more than perceptible, and perhaps barely so. It is an insufficient basis for more than no-regrets policies, particularly when further steps involve intolerable and politically unacceptable increases in energy costs, major losses in markets, aggravated debt, and inevitably large-scale unemployment.

There remains a strong argument for deferring major commitments for 30 years, allowing time for the development of economically competitive and ecologically safe and acceptable renewable technologies, while benefiting from inexpensive no-regrets policies and overlooked elementary benefits. The world will not face calamitous outcomes within 30 years, and yet will be better equipped for the major challenge should it come. By 2030, it will be clearer as to whether or not we are within a genuine warming period; and whether variations in human activities could possibly influence the situation.

It should be noted also that much of the developing world was excluded from the Kyoto commitments; a fact already noted by the US Senate. Countries that have no hesitation in brandishing nuclear weapons, plead poverty when it comes to participation in world events, continuing to seek official aid as Third World countries.

Box 9.11 summarizes the views of *The Economist* on the entire issue of global warming.

Further Reading

Bolin, B. (1989) Changing climates. *The Fragile Environment,* The Darwin College Lectures, Cambridge University Press, Cambridge.

Brydges, T.G. and Wilson, R.B. (1991) Acid rain since 1985–times are changing, *Proceedings of the Royal Society of Edinburgh,* **97b**: 1–16.

Intergovernmental Panel on Climate Change (1995) *The Second Assessment Report,* comprising three volumes: *Vol. 1 The Science of Climate Change* (the contribution of Working Group I); *Vol. 2 Impacts, Adaptations and Mitigation of Climate Change; Scientific and Technical Analyses* (the contribution of Working Group II); *Vol. 3 Economic and Social Dimensions of Climate Change* (the contribution of Working Group III), Cambridge University Press, Cambridge.

Jones, P.D., Wigley, T.M.L. and Wright, P.B. (1986) Global temperature variations between 1861 and 1984, *Nature,* **322** (6078): 430–4.

Larsen, B. and Shah, A. (1995) Global climate change, energy subsidies, and national carbon taxes in *Public Economics and the Environment in an Imperfect World* (eds Bovenberg and Cnossen), Kluwer, Dordrecht, pp. 113–32.

Lave, L.B. and Seskin, E.P. (1977) *Air Pollution and Human Health,* Johns Hopkins Press, Baltimore.

Nordhaus, W.D. (1994) *Managing the Global Commons: The Economics of Climate Change,* MIT Press, Cambridge.

United Nations (1995) *Report of the Conference of the Parties to the Framework Convention on Climate Change,* held in Berlin, 28 March–7 April, UN, New York.

World Health Organization (1990) *Potential Health Effects of Climatic Change,* WHO, Geneva.

10
Population and the Quality of Life

10.1 Introduction

This chapter addresses the central question of the world' s population, its size and rate of growth, and the carrying capacity of the Earth. While food supplies have kept up with the growth of population in most places at most times, there is evidence of strain in the system, particularly in respect of fisheries.

The rapid growth of population since about 1750 has been due to improving conditions, reduced infant mortality rates, better diets, better health care, the success of immunization and vaccination programmes, and the control of plagues and epidemics, with consequent extensions in the expectation of life in virtually all countries, developed and developing. World population increased from about 890 million in 1750 to 5771 million in 1996. Population may peak at about 10.6 billion in 2080 and then decline. The density of population per square kilometre may reach 40 or more, up from 6.5 per km^2 in 1900.

The process of urbanization will intensify to a stage when over half the total urban population will live in cities with populations of more than one million, with many cities having more than 10 million inhabitants. Cities never fail to attract, despite the problems and congestion there.

The chapter also addresses water quality and the crucial issue of sanitation. Measures of progress are examined. The chapter also addresses the blessings and curses of the automobile. Official aid is also examined.

10.2 World Population

The human population of the world remained remarkably stable over many centuries, certainly up to the beginnings of the agricultural, commercial, industrial and scientific revolutions. From the 17th century onwards, population growth boomed and has become an overwhelming flood. This spectacular growth has many explanations, such as falling death rates in the early and later years of life, improved general health and well-being, more food and better diets, progressively improving health care and sanitation, improved levels of education, increased real incomes and standard of living, improved public infrastructure in respect of water supplies and drainage, improved solid waste management, vaccinations and inoculations, the emergence of germ theory, the control of epidemics, food inspection and analysis, overcrowding abatement, air pollution control, safer working conditions, rodent control, the reduction of malnutrition, and the eradication or containment of many hitherto lethal diseases.

By 2000 the world population will stand at around six billion, an increase of one billion since 1987 when the population stood at five billion. If the growth rates of the last decade were to continue, the world population would double again within the next 50 years. During 1996, 140 million babies were born, 126 million (90%) of these in developing countries. Each day, world population increased by 240 000 (the result of 383 000 births and 143 000 deaths overall). The 1996 total amounted to some 90 million more than the previous year. Table 10.1 reveals that during the past 25 years the expectation of life at birth has increased by 5 years in the case of the richer countries and no less than 13 years in the poorer countries.

It is not surprising that many commentators have expressed concern about this explosion of population over recent years (though not all share the views of these pessimists: see Boxes 10.1 and 10.2).

However, the rate of growth of populations in general has eased slightly. In 1990–94, population growth averaged 1.57% a year, compared with 1.73% over the previous 15 years.

Box 10.1 Ehrlich and Ehrlich on Population

The emphasis now needs to be put on reducing the scale of the whole human enterprise by shifting to sustainable development in *both* rich and poor nations, including moving towards population shrinkage in both. Growth in the physical economies in rich nations – growth that in the past greatly improved the lives of their citizens – must now be recognized as the disease, not the cure.

Source: P.R. Ehrlich and A.H. Ehrlich (1991) Healing the Planet: Strategies for Resolving the Environmental Crisis, Addison-Wesley, Massachusetts, p. 9.

Box 10.2 Gro Harlem Brundtland on Population Growth

Population growth is one of the most serious obstacles to world prosperity and sustainable development. We may soon be facing new famines, mass migration, destabilization, and even armed struggle as peoples compete for ever more scarce land and water resources. In the more developed countries, today's generations may be able to delay their confrontation with the imminent environmental crisis, but their children will be facing the ultimate collapse of vital resource bases.

Source: G.H. Brundtland (1994) The solution to a global crisis, Environment, 36(10): 17–20.

Table 10.1 *Life expectancy at birth in years since the UN Conference on the Human Environment 1972, male and female*

<table>
<tr><th></th><th colspan="2">1972</th><th colspan="2">1995</th></tr>
<tr><th></th><th>Male</th><th>Female</th><th>Male</th><th>Female</th></tr>
<tr><td>Australia</td><td>67.9</td><td>74.2</td><td>74.5</td><td>80.8</td></tr>
<tr><td>Austria</td><td>66.5</td><td>73.3</td><td>72.9</td><td>79.4</td></tr>
<tr><td>Bangladesh</td><td colspan="2">45</td><td>57.0</td><td>57.0</td></tr>
<tr><td>Belgium</td><td>67.7</td><td>73.5</td><td>72.4</td><td>79.1</td></tr>
<tr><td>Britain</td><td>68.7</td><td>74.9</td><td>74.4</td><td>79.7</td></tr>
<tr><td>Canada</td><td>70.8</td><td>75.2</td><td>74.7</td><td>81.7</td></tr>
<tr><td>China</td><td colspan="2">59</td><td>69.1</td><td>72.4</td></tr>
<tr><td>Czech Republic</td><td colspan="2">na</td><td>68.9</td><td>76.6</td></tr>
<tr><td>Denmark</td><td>70.6</td><td>75.4</td><td>72.5</td><td>77.8</td></tr>
<tr><td>Finland</td><td>65.4</td><td>72.6</td><td>71.7</td><td>79.4</td></tr>
<tr><td>France</td><td>68.0</td><td>75.5</td><td>73.1</td><td>81.3</td></tr>
<tr><td>Germany</td><td>67.6</td><td>73.6</td><td>73.2</td><td>79.8</td></tr>
<tr><td></td><td colspan="2">(West)</td><td></td><td></td></tr>
<tr><td>Greece</td><td>67.5</td><td>70.7</td><td>74.6</td><td>79.8</td></tr>
<tr><td>India</td><td>41.9</td><td>40.6</td><td>60.4</td><td>61.2</td></tr>
<tr><td>Indonesia</td><td colspan="2">47</td><td>62.0</td><td>65.0</td></tr>
<tr><td>Ireland</td><td>68.1</td><td>71.9</td><td>71.0</td><td>76.7</td></tr>
<tr><td>Italy</td><td>67.2</td><td>72.3</td><td>73.6</td><td>80.2</td></tr>
<tr><td>Japan</td><td>69.1</td><td>74.3</td><td>76.6</td><td>83.0</td></tr>
<tr><td>Korea (South)</td><td>51.1</td><td>53.7</td><td>68.0</td><td>76.0</td></tr>
<tr><td>Malaysia</td><td colspan="2">62</td><td>69.0</td><td>73.0</td></tr>
<tr><td>Netherlands</td><td>71.0</td><td>76.4</td><td>74.0</td><td>80.0</td></tr>
<tr><td>New Zealand</td><td>68.4</td><td>73.8</td><td>73.4</td><td>79.1</td></tr>
<tr><td>Norway</td><td>71.0</td><td>76.0</td><td>74.2</td><td>80.3</td></tr>
<tr><td>Pakistan</td><td>53.7</td><td>48.8</td><td>62.0</td><td>64.0</td></tr>
<tr><td>Philippines</td><td colspan="2">57</td><td>66.0</td><td>69.0</td></tr>
<tr><td>Poland</td><td>66.9</td><td>72.8</td><td>67.4</td><td>76.0</td></tr>
<tr><td>Portugal</td><td>60.7</td><td>66.4</td><td>70.8</td><td>78.0</td></tr>
<tr><td>Russia</td><td>65.0</td><td>74.0</td><td>57.7</td><td>71.1</td></tr>
<tr><td></td><td colspan="2">(USSR)</td><td></td><td></td></tr>
<tr><td>Singapore</td><td colspan="2">68</td><td>74.4</td><td>78.5</td></tr>
<tr><td>Spain</td><td>67.3</td><td>71.9</td><td>73.2</td><td>81.1</td></tr>
<tr><td>Sri Lanka</td><td colspan="2">64</td><td>70.0</td><td>75.0</td></tr>
<tr><td>Sweden</td><td>71.9</td><td>76.5</td><td>75.1</td><td>80.6</td></tr>
<tr><td>Switzerland</td><td>68.7</td><td>74.1</td><td>74.7</td><td>81.4</td></tr>
<tr><td>Taiwan</td><td>65.8</td><td>70.4</td><td>71.6</td><td>77.6</td></tr>
<tr><td>Thailand</td><td>53.6</td><td>58.7</td><td>66.0</td><td>71.0</td></tr>
<tr><td>USA</td><td>66.6</td><td>74.0</td><td>67.4</td><td>75.5</td></tr>
</table>

Sources: Britannica Year Books, World Bank Atlas, and UN Demographic Statistics.

Many countries in Asia and Africa are seeing lower total fertility rates (the number of children a woman will bear during her lifetime). Kenya's rate has dropped from more than 8 children per woman in the late 1970s to 6.3 in 1990–95, while in Algeria the fall has been from 6.3 children in 1980–85 to 3.8 in 1990–95. However, in other African countries, the rate shows no sign of dropping. In the developed world generally, the total fertility rate is less than 2, whilst throughout almost the whole of Africa and the Middle East, the rate is 5 or more. Many countries now discourage large families and encourage wise family planning; China has a very restrictive policy on the size of families.

In 1996, Latin America's population totalled 486 million, with an annual growth rate of 1.9%; the total fertility rate remained at 3.1, ranging from 1.5 in Cuba to 5.2 in Honduras.

During the same period, Europe has continued to report a negative rate of natural increase, with continued low fertility in Western Europe, and the collapse of the birth rate in the former Communist countries. In Russia, life expectancy for males dropped to a level similar to that found in industrialized countries at the beginning of the 20th century (i.e. around 57 years). In contrast, the highest male life expectancy was in Japan at 76 years. Japan also had the lowest infant mortality rate in the world at 4.2 infant deaths per 1000 live births.

As circumstances change, the elements encouraging and indeed demanding large families also change. The development of social services, falling infant mortality rates, the need to have at least one surviving son on reaching retiring age, and social and cultural attitudes, produce a slow and gradual change. Even so, many remain anxious about the effects of unrestricted population growth, the ability to feed forever a growing number of mouths, and the effects of the resource base underlying all activity such as soil, water and other resources. There is also concern about the quality of life that can be offered to so many people. Table 10.2 summarizes the growth of population since AD 14, while Box 10.3 examines the effects on density of growing world numbers. The views of Lutz (1996) are embodied in Box 10.4.

According to the WHO *World Health Report 1998*, the world average life expectancy of babies born in 1998 was 66 and is likely to reach 73 by 2025. The world's population is expected to reach eight billion by 2025. By then some 800 million people, or over 10% of the population, will be over 65. The gap between the richest and poorest countries is closing gradually.

By 2025, those born in that year in some 26 mainly Western countries may expect to live to 80 years or more. African countries rate lowest but with an expectation of continuous improvement. Life expectancy in Sierra Leone is predicted to reach only 51 by 2025, but that is still an improvement on the present life expectation of 38.

The UN Population Division makes varying assumptions about fertility and mortality to arrive at high, medium and low estimates of future world population figures. The UN medium variant assumes that mortality will fall globally to life expectancies of 82.5 years for males and 87.5 years for females between the years 2045 and 2050. This estimate assumes not only modest declines in mortality, but adequate food supplies, water resources, and continued medical, scientific and technological advances.

Table 10.2 *World population*

Year	World population (millions)
14	256
350	254
600	237
800	261
1000	280
1200	384
1340	378
1500	427
1600	498
1650	516
1700	641
1750	731
1800	890
1920	1 968
1930	2 145
1940	2 340
1950	2 500
1960	3 000
1966	3 288
1974	4 000
1987	5 000
1994	5 600
1996	5 771
2010 projected	6 918
2200 projected	11 500

Sources: Colin Clark (1977) Population Growth and Land Use, Macmillan; UN Demographic Yearbooks; Britannica World Data.

During the same period, the UN medium projection indicates the world's total fertility rate (TFR) levelling to an average of 2.1 births per woman, meaning that each generation replaces itself and no more. Replacement fertility sustained for about 70 years would yield a constant population with a stable age distribution. This particular scenario indicates zero population growth, or stabilization. The total world population would have reached 9.8 billion. The annual growth rate would be 1.0%, falling to zero as total world population stabilizes at just over 11 billion in or after 2100.

The UN high projection assumes a TFR of 2.60 children per woman and a total population of 11.9 billion by 2050. The most likely outcome is the medium projection, for population growth rates and TFRs are declining nearly everywhere.

The UN goal of stabilization and zero population growth by the end of 2100 may be the final stage in the demographic transition which began in Europe, spreading to Latin America, Asia and Africa. Increasingly, the future population of the globe will be what couples decide it should be.

Box 10.3 Population: The Effects on Density of Growing World Numbers

Area of the Earth		511 000 000 km^2
Water		361 000 000 km^2
Land		150 000 000 km^2
Population	1900	1 000 000 000 persons
(UN est.)	1950	2 500 000 000 persons
	1995	5 800 000 000 persons
	2000	6 000 000 000 persons
(Lutz est.)	2100	10 350 000 000 persons

Density of population per square kilometre

1900	6.5
1950	17.0
1995	39.0
2000	40.0
2100	69.0

Present density of selected countries (per square kilometre)

China	126.0
India	296.0
Australia	2.3
Britain	240.0
USA	27.6
Japan	331.8

Box 10.4 Lutz on World Population

- The world's population will peak at about 10.6 billion in 2080 and then decline.
- The main reason for the slowdown is that birth rates world-wide are falling faster than was expected.
- By 2100, world population will have fallen to 10.35 billion.
- The world-wide drop in fertility rates is contributing to the outcome.
- To replace people, the fertility rate should be 2.1; in China the fertility rate has already dropped below 2.0.
- The conclusions are based on a weighted assessment of 4000 different possible scenarios.

Source: W. Lutz (1996) The Future Population of the World: What Can We Assume Today? Earthscan, London, on behalf of the International Institute for Applied Systems Analysis, Laxenburg, Austria.

Box 10.5 Australia: A Choice of Population

Australia's population remains modest at just over 18 million (1996); with fertility just below replacement level, any growth in population will be driven by immigration alone.

If fertility remains at its current level with immigration held to the low end of the post-war range (50 000 net persons per year), the population of Australia in 2040 will be 23 million and almost stationary (Australian Academy of Science 1995). On the other hand, with a slow increase in life expectancy and fertility rising to replacement levels, coupled with immigration at a relatively high level of say 170 000 people per annum, the population in 2040 could be 37 million, and fast-growing.

The differences between these two scenarios are very great, the higher population requiring four more cities the size of Melbourne and Sydney, or a doubling of the size of the ten main cities of Australia.

Within these possible targets, regard must be given to Australia's natural resources and assets such as water (Australia is the driest continent on Earth), and to degraded soil and ecological systems. While the area of Australia is only slightly less than that of the United States (without Alaska), there is a lack of potential for population growth beyond relatively narrow limits. It is unlikely that the population of Australia will ever exceed that of New York State.

The contrast is remarkable for the density of population in Australia is currently no more than 2.3 persons per square kilometre (compared with 27.6 in the US; 126.0 in China; 296.0 in India; and 331.8 in Japan). The average for the world is 40.0 per square kilometre.

A limited capacity for population growth does not restrict the potential for economic growth, however. Including natural assets, the World Bank has classified Australia as the richest country on Earth, per capita.

It has been argued that the stabilization of Australia's population by the mid-21st century would align Australia with global policies endorsed by international bodies. The superficial view that Australia could ease overcrowding problems elsewhere in the world, must be seen in the context that the world's population increases at a rate of over 110 000 a day.

Box 10.5 is an interesting reflection on the population problem of Australia, which has a geographical area similar to that of the USA and a current population of little more than 18 million. Paul Ehrlich has suggested that a population of no more than 10 million is suitable for Australia, making it easier to get to Bondi Beach. How to get rid of 8 million people has yet to be explained!

10.3 Mortality and Morbidity

The infant mortality rate reflects the annual deaths of infants aged under one year, per thousand live births. An early crude indicator of social welfare, it continues to be a valuable indicator of those aspects of the quality of life impinging on health and the development of health services generally. The infant mortality rate

Table 10.3 *Infant mortality rates: the annual deaths of infants aged under one year per thousand live births*

	1970–75	1990–95
Africa	131	93
Asia	98	65
Europe	25	12
North and Central America	35	19
Oceania	41	27
South America	84	48
World	93	64
Individual cases:		
Australia	17	7
Austria	24	7
Britain	17	7
Canada	16	7
Finland	12	5
France	16	7
Germany	21	6
Japan	12	4
Singapore	19	6
Sweden	10	5
USA	18	9

Source: UN Population Division and the World Bank.

in developed countries is now about seven deaths per thousand, or less, and in poorer, developing countries about 20–140 deaths per thousand (see Table 10.3).

In recent years, emphasis has also been placed on the under-five mortality rate per 1000 live births. Between 1980 and 1996, India almost halved its under-five rate, with similar achievements in Egypt, Indonesia, Turkey, the Philippines, Brazil, China, Thailand and Mexico.

Infant mortality rates have generally declined throughout the world (by as much as 25%) during the last decade. However, evidence of solid progress does not disguise the fact that each year, three million babies born in developing countries do not survive for more than one week. Many people remain at the bottom of the health pyramid, with low incomes, and limited access to safe drinking water or adequate sanitation. Most urban areas, in both developing and developed countries, embrace large numbers of people who are unemployed on a long-term basis, elderly, and beyond the reach of social and welfare services. Many countries have very limited social security systems, or none.

In 1993, over 50 million people died world-wide. The main causes of death were infectious and parasitic diseases, and diseases of the circulatory system. Table 10.4 summarizes the major environmentally induced pathologies. Box 10.6 summarizes the views of the UN Secretariat on the condition of humanity, while Box 10.7 analyses the characteristics of poverty.

Box 10.6 The UN Secretariat on the Condition of Humanity

For the mid-1980s, it has been estimated that every year 17 million people (including 10.5 million children under the age of five) in the developing countries died as a result of infectious and parasitic diseases, as compared with about half a million in the developed countries. In the developing countries malnutrition, inadequate water supply and sanitation, poor hygienic practices and overcrowded living conditions all contribute to the high incidence of diarrhoeal and infectious diseases. In 1990, despite dramatic improvements in the standards and levels of services in drinking water supply and sanitation achieved over the previous two decades, especially in the rural areas of developing countries, 1.2 billion people did not have access to clean water and 1.7 billion people were not served by adequate sanitary arrangements. In fact, most towns and settlements in Africa and Asia have no sewerage system at all, including many cities with one million inhabitants or more; and 90 per cent of the sewage collected is discharged without treatment, thereby polluting the soil and the water in the area. Refuse collection services are inadequate or non-existent in most urban residential areas, an estimated 30–50 per cent of solid wastes generated within urban centres being left uncollected ... in cities that have sewerage and refuse collection services, rarely more than a small proportion of the population is served, typically those living in richer residential districts. In many rural areas, the fetching of water and fuelwood take up a substantial amount of the time of women and children, while such water is often contaminated and a shortage of fuelwood limits the number of hot meals each day. Children are kept out of school to help which explains high fertility preferences. Air pollution levels also damage health.

Source: UN Secretariat (1992) Population and the environment: an overview, in Population, Environment and Development: Proceedings of the UN Expert Group Meeting, convened as part of the substantive preparations for the International Conference on Population and Development 1994.

Box 10.7 Characteristics of Poverty in Developing Countries

Poverty: people living on less than US$1.00 per day with little access to skilled jobs, health care, adequate nutrition, clean and safe drinking water, sanitation and schools, with low life expectancy, and only one meal a day

Percentage of population living in poverty

Over 60%

Guinea-Bissau (a life expectancy of 38 years)
Niger
Zambia (a life expectancy of 46 years)

Over 40%

India (more than half the population)
Kenya
Nepal

Continued on page 278

Continued from page 277

Peru
Rwanda
Uganda
Zimbabwe

Over 20%

Brazil
China
Ethiopia
Nigeria
Philippines
South Africa

Less than 20%

Chile
Colombia
Cote d'Ivoire
Czech Republic
Dominican Republic
Egypt
Indonesia
Jamaica
Kirgistan
Malaysia
Mexico
Morocco
Pakistan
Poland
Russia
Sri Lanka
Venezuela

NB Since the mid-1970s poverty in East Asia has declined from 6 out of 10 households to 2 out of 10 households. Poverty has been largely eradicated in the economies of Hong Kong, South Korea, Singapore and Taiwan, with sharp declines in Indonesia and Thailand. The decline in poverty has been described by the World Bank as completely unprecedented in human history. Concern has shifted to growing inequalities.

Source: World Bank and Britannica 1997.

According to the *World Health Report 1998*, average life expectancy had reached 66 years by 1998 – arise of more than a third since 1955. Childhood mortality had fallen dramatically. In 1955, over 1 in 5 children died before their fifth birthday; now fewer than 1 in 12 die. Infectious diseases such as AIDS, tuberculosis, malaria and others were, however, still responsible for a third of

Table 10.4 Major environmentally induced pathologies

Pathology	Environmental cause(s)	Estimated global burden
Respiratory diseases	Indoor and outdoor air pollution	Several hundreds of millions
Diarrhoeal diseases (cholera, typhoid, dysentery, etc.)	Contaminated food and water	3.1 million infant and child deaths per annum (1995)
Intestinal parasites	Unsafe food	Several hundreds of millions; 135 000 deaths (1995)
Malaria	Natural endemicity	2.1 million deaths per annum (1995); 267 million cases per annum (1995)
Tuberculosis	Poor hygiene, overcrowding, malnutrition	3.1 million deaths per annum, among 20 million cases (1995)
Malnutrition	Food shortages, poor diets, social factors	Several hundreds of millions
Smallpox		Eradicated, with six other diseases including polio targeted for eradication within a few years

Sources: WHO and WHO Commission on Health and Environment.

the 52.2 million deaths in 1997. Cancer and cardiovascular disease killed more people in developing countries than in developed countries, but were responsible for only a third of the deaths in poor countries, compared with two-thirds in the rich countries.

10.4 Urbanization

Urbanization is a process characterized by the movement of people from rural to urban areas or cities, where people are permanently concentrated in densely populated, often relatively small, districts. The UN has recommended that countries should regard all places with more than 20 000 inhabitants living close together as urban. In fact, nations compile their statistics on the basis of many different standards. In the USA, for example, an urban place means any locality where more than 2500 people reside. But whatever the definition, the course of human history has been marked by a process of accelerated urbanization.

However, it was not until the time of classical antiquity that cities with more than 100 000 inhabitants could be found, and even these did not become common until the sustained population growth and industrialization of the last two hundred years. Indeed, even in 1800 less than 3% of the world's population was living in cities of 20 000 or more; this had increased to about 25% by the mid-1960s and to about 40% by 1980.

The greatest city of antiquity was Rome which ultimately covered some 10 km^2 and had at least 800 000 inhabitants. The heart of the Roman Empire, Rome became the site of grandiose palaces, temples, public baths, theatres, stadiums such as the Colosseum, and many other public buildings. A system of aqueducts brought drinking water from the hills up to 70 km away, while within the city water was distributed to individual houses and many fountains. The ultimate decline of the Empire saw Rome shrink to some 30 000 people.

Medieval towns were rarely as large as Rome, even London having fewer than 100 000 citizens during that period. However, the Industrial Revolution, beginning in the late 18th century gave fresh impetus to the process of urbanization. The new factories needed large pools of labour immediately to hand and many specialized workers. Immigrants from impoverished rural districts poured into the growing towns and cities, often leaving rural slums to live in crowded city slums awash with refuse, disease, rodents and other insanitary conditions.

Global urbanization is set to continue, with increasing tempo in the developing world. Almost half the population of the world is already urban. In the 21st century it is expected that the majority of the world's population will live in urban centres and megacities. Indeed, by 2025, it is expected that 8 out of 10 residents of the industrialized countries will live in cities and towns, though the larger complexes may have stabilized or declined.

Urbanization processes have been quite rapid in the developing countries. In 1950, only some 17% of the total population lived in urban areas. This proportion had grown to 26% per cent by 1970 and 34% by 1994. By 2025, over half the population of developing countries will live in urban areas (see Table 10.5 and Box 10.8).

Box 10.8 The UN Centre for Human Settlements on City Growth

. . . from both an environmental and a socio-economic point of view, migration to urban settlements is the most desirable option. Although the high density of human activity . . . leads to special environmental problems, such as congestion and pollution, it is also true that cities are better places than remote and unplanned rural areas to mitigate and control the environmental effects of their development. City-dwellers have much more leverage than isolated rural-dwellers to exert pressure on the Government, local authorities and industry to protect their own health and to limit the impact of activities that are damaging to the environment.

Source: The UN Centre for Human Settlements (1994) Cities in the developing world, in Population, Environment and Development: Proceedings of an Expert Group Meeting, UN, New York.

Table 10.5 *Urban population, developed and developing countries, 1950–2025 (millions)*

Year	Developed countries		Developing countries	
	Urban	Percentage of total population	Urban	Percentage of total population
1950	448	53.8	285	16.9
1960	571	60.5	459	22.1
1970	699	66.6	675	25.5
1980	798	70.2	972	29.3
1990	876	72.7	1385	33.9
2000	945	74.8	1972	39.5
2010	1003	76.8	2733	46.5
2020	1050	78.4	3599	53.5
2025	1068	79.0	4051	56.9

Source: UN Centre for Human Settlements (1994) Nairobi.

In 1960, more than one-third of city dwellers world-wide lived in cities with fewer than 100 000 inhabitants, while more than one-third lived in cities with populations of more than one million. By the year 2000, almost half the total urban population of the world lived in urban agglomerates of a million inhabitants or more. By the year 2000, 18 of the 24 world cities of over 10 million inhabitants were in developing countries. Further, the largest agglomerations in the developing world will continue to grow. Research has indicated no relationship between quality of life and city size (UN Centre for Human Settlements 1994). In both urban and rural areas, the poor have no access to safe drinking water, elementary sanitation, and primary health-care services. Informal urban settlements are often located in hazardous, flood-prone and unhealthy sites, where overcrowding is compounded by exposure to communicable diseases, and garbage dumps constitute a permanent health hazard as well as a source of income.

Table 10.6 lists the world's 25 largest metropolitan areas; note that London and Rome no longer make it onto the list. Box 10.9 quotes Suzuki and McConnell on the social fabric, while Box 10.10 delivers another quote from *The Economist*.

10.5 World Food Production

Throughout the 1960s, world food production outpaced the growth of population, yielding on average in most places an increased consumption per head. Higher crop yields were achieved through the adoption of new varieties of rice, wheat, and other grains associated with heavy applications of fertilizers and improved irrigation. In fact, during and following the 1960s, yields in several countries increased dramatically, the benefits being felt in India and Pakistan, the Middle and Far East, North Africa and Latin America. However, events of the early

Box 10.9 Suzuki on Population

For thousands of years, small communities of people ensured relative tranquillity while providing for the social needs of their communities. The explosive rate at which our species has been converted to an urban creature has been accompanied by a deterioration of the social fabric that held people together.

Source: D. Suzuki and A. McConnell (1997) The Sacred Balance, Allen and Unwin, Australia, p. 174.

Box 10.10 The Developing World

A World Bank study last year put the cost of air and water pollution in China at $54 billion a year, equivalent to an astonishing eight per cent of the country's GDP. Another study estimated the health costs of air pollution in Jakarta and Bangkok in the early 1990s at around ten per cent of these cities' incomes. These are no more than educated guesses, but whichever way the sums are done, the cost is not negligible.

Another recent World Bank study found that across a range of Asian countries, including Thailand, Indonesia and China, the cost of a simple set of environmental measures – such as phasing out lead in petrol and investing in clean water supplies – would be much lower than the value of the benefits to human health.

Source: Dirt poor: a survey of development and the environment, The Economist, 21 March 1998.

1970s shook the conviction that the threat of Malthusian-type disasters had finally been overcome.

In 1972, there were serious climatic and economic setbacks for much of the world. Severe droughts occurred in Russia, India, South-east Asia, Australia, Central and South America and the Sahel region of Africa.

An El Niño event devastated Peru's protein-rich anchovy industry. The reduction of the cool upwelling water off the western coast of South America reduced the availability of plankton; when the plankton decrease so do the anchovies.

Grain supplies from many major grain-producing areas were depleted. The resulting famines debilitated or killed tens of millions of people, the total number of deaths in India and Bangladesh alone attributable to the bad harvests was at least a million. The shortfalls in production led to a 3% drop in global grain production. When combined with a growing need for food and an annual population growth of 2%, this superficially small drop in production had grave consequences.

Total grain yields began dropping in 1973, down by roughly 4.5%, while harvested areas increased only slightly. Consequently 1974 also saw reductions in world-wide grain production; this occurred when grain stocks were the lowest

Table 10.6 *The world's 25 largest metropolitan areas by population, 1995*

	Estimated population (millions)
Tokyo, Japan	27.9
São Paulo, Brazil	16.4
New York City, USA	16.3
Mexico City, Mexico	15.6
Bombay, India	15.1
Shanghai, China	15.1
Los Angeles, USA	12.4
Beijing, China	12.4
Calcutta, India	11.7
Seoul, South Korea	11.6
Jakarta, Indonesia	11.5
Buenos Aires, Argentina	11.0
Tianjin, China	10.7
Osaka, Japan	10.6
Lagos, Nigeria	10.3
Rio de Janeiro, Brazil	9.9
Delhi, India	9.9
Karachi, Pakistan	9.9
Cairo, Egypt	9.7
Paris, France	9.5
Manila, Philippines	9.3
Moscow, Russia	9.2
Dhaka, Bangladesh	7.8
Istanbul, Turkey	7.8
Lima, Peru	7.5

Source: UN Population Division; Britannica Year Book 1997.

in decades. However, near the end of the 1970s, a number of years of good weather returned. The year 1980 proved disappointing, with hot and dry weather in the USA, and poor harvests in Russia as well. The climatic events of the 1970s and the 1980s reveal the vulnerability of the human race to the adverse effects of abnormal weather; climate is not a constant.

Also certain regions present a continuing problem. In the Sahel, the semi-arid region of western Africa extending from Senegal to the Sudan and bordering the edges of the arid Sahara Desert, the loss of human life by starvation and disease has been calamitous. The Sahel has natural pasture with short grass and tall, herbaceous perennials, providing forage for animals including camels, grazing cattle, and sheep. The area has none the less suffered overstocking and overfarming. At least eight months of the year are normally dry, while intermittent severe droughts worsen the situation. Severe famine and drought afflicted the Sahel in 1970–73 and again in 1983–85, while malnutrition remains endemic.

In 1996, a world food crisis was narrowly averted with the recovery of grain production. The increase in grain production was widespread among developing nations. The year 1996 was also noteworthy for the World Food Summit where nations committed themselves to efforts to reduce the number of undernourished people in the world by half by the year 2015.

World agriculture in the 1990s was also influenced by underlying market forces; world markets for food and animal feed had become more integrated. Domestic agricultural policies underwent change and trade barriers were reduced.

The World Trade Organization (WTO), a body created by the Uruguay Round of the General Agreement on Tariffs and Trade (GATT), to give expression to and co-ordinate that agreement towards freer trade, came into operation in 1995. Producers and consumers in many nations began to respond more quickly to world market forces. Large areas of cropland were transferred from other uses to grain production in 1996.

Markets began to grow, especially for meat, fresh fruits and vegetables. China experienced a growing demand for meat, which expanded the demand for feed grains and oilseed meal.

However, a survey by the UN Food and Agriculture Organization (FAO) showed that in the early 1990s there were about 850 million people with inadequate access to food, down slightly from the 900 million people of 20 years earlier, even though the populations of the developing countries had increased by 1.5 billion (1500 million) over that period.

The world's population, expected to increase by 50% by the year 2030, faced a contracting per head supply of tillable land and fresh water.

However, the 1990s witnessed the beginning of a new green revolution: genetically modified food, plants and bacteria. The original idea promoted by Monsanto was to develop genetically modified crops that are not harmed by large doses of glyphosate-based herbicide, or Roundup, a Monsanto product. Monsanto's first mutated crop was soya beans, sprayed with Roundup. The beans thrived while the weeds died. Once Monsanto had paved the way the other large agrichemical corporations such as Novartis, AgroEvo, Dupont, Zeneca and Dow joined in with their own genetically engineered products. In 1997, some 12 million ha were planted with mutated soya, corn, cotton and rapeseed, which was a substantial increase over the area planted in the previous two years. The area doubled again in 1998, while genetically modified versions of most other major crops were on the way. As agriculture accounts for some 65% of the global economy, an economic revolution appears to be at hand. The creation of herbicide-resistant crops boosts yields, without any apparent environmental side-effects. Nevertheless, many consumers have reservations about eating genetically modified food, although there might be little option in the future.

Some European countries are moving towards the labelling of genetically modified foods, and there is little doubt that the WTO will become involved. The USA is also grappling with this problem. Whatever the outcomes, it is likely that a new green revolution is to hand to keep world production ahead of world population.

The FAO World Food Surveys (FAO 1996, 1997) confirm that the per capita dietary energy supplies have continued to increase in the developing countries as a whole, with the result that during the two decades from 1969–71, the prevalence of food inadequacy declined. Some 20% of the total world population had inadequate access to food in 1990–92, compared with 35% two decades ago, notwithstanding the addition of 1.5 billion people to the population of developing countries during this period. The number of people with an inadequate food supply declined from 918 million in 1969–71 to 906 million in 1979–81 and to 841 million in 1990–92. However, currently, 1 in 5 people in the developing world face an inadequate supply of food.

Globally, there is an indication that the prevalence of weight deficiency among children under five is declining over time in a number of countries. The pattern is strongest in South Asia and Latin America, although the prevalence of weight deficiency is actually increasing in sub-Saharan Africa. The experience in other regions is mixed. At the start of the present decade, 2 in 5 children under the age of five in the developing world were stunted, and 1 in 3 were underweight. In absolute numbers, this meant that about 200 million children under five in the developing world were stunted in growth, with 180 million being underweight; almost 50 million were wasted. Hence the magnitude of the problem remains daunting. This must be compared with the incidence of obesity in the USA, and in many developing countries in the higher income groups. The aim of FAO remains the guarantee of food security for everyone, everywhere.

Comments on the world food supply by *The Economist* appear in Box 10.11 and again by Partha Dasgupta in Box 10.12. Box 10.13 records the response of Simon and Kahn to *Global 2000: A Report to the President*, demolishing the findings of that report.

10.6 Fisheries

An estimated 950 million people, nearly a fifth of the world's population, depend on fish as their primary source of protein (World Resources Institute 1996). Most

Box 10.11 World Food Production

The facts on world food production are truly startling for those who have heard only the doomsayers' views. Since 1961, the population of the world has almost doubled, but food production has more than doubled. As a result food production per head has risen by 20 per cent since 1961. Nor is this improvement confined to the rich countries. According to the Food and Agriculture Organization, calories consumed per capita per day are 27 per cent higher in the third world than they were in 1963. Deaths from famine, starvation and malnutrition are fewer than ever before.

Source: The Economist, 20 December 1997, p. 20.

Box 10.12 Economic Growth and Economic Analysis

... since the end of the Second World War, income per head has grown in nearly all poor countries. In addition, growth in world food production since 1960 has exceeded the world's population growth by an annual rate of about 0.6 per cent. This has been accompanied by improvements in a number of indicators of human well-being, such as the under-5 survival rates, life expectancy at birth, and literacy. In poor regions, this has occurred in a regime of population growth rates substantially higher than in the past ... but statistics on past movements of gross world income and agricultural production say nothing about the environmental resource-base, that is, the depletion of natural capital and soil.

Source: P. Dasgupta (1996) The Economics of the Environment, Beijer Discussion Paper No. 80, Stockholm.

Box 10.13 Global 2000: A Report to the President

A report to the President of the United States in 1980 prepared by a committee of the 'great and the good'. Global 2000 predicted that population would increase faster than world food production, so that food prices would rise by between 35 per cent and 115 per cent by the year 2000. By the close of the century, the world food commodity price index had fallen by 50 per cent.

The predictions stimulated a response by Simon and Kahn (Blackwell, 1984) entitled *The Resourceful Earth: A Response to Global 2000*, completely demolishing the findings of that report.

Sources: Based on comments by The Economist, 20 December 1997, p. 20; Pearce and Turner (1997) The historical development of environmental economics in *Environmental Management: Readings and Case Studies* (eds L. Owen and T. Unwin), Blackwell, Oxford.

consumers live in developing countries. According to FAO, 1994 produced the highest ever world fish catch, a 7.3 million tonne increase over the previous year, to reach 110 million tonnes. Three-quarters of this was for human consumption. China topped the list of fish-producing nations, followed by Peru, Chile, Japan, the USA, India, Indonesia, Russia, Thailand, South Korea, Norway and the Philippines, with 17 other significant producers.

World fish production comes from three sources: the marine catch harvested on the high seas and coastal waters; the inland catch from lakes and rivers; and aquaculture, both freshwater and marine. Aquaculture or fish farming comprises the rearing of fish, shellfish, and some aquatic plants under controlled and managed conditions. Production from this source constantly increases. Of the total global fish harvest, almost 80% is caught by marine fishing fleets, and 16% through aquaculture.

Between 1950 and 1989, the marine harvest increased nearly fivefold. In 1989, the marine harvest peaked though the global catch continued to increase due to the development of aquaculture. The growth of the marine harvest over those years reflected the steady growth in the number and size of fishing vessels, the sophistication of their fishing gear, and the increasing demand of a growing world population.

In 1993, the FAO estimated that more than two-thirds of the world's marine fish stocks were being fished at or beyond their level of maximum productivity. Indeed, a number of species were either already depleted from over-fishing or in imminent danger of serious depletion due to overharvesting. Many fish stocks were being fished at their biological limits.

Over-exploited species included halibut, haddock, cod, hake, redfish, orange roughy, bluefin tuna, albacore and swordfish. Over-fishing has been more severe in some regions than in others. For example, the North Atlantic has been subject to considerable fishing pressures causing steep declines in some fish, such as cod. The north-western Pacific off the Asian coast is another site of very heavy fishing. The effect has been a decline in high-value fish and a growth of catches of lower value, though many of these serve the aquaculture industry.

Over-fishing is not due solely to large-scale industrial fishing fleets but also due to small-scale fleets and subsistence fishing in the developing world, much fishing being done close to shore.

While over-fishing is the prime cause of decline, this emerges from the fact that most waters are open-access resources, open to anyone with a boat and some fishing gear. Fishing may continue to be profitable after the resource base has begun to erode. Fish only need to cover operating costs. The problem has also been aggravated through many nations subsidizing the construction of fishing vessels and fish-processing facilities. The global fishing fleet is probably 30% larger than it needs to be. This overcapacity has tended to aggravate the problem. The FAO has estimated that the total expenses of the world fleet exceed total revenues by nearly US$50 billion a year. A variety of subsidies tend to fill the gap. Reducing or adjusting subsidies to discourage new entrants and retire older vessels will be necessary to shrink the inflated fishing fleets. Individual transferable quotas (ITQs) are being tried around the world with a measure of success. In 1992, the UN successfuly banished the use of long drift nets.

In 1982, the UN Convention on the Law of the Sea became the legal framework for all fishing nations. In 1995, the UN Convention on Straddling Stocks and Highly Migratory Fish Stocks was formally adopted; it seeks to protect fish stocks and improve monitoring.

In 1995, FAO adopted the voluntary Code of Conduct for Responsible Fisheries, which outlines the principles and standards for fisheries management and development. The Code addresses fisheries management, fishing operations, aquaculture development, integration of fisheries into coastal area management schemes, post-harvest practices, and trade and fisheries research. The expression 'plenty more fish in the sea' is no longer apt. The condition of the world fisheries in the 1990s has left little doubt that the fisheries are under great stress.

The social consequences of a further decline in marine fisheries could be severe. The fishing industry employs, directly or indirectly, world-wide, some 200 million people. In eastern Canada, the closure of the cod fishery in 1992 meant some 40 000 workers losing their jobs. While the task of reducing the world's fishing fleets and regulating access to fish stocks may require painful economic and social adjustments, the cost of failing to do so will be much higher.

Box 10.14 sets out eight rules for the marine environment based on *The Economist* survey, *The Deep Green Sea.*

Faced with depleting fish stocks and alarmed at the potential effects on economies, in 1998, some 80 nations reached a voluntary agreement to limit fishing world-wide coupled with protection for sharks and seabirds. The accord received the support of all key fishing nations, including the US.

Species at greatest risk include Atlantic cod and redfish, haddock, anchovy, swordfish, tunas such as bluefin, bigeye and albacore, and sharks.

Box 10.14 Eight Rules for the Marine Environment

- Stop subsidizing the destruction of the marine environment. A study by the World Bank has estimated subsidies as worth up to US$16 billion a year, by way of direct aid to build boats and tax concessions.
- Limit access to marine resources to prevent over-fishing. The FAO estimates that the world's catch could be 10–20% greater with good fisheries management.
- Where possible, put marine resources into private hands by, for example, quotas. Transferable quota systems have been tried in New Zealand, Australia, Chile, Iceland, the Netherlands and the USA.
- Recognize that, except in emergencies, governments and international organizations make poor administrators. Follow the FAO Code for responsible fisheries.
- Seek independent assessment of stocks through the International Council for the Exploration of the Sea.
- Create reserves to preserve important habitats; reserves are now seen as a way to preserve fish stocks around the world. There is a movement in the USA to get 20% of coastal waters set aside as reserves, within 20 years.
- Limit the effects of pollution from mining, exploration, shipping, sewage effluent, oil and gas production.
- Promote stewardship of the oceans, with all the privileges and responsibilities that implies.

Source: Adapted from The Economist survey, The Deep Green Sea, 23 May 1998.

10.7 Water Quality and Sanitation

Water resources embrace the entire range of natural waters that are found on Earth and are of potential benefit to humanity. The resources readily available for use are the waters of the oceans, rivers, lakes, precipitation, groundwater and aquifers, glaciers and snowfields. Groundwater accounts for over 95% of the Earth's usable freshwater resources and plays an important part in maintaining soil moisture, stream flow and wetlands. Over half the world's population depends on groundwater for drinking water supplies. In fact, the contamination of groundwater supplies is often not detected until noxious substances appear in drinking water. Most of these contaminants are derived from agricultural, urban and industrial uses, including fertilizers, pesticides, septic tank systems, street drainage, and air and surface-water pollution. Effective conservation of groundwater supplies requires the integration of land use and water management.

Water is plentiful around the world, though not always in the right places. It is vital to life, participating in virtually every process that occurs in plants and animals, and in the environment. As a liquid, water is colourless, tasteless and odourless, with a marked ability to dissolve many other substances. Water is the working fluid of steam systems, and a heating and cooling medium. Impurities and pollutants may impair the efficient use of water, and threaten public health.

A water-resource strategy is a spectrum of measures to conserve surface water and groundwater; to ensure the continued availability of water for growing domestic, commercial and industrial uses; to ensure sufficient water for natural ecosystems; and to promote sustainable development. Notwithstanding the ready availability of water in most countries through most seasons, some areas suffer acute shortages, notably in areas of drought, and in arid and semi-arid areas, where rainfall may be quite inadequate and desalination plants are required.

The human use of natural waters has increased steadily over the centuries, due to a progressive increase in population and a progressive increase in consumption per head of the population. While a substantial proportion of the world's population spends much time obtaining water for home consumption, others with ready access to piped and treated water enjoy almost unlimited access for domestic and industrial use. According to the World Bank, the USA uses 1870 m^3 of water per person per year. Likewise, Canada, with a large farm sector, is the second largest consumer, using 1602 m^3 per head per year. In descending order are Italy, Australia, Belgium, Mexico, Russia, Spain, Japan, France, India, Germany, Greece, Holland and China. Britain is well down the list at 21st, with a consumption per head of only 205 m^3 per head. However, Britain uses only 16.6% of its total available supply. Assessing consumption in countries without piped supplies presents greater difficulties, while the threats to public health are much greater.

In 1977, the Global Water Quality Monitoring Program was introduced by UNEP, WHO, WMO and UNESCO, as an important arm of the UN Global Environmental Monitoring System (GEMS), one of the components of Earthwatch.

GEMS comprises five closely linked major programmes, each containing various monitoring networks relating to pollution (air and water), climate, health, ecology and oceans.

The Water Quality Monitoring Program was the first of its kind to address global issues of water quality through a network of monitoring stations in rivers, lakes, reservoirs and groundwater on all continents. Technical co-operation with developing countries was the principal focus of the first phase, resulting in the expansion of monitoring networks into many countries. Phase 2 of the programme (1990–2000) concentrates on data interpretation, assessment of global and regional water quality issues, and water quality management issues. The number of monitoring stations is being increased, baseline monitoring of pristine water is being introduced, as well as trend stations to follow long-term changes in water quality in relation to industrial, agricultural and municipal pollution and various land uses; and global river flux stations at the mouths of major rivers to determine the annual cycles of critical pollutants and nutrients from river basins to the oceans. The long-term objectives of Phase 2 include providing governments, the scientific community and the public with water quality assessments and information relating to the health of human populations, aquatic ecosystems, and the global environment. It is also intended to strengthen water quality monitoring networks in developing countries, including the improvement of their analytical capabilities and the quality of the data they produce. Findings so far reinforce the findings of the UN Secretariat.

The decade 1980–90 was declared by the UN to be the International Drinking Water Supply and Sanitation Decade, the aim being to bring clean and safe water and adequate sanitation to everyone by 1990. This was to involve bringing a water supply to 1800 million people, and providing sanitation for 2400 million people. The programme was to cost more than US$600 billion at 1978 prices. Although the goals for the decade were not met, the number of urban people in the developing world with access to adequate water increased by about 80%, and the number of people with adequate sanitation facilities increased by 50% (World Resources Institute 1996). However, the rapid rise in urban populations offset these gains, and in 1994 more than 220 million urban dwellers (13% of the developing world's urban population) still did not have access to a safe and reliable water source, while more than 420 million (25% of the developing world's urban population) did not have access to sanitation services.

The decade made clear that the high-cost water and sanitation systems adopted widely throughout the developed world would not work in the developing world. Piped water and flush toilets inside homes are beyond reach, and lower-cost alternatives have been adopted. Pit latrines with concrete caps have come into use. Such systems cost about one-tenth of conventional systems. Modified drainage systems with septic tanks and houses connected to each other have come into vogue. But problems abound, including pricing policies and cost recovery.

In 1992, UNEP inaugurated the International Environment Technology Centre, dedicated to the transfer of technologies relevant to conservation of the environment. The centre has two offices in Japan, including one in Shiga focusing on the management of freshwater lakes and reservoirs and their basins (UNEP 1994a).

10.8 The Risk Transition

Most developing nations are experiencing substantial reductions in some kinds of risk to human health and welfare such as infectious diseases, intestinal infestations, malnutrition, respiratory diseases, malaria and other tropical diseases. Such declines are associated with lengthened life spans, declining infant mortality rates and deaths under five, and a fall in the incidence of ailments at all times of life. The downward trends are at different rates in different places. At the same time, industrialization and urbanization bring increases in the number of modern hazards. Figure 10.1 illustrates the decline in traditional afflictions and the emergence of modern risks. These modern risks include cardiovascular problems, cancers, the risk of death on the roads, homicides and suicides, poisonings, and degenerative conditions. However, modern risks do not approach the levels of traditional risks; the quality of life has dramatically improved (Smith 1988). The ages of pestilence and famine, and life expectancies of 20–40 years, have been substantially though not entirely eliminated.

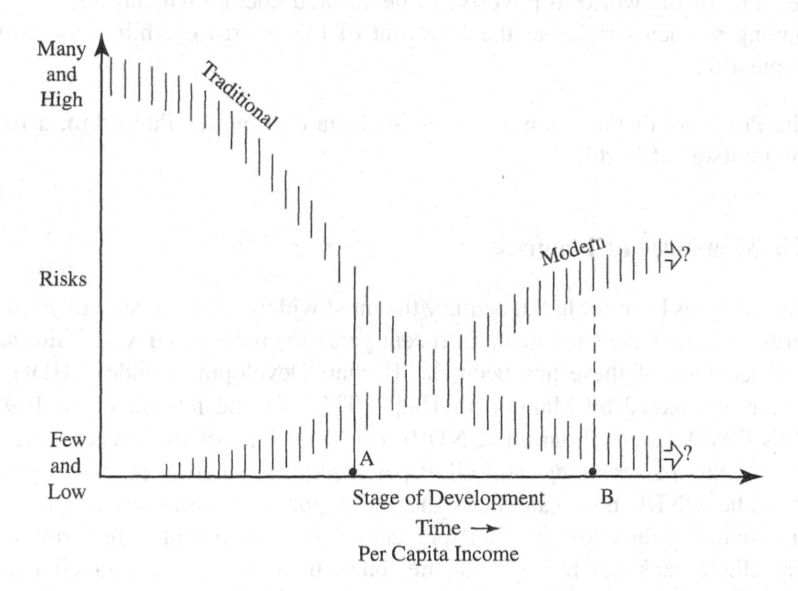

Figure 10.1 *The risk transition: at point A, traditional risks are dropping rapidly, and modern risks growing; at point B, traditional risks have been largely eliminated*

A stage may be reached where traditional risks are all but squeezed out, but modern risks keep rising with extended life spans. People in many developed and some less developed countries look forward to life spans of over 80 years.

10.9 The Independent Commission on Population and Quality of Life

The Independent Commission on Population and Quality of Life was launched on the initiative of several organizations that had been involved for many years in the field of population studies and family planning. Two preparatory meetings were held which led to the creation of the Commission in 1992. The purpose of the Commission was to develop and disseminate a fresh vision of international population matters. The commission met every six months in various centres; it also held regional meetings in southern Africa, western Africa, North America, southern Asia, Latin America, South-east Asia and Eastern Europe. Its report, *Caring for the Future: Making the Next Decades Provide a Life Worth Living* (ICPQL 1996), proposed the following:

1. targets and timetables to improve the standards of health care and education throughout the world;
2. a tax on international financial transactions to raise the necessary funding;
3. a rejection of the overreliance on free-market economics that condemns sections of the world to poverty, ill health and unemployment; and
4. putting women's rights at the forefront of the effort to stabilize the world's population.

The President of the Commission was Maria de Lourdes Pintasilgo, a former prime minister of Portugal.

10.10 Measures of Progress

While GDP has been for half a century the most widely used measure of economic progress, efforts have been made in recent years to create and develop alternative yardsticks. One of these has been the Human Development Index (HDI). The HDI was pioneered by Mahbub ul Haq (1934–98) and introduced in 1990 by the UN Development Program (UNDP). It is a cocktail of life expectancy, adult literacy, years of schooling, and GDP per capita measured at purchasing-power parity. The UNDP now calculates the index for most countries in the world. Some countries rank low in GDP per capita, but much higher in terms of the index; others rank much lower on the index than their per capita GDP might suggest. For rich countries, the differences in ranking tend to be smaller, there being little difference in the cases of Canada, Japan, France, the USA and Britain.

The poorer countries reveal much greater diversity. While the HDI represents a distinct improvement over GDP alone as a measure of welfare, it does not take account of environmental degradation. Hence the quality of life could improve while the long-term resource base deteriorates (UNDP 1990). In quite recent years, the HDI has attempted to embrace gender issues and the struggle for women's rights. On these issues, the Nordic countries score the highest.

An alternative index is the Daly–Cobb Index of Sustainable Economic Welfare (Daly and Cobb 1989). It is a comprehensive indicator of well-being that takes into account not only GDP but distributional inequality and environmental degradation such as the depletion of non-renewable resources and loss of farmland, loss of wetlands, global warming and the costs of air and water pollution. However, the index has been calculated only for the USA; in any event few countries have enough information available. The index remains the most sophisticated indicator of progress, although of restricted application.

Recently, the World Bank has suggested new indicators, one of which relates to wealth. However, the Bank's estimates of wealth go far beyond the sources of wealth traditionally considered. They include estimates of natural capital such as the value of land, water, timber, subsoil assets, human capital and economic capital. Such an assessment is in effect, a balance sheet of a nation's primary assets.

The importance of including natural and human capital is that these quantities may be significantly larger than produced assets. Natural assets exceeded the produced assets in half the countries assessed. Australia emerged as the richest nation in the world, much to the surprise of Australians who have long recognized their place as about half-way down the OECD scale of material welfare (World Bank 1995a). The Bank is also developing an index of genuine savings.

In 1998, efforts to improve the measurement of real progress were augmented by the CSIRO publication, *Measuring Progress: Is Life Getting Better?* (Eckersley 1998). This work comprises contributions from 25 leading economic, social and environmental scholars. In Australia, as elsewhere, progress has been defined in material terms as growth in GDP. However, progress in these terms has been matched by a growing belief that things are getting worse for many people, with increasing rates of crime, drug abuse, suicide, depression, the increasing variations in income between rich and poor, and the all-pervading sense of job insecurity. Today, corporate success is measured by profits on the one hand and the progressive retrenchment of the workforce on the other.

This study has suggested yet another index, a Genuine Progress Index (GPI). The GPI is calculated by adjusting the basic economic measure GDP for a variety of environmental factors and the growing disparity of income. It appears that in Australia, the GPI has stopped rising, although the GDP continues to climb. However, the value judgements that go into this calculation must be recognized. It is indeed difficult to reduce to a few figures social factors of a vague, intangible and subjective nature.

While GDP may remain a rough guide to material progress, it can never indicate who gains and loses, nor attempt to measure happiness or social well-being.

Box 10.15 conveys the views of John Maynard Keynes on the standard of living over the last 4000 years, with particular reference to the last 200 years. Box 10.16 quotes from *The Communist Manifesto* of 1848 in which Marx and Engels recognize the achievements of capitalist society. The quotations provide a perspective within which subsequent developments may be placed.

Box 10.15 Keynes on the standard of living over 4000 years

> From the earliest times of which we have record – back to say two thousand years before Christ – down to the beginning of the eighteenth century, there was no very great change in the standard of life of the average person living in the civilised centres of the earth. Ups and downs certainly. Visitations of plague, famine, and war. Golden intervals. But no progressive, violent change ... From the sixteenth century, with a cumulative crescendo after the eighteenth, the great age of science and technical inventions began, which since the beginning of the nineteenth century has been in full flood ... and the growth of capital has been on a scale which is far beyond a hundred-fold of what any previous age had known.

Source: John Maynard Keynes (1883–1946) in Essays in Persuasion (1951) Rupert Hart-Davis, London, p. 360.

Box 10.16 The Communist Manifesto 1848

The Communist Manifesto was an exposition by Karl Marx (1818–1883) and Friedrich Engels (1820–1895) on the failings of the capitalist system and the inevitability of a socialist society established through the dictatorship of the proletariat (the working classes). It is perhaps the best known piece of communist literature. It blames the bourgeoisie (the entrepreneurial or capitalist class) for the promotion of naked self-interest as the governing influence in society, everything being reduced to an exchange value. Yet the revolutionary role of the bourgeoisie in the framework of history received full recognition:

> The bourgeoisie, during its rule of scarce one hundred years, has created more massive and more colossal productive forces than have all preceding generations together ... It has been the first to show what man's activity can bring about. It has accomplished wonders far surpassing Egyptian pyramids, Roman aqueducts, and Gothic cathedrals; it has conducted expeditions that put in the shade all former Exoduses of nations and crusades.

This tribute to capitalism, a system dominating much of the world today, remains a striking feature of a short work dedicated to the overthrow of that system.

10.11 The Automobile

Of all the means of personal transport available today by road, air, rail and water, the car emerges as the most controversial. From the early days the car has been the dream of millions, the key to personal freedom and mobility. To have a small house with a lock-up garage remains a high priority for many. Indeed there are today over 550 million vehicles in the world, a figure which could double in the next 30 years. At one time, in say the 1930s, proximity to a railway station or bus stop ranked high in the list of needs, with the bicycle still in vogue and the legs doing a great deal of walking. Gradually the world changed for many with the acquisition of a small car: the ability to live further out of town, independence of other people's timetables, immediacy and comfort all became important. The absence of a vehicle produced a sense of inferiority, of not being quite as successful in life. The automobile industry expanded immensely in the industrial world, yet as Box 10.17 reveals, the industry, in mass production terms, is just 100 years old. It helped to create that transformation of society which we know today. Box 10.18 gives a Friedman perspective.

Box 10.17 History of the Automobile

1769 Nicolas-Joseph Cugnot of France designed and built the world's first true automobile, a heavy steam-powered tricycle, capable of 3.6 km h^{-1}
1876 Niklaus August Otto developed the four-stroke, gasoline-powered engine
1877 The Duryea Motor Wagon produced
1885 Carl Benz builds his first automobile
1889 The Daimler-Maybach car was developed by Gottlieb Daimler
1890 First American battery-powered car
1892 Levassor cars in production in France
1895 First motor race held, from Paris to Bordeaux
1896 Peugeot established
1898 There were 50 car manufacturers in the USA
1901 First Oldsmobiles marketed
1904 Studebaker cars; Rolls Royce established
1906 The Vanderbilt Cup Race; British Austin and Wolseley cars in production; Fiat incorporated in Italy
1907 Peking to Paris motor race
1908 Henry Ford markets the Model T-Ford, sales reaching 10 000; General Motors Corporation founded
1913 Ford moving assembly line; Morris Oxford appears in Britain
1919 First Citroën car produced
1920 Bentley Motors established in Britain
1925 Chrysler Corporation established, with Walter Chrysler as president
1926 Daimler-Benz formed (first company to use diesel engines in cars); Volvo established

Continued on page 296

Continued from page 295

1933 Toyota Motor Corporation established
1937 Volkswagen introduced
1948 Honda Motor Company founded
1950 Porsche sports car; Dunlop announced the disc brake
1951 Buick and Chrysler introduced power steering
1952 Rover's gas turbine car sets a world speed record
1953 Haagen-Smit identifies motor vehicle exhaust gases as primary source of photochemical smog in Los Angeles basin
1954 Campbell achieves land speed record; Bosch introduced fuel injection for cars
1963 Positive crankcase ventilation introduced in the US
1965 Ralph Nader publishes *Unsafe at any Speed,* marking the beginning of safety standards
1966 California introduces legislation to reduce pollution by cars
1967 Energy-absorbing steering columns
1972 Safety tyres introduced; seat belts required in the US
1980 Lean-burn engines introduced
1985 Lead-free gasoline introduced for use with catalytic converters
1990 Los Angeles pollution much reduced but still too high
1996 2000 classic American cars parade in Detroit to celebrate the 100th anniversary of the American motor industry
1997 500 million cars serving 5.6 billion people
1998 50 million cars roll off the assembly lines throughout the world
2020 Projected world car population of one billion

Box 10.18 The Friedmans on Environmental Achievement

If we look not at rhetoric but at reality, the air is in general far cleaner and the water safer today than one hundred years ago. The air is cleaner and the water safer in the advanced countries of the world today than in the backward countries. Industrialization has raised new problems, but it has also provided the means to solve those problems. The development of the automobile did add to one form of pollution; but largely ended a far less attractive form.

Source: M. Friedman and R. Friedman (1980) Free to Choose: A Personal Statement, Macmillan, Australia, p. 218.

Another surprising feature of the car is its distribution. While in the USA, 60% of households own two or more cars, and 20% own three or more, the average ownership for OECD countries is about 400 per thousand residents, a more modest figure. However, in the developing world there are about 70 cars per thousand residents in Latin America, 29 cars per thousand people in Asia and the Pacific, and about 15 cars per thousand in Africa. Further details of this maldistribution are revealed in Table 10.7: there are one or two people per car in

Table 10.7 *Motor vehicles: an unequal distribution*

	No of vehicles (millions)*	Persons per vehicle
United States	210	1.2
Canada	18	1.6
Mexico	7.5	7.2
Britain	23	2.5
Germany	43	1.9
France	33	1.9
Italy	32	1.8
Spain	17	2.3
Sweden	4	2.3
Poland	9	4.3
Turkey	4	16
Hungary	3	4.0
Rumania	2.5	9.5
Russia	2.5	9.5
Brazil	15.5	10
China	9	131
India	6	163
Saudi Arabia	3	6
Japan	64	2
South Africa	5.5	7
Indonesia	4	51
Malaysia	3	7
Egypt	1.5	38
Nigeria	1.3	66
South Korea	8	6
Taiwan	5	4.5
Thailand	4	15
Philippines	2	32
Australia	11	1.7
New Zealand	2	1.8
World	553	

*Includes automobiles, trucks and buses.
Source: Britannica Year Book.

the most advanced countries, and around 150 people per car in China and India. One knows that in China the bicycle thrives and in Taiwan the motor-scooter is paramount, but in emerging countries such as Thailand and South Korea, the traffic jams of cities such as Bangkok and Seoul are now legendary.

The car everywhere remains a symbol of ambition, and as Box 10.19 seeks to demonstrate, the attractions of the car are numerous even recognizing current and emerging problems. The essential problem with the car is that it does not carry a full share of its social costs, and improvements in such characteristics as its energy consumption, noise control, pollution control and safety features tend to be offset by increasing numbers. In 1996, 50 000 000 new vehicles were produced. The disadvantages of the car are set out in greater detail in Box 10.20.

Some indication of the lines of future policy are given by Ricardo Neves in Box 10.21. The confidence in the hypercar expressed by David Suzuki and Amanda McConnell in Box 10.22 should be modified by the fact that it needs to be tied down in a slight breeze and does not carry children and groceries. However, the vision of a carless society takes society back say to the 1920s. There are great problems about receding standards of living and there is no evidence that people would accept it in any country. The essential problem is to modify the less attractive aspects of the car, and promote the zero-emission vehicle as an increasing component of the total vehicle stock. See Box 10.23 on the fuel cell and the zero-emission car.

10.12 Official Aid

The Independent Commission on International Development issues was set up in 1977, under the chairmanship of Willy Brandt following his resignation from the German Government. The Commission, prestigious and independent, was hence known as the Brandt Commission.

Box 10.19 The Advantages of the Car

- Cars offer freedom and mobility
- The car is the only modern form of transport that will move precisely from point A to point B carrying passengers, children and groceries, at a time of the driver's convenience
- Cars are a cheap, economical, comfortable and relatively safe form of transport
- Cars offer the ability to make countless calls in a relatively short period of time
- Cars offer the ability to travel long distances at speed with few refuelling stops
- Cars offer the ability to go for a run, simply for the pleasure of driving around
- Cars can be a joy to look at and fun to drive
- Models can 'make a statement' about the driver
- Cars offer access to shopping malls with parking facilities
- Road-pricing policies, using electronic techniques, can do much to relieve congestion
- In 1994, the Program for a New Generation of Vehicles brought together the US government and the three major US manufacturers to share knowledge and produce new designs of vehicles with higher economy and virtually zero emissions, at a competitive price
- Considerable effort is being devoted to the development of satisfactory electric cars, in the US and Europe
- Fuel-cell cars, the most ambitious of the hybrid models, are also under development
- Public transport alternatives are unsatisfactory, poor or non-existent, and in any event do not take the passenger door-to-door, but only stop-to-stop

Box 10.20 The Disadvantages of the Car

- Cars and other vehicles are seen as the worst polluters of urban air and the largest producers of carbon dioxide, the principal factor in global warming
- As the number of cars progressively increases, delays, congestion and traffic jams ensue
- The benefits of lean-burning devices and catalytic converters, which reduce pollution markedly from individual vehicles, are collectively offset by increasing vehicle numbers
- Noise levels worsen
- Road-construction programmes split communities, absorb enormous amounts of land, and disrupt greenery
- There is evidence of respiratory diseases being caused by cars and trucks emitting fine particles and volatile compounds
- The number of cars is likely to double by the year 2030
- Space on the road is allocated largely without regard to market mechanisms, and clean air is supplied free to the polluting motorist
- There is no clear relationship between the benefits of a car trip and the social costs (pollution, noise, wear and tear, frustration to pedestrians, risk of accidents, road repairs, overpasses)
- Taxes paid by motorists are paid on the vehicle, not the distance travelled; the true cost of each trip is avoided
- Studies by the World Resources Institute put the social cost of driving in the US at $300 billion a year or 5.3% of GDP. This estimate covers the cost of building and repairing roads, loss of economic activity due to congestion, the cost of illnesses caused by air pollution, and medical care for the victims of two million accidents a year
- Economic growth promotes car ownership and travel
- The fuel-cost of motoring has fallen dramatically in real terms in the last 10 years
- Britain's Department of Transport studies have shown that a 10% rise in fuel prices would cut consumption by only 3%, i.e. demand is inelastic
- Employers subsidize company cars and parking facilities
- Off-road, four-wheel-drive models are strongly in vogue but often damage the environment
- Biomass ethanol and methanol, natural gas and liquified petroleum gas have all been tried since 1973; they may cut emissions, but cost a lot more

The Brandt Commission had been set up at the instigation of, among others, the World Bank. The 20 members of the Commission were drawn from richer, northern industrial countries and poorer, southern countries. The prime purpose of the Commission was to recommend an emergency programme in respect of international trade imbalances and longer-term reforms.

The 1980 report of the Commission, *North–South: A Program for Survival*, envisaged a large transfer of resources to the less developed countries to reach 0.7% of the GDP of the rich countries by 1985, and 1.0% by the year 2000, together with other measures by way of increased lending and major reforms in the international economic system. The report recommended a new institution,

Box 10.21 Neves on Transportation

Cities need to begin to develop programs that will curtail car use and promote an integrated, environmentally sustainable urban transportation system with a clearly defined place for nonmotorized vehicles. Transferring the real cost of driving to car users instead of continuing to subsidize car ownership is an important concept to consider. In addition instead of continuing to expand road networks to meet the spiralling demand, cities need to find ways to reduce existing as well as future travel demand.

Source: Ricardo Neves, President of the Institute of Technology for the Citizen in Rio de Janeiro, Brazil, as quoted in World Resources 1996–97, p. 96.

Box 10.22 Suzuki and McConnell on the Car

Hypercars are capable of travelling 150 kilometres on a litre of gas and could allow us to use vehicles with greatly reduced ecological impact. This could buy time for the design and construction of living spaces for most of humanity that eliminate the need for cars altogether.

Source: D. Suzuki and A. McConnell (1997) The Sacred Balance, Allen and Unwin, Australia, p. 122.

Box 10.23 The Fuel Cell and the Zero-Emission Car

Fuel cells produce electricity by hydrogen and oxygen reacting together electrochemically, rather than by combustion. The only product of this reaction is water; there are no noxious pollutants such as carbon monoxide, oxides of nitrogen or carbon dioxide which contribute to the greenhouse effect. Fuel cells if applied to vehicles could make an invaluable contribution to ending the greenhouse effect and global warming. The fuel cell was developed by William Grove in 1839, but has been utilized only in expensive roles such as space travel, providing electricity and water to astronauts.

A fuel cell consists essentially of a fuel supply (hydrogen), an oxidant (oxygen, say from the air) and two electrodes (the anode and the cathode) on either side of an electrolyte. The latter is a material that conducts electricity by the passage not of electrons, but of electrically charged atoms, or ions. There are five types of fuel cells, but only two are near to becoming practicable propositions: these are the phosphoric-acid cell which is possibly suitable for commercial electricity generation, and the proton-exchange-membrane (or PEM cell) which may become the main way of powering automobiles. PEM cells go back to the 1950s, but until recently required very expensive platinum catalysts; these have now become very much cheaper through better deployment.

Continued on page 301

Continued from page 300

Both Daimler and Toyota have released small cars with prototype fuel-cell engines. The Toyota vehicle has a range of 500 km. Both vehicles use methanol rather than hydrogen; the hydrogen is produced on board by a special reactor. However, cells remain expensive to make compared with conventional petrol engines, and would need mass production to dramatically reduce costs. Both prototypes appeared on show at the 1997 Tokyo Motor Show. Future models may be hybrids, using petrol on the open road and electricity for city driving. It is claimed that competitive vehicles may emerge by 2005. A fuel-cell vehicle using methanol (from natural gas) might well reduce carbon dioxide emission from vehicles throughout the world by 50%. Only hydrogen offers zero-emissions.

a World Development Fund, with universal taxes flowing into its coffers. This was not the first attempt to establish channels of assistance to poorer countries, as individual countries had attempted to do so since the Second World War, but the Commission sought a world-wide commitment of a formal nature.

In 1992, at the UN Conference on Environment and Development, the developed countries reaffirmed their commitment to reach the target of 0.7% of GDP for foreign aid as soon as possible; this proposal was incorporated into Agenda 21. It was also decided that the UN Commission for Sustainable Development would regularly monitor and review progress towards this target.

The actual performance of the richer nations has fallen well short of the nominated target. Box 10.24 indicates the actual performance of nations, the total reaching 0.3% of combined GDP, the lowest ratio since 1992. This was, however, eclipsed by the flows of private capital. Between 1992 and 1996 net official aid to developing countries from developed countries fell by 16% in real terms.

The US Agency for International Development manages the US foreign aid programme. The total aid disbursements in recent years for the US have been about 0.2% of GDP, below that for the OECD countries as a whole. On the other hand, some 80% of US aid is untied, i.e. recipients are not obliged to purchase from the US. This is one of the lowest percentages in the world.

China is still the largest recipient of foreign aid, receiving in 1996 some $2.6 billion. The only other countries to receive similar amounts were Egypt and Israel. In 1996, foreign aid to Mozambique amounted to 72% of its GDP; it is also important to Rwanda, Tanzania and Zambia.

The UNDP has suggested that 20% of government expenditure in developing countries should be allocated to helping the poor meet their needs for food, water, sanitation, basic health care, family planning and education for their children. At the same time, industrialized countries were asked to allocate 20% of their development aid to meet these priority needs, as a minimum.

Box 10.25 sets out the views of Jeffrey Sachs, Professor of International Trade at Harvard and Director of the Harvard Institute for International Development.

Box 10.24 Official Aid as a Percentage of GDP 1995

Between 0.5 and 1.0%

Denmark
France
Netherlands
Norway
Sweden

Up to 0.5%

Australia
Austria
Belgium
Britain
Canada
Finland
Germany
Italy
Ireland
Japan
Luxembourg
New Zealand
Portugal
Spain
Switzerland
USA

In 1995, the rich countries of the OECD contributed a total of US$59 billion in foreign aid to poorer countries; the aid was worth 0.3% of their combined GDP, the lowest ratio since the UN announced a global target of 0.7% in 1972 and 1992. The total of official aid, export credits, and private-sector flows to developing countries rose to US$253 billion, an all-time high, with private flows of US$170 billion being by far the largest component.

Sources: OECD and The Economist.

10.13 Summary

The rapid growth of population since about 1750 has been due to improving conditions, reduced infant mortality rates, better diets, better health care, the success of immunization and vaccination programmes, the control of plagues and epidemics, the elimination of smallpox and the much reduced incidence of poliomyelitis and tuberculosis and childhood diseases generally, with consequent increases in life expectation in most countries, developed and undeveloped. Food supplies have kept apace though there is evidence of strain, particularly in respect

Box 10.25 Sachs on Official Aid

Aside from humanitarian emergencies ... the main function of aid should be to respond to the fact that private markets do not attend to many international public goods – goods that heavily influence the success or failure of the poorest countries. Much of the developing world is burdened by disease and environmental stresses that are deeply debilitating the lives of millions of people. Malaria, for instance, afflicts around 500 m a year, but global public spending on a vaccine has been less than ten cents a case in recent years ... similarly hookworm undermines the health of millions of people throughout the tropics, but almost no public money is spent on basic epidemiology or vaccine research. The developing world lacks basic scientific and technical means to address the environment, health, population and agriculture.

Source: J. Sachs (1998) Global capitalism: making it work, The Economist, 12 September, pp. 18–23.

of fisheries. An easing of the growth of the world's population is clearly essential. The population now approaches six billion, with a prospect of growing to 11 billion by the year 2080. The density of population per square kilometre of the land area of the Earth may reach 40 or more.

The secular process of urbanization will intensify, with many cities having more than 10 million inhabitants, and some exceeding 15 million. Some cities, such as Paris and London, once considered large, now appear quite modest in size.

However, poverty remains a persistent problem, despite solid progress. Although the proportion of those living in poverty declines every year, the absolute number of those in poverty remains stubborn. There is also the problem of the maldistribution of resources, nationally, regionally and internationally. Further, the gap between rich and poor tends to widen, rather than to narrow. In all countries, increases in wealth seem to gravitate to the top 10% of the population. The level of official aid appears to languish well below the level promised at international conferences. One explanation is that contributing nations have lost faith that aid reaches the victims, instead gravitating to the richest families and vested interests. However, progress in many essentials remains in evidence.

The automobile receives special attention in this chapter. Sir Winston Churchill once referred to the 'infernal combustion engine' as one of the prime evils of society. Yet almost everybody wants a car and regards public transport as the means of transport for those who have not quite made it. Reserved parking in the inner city has become the *sine qua non* of social success and personal grandeur. The zero-emission car is contemplated.

Some attention is devoted to the risk transition in evolving society, the Independent Commission on Population and the Quality of Life, measures of progress, and the choice of population for Australia. Australia, a very large and somewhat barren land, has a small population of barely over 18 million; population growth is governed largely by immigration. Current debate centres on the question of the optimal size of population and its distribution, and the abolition of the multicultural society that has emerged, reverting to a white-skinned ageing and declining population on the southern fringe of Asia, where China is an emerging superpower.

Further Reading

Cipolla, C.M. (1967) *The Economic History of World Population*, Penguin, Harmondsworth.

Clark, C. (1977) *Population, Growth and Land Use* (2nd edn), Macmillan, London.

Dasgupta, P. (1995) The population problem: theory and evidence, *Journal of Economic Literature*, **XXXIII** (December), pp. 1879–902.

Independent Commission on Population and Quality of Life (1996) *Caring for the Future: Making the Next Decades Provide a Life Worth Living: The Report of the ICPQL*, Oxford University Press, Oxford.

Lutz, W. (ed.) (1994) *The Future of World Population*, Earthscan, London.

Lutz, W. (1996) *The Future Population of the World: What Can We Assume Today?* Earthscan, London.

Myrdal, G. (1970) *The Challenge of World Poverty: A World Anti-poverty Program in Outline*, Pantheon, New York.

Myrdal, G. (1973) The economics of an improved environment, in *Who Speaks for Earth?* Norton, New York.

Ness, G. and Gholay, M. (1996) *Population and Strategies for National Sustainable Development: A Handbook on Population Appraisal and Planning Processes* (for IUCN and UNFPA), Earthscan, London.

Ramphal, S. and Sinding, S.W. (1996) *Population Growth and Environmental Issues*, Praeger, Westport and London.

UNDP (1998) *Human Development Report*, Oxford University Press, Oxford.

UN Group of Experts on the Scientific Aspects of Marine Pollution (1990) *The State of the Marine Environment*, Blackwell, Oxford.

UN Population Division (1997) *World Population Prospects*, UN, New York.

Weber, M. and Gradwohl, J. (1995) *The Wealth of Oceans*, Norton, New York.

Bibliography

Adams, J. (1995) *Cost–Benefit Analysis: Part of the Problem, Not the Solution*, Green College Centre for Environmental Policy and Understanding, Oxford.

Ahmad, Y.J., El Serafy, S. and Lutz, E. (1989) *Environmental Accounting for Sustainable Development*, A UNEP–World Bank Symposium, World Bank, Washington, DC.

Anderson, K. and Drake-Brockman, J. (1995) *The World Trade Organization and the Environment*, paper presented to the 2nd Environmental Outlook Conference, March, Australian Centre for Environmental Law and Department of the Environment, Sport and Territories, Sydney.

Anderson, M.S. (1994) *Economic Instruments and Clean Water: Why Institutions and Policy Design Matter*, OECD Public management meeting on alternatives to traditional regulation, May, Paris.

Arrhenius, S. (1896) On the influence of carbonic acid in the air upon the temperature of the ground, *The London, Edinburgh and Dublin Philosophical Magazine and Journal of Science*, series 5, **41** (251): 237–76.

Arrow, K. (1994) Discounting and Climate Change Prospects, a communication to the IPCC working group III, March, Montreux, Switzerland.

Arrow, K. (1996) Foreword, in *Rights to Nature* (eds S. Hanna, C. Folke and K.-G. Mahler), Island Press, Washington, DC.

Arrow, K., Bolin, B., Costanza, R., Dasgupta, P., Folke, C. *et al.* (1995) Economic growth, carrying capacity, and the environment, *Science*, **268** (April): 520–1.

Arrow, K.J. *et al.* (1996) *Benefit–Cost Analysis in Environmental Health and Safety Regulations*, AEI Press, Washington, DC.

Asian Development Bank (1994) *Climate Change in Asia*, Manila.

Australian Academy of Sciences (1995) *Population 2040: Australia's Choice*, Proceedings of an AAS symposium 1994, Canberra.

Australian Bureau of Agricultural and Resource Economics (1997) *The Economic Impact of International Climate Change Policy*. Research Report 97.4, Commonwealth of Australia, Canberra.

Barber, M.S. (1993) Inadequate IA threatens life, health and property: the Three-Gorges water control project feasibility study, in *Development and the Environment*, Proceedings of the 13th Annual Meeting of the International Association for Impact Assessment, 12–15 June, Shanghai, China, IAIA.

Barbier, C.P. (1963) *William Gilpin: His Drawings, Teachings and Theory of the Picturesque*, Clarendon Press, Oxford.

Barbier, E.B. (1989) *Economics, Natural Resource Scarcity and Development*, Earthscan, London.

Barde, J.P. and Pearce, D.W. (1991) *Valuing the Environment: Six Case Studies*, Earthscan, London.

Barney, G.O. (1980) *The Global 2000 Report to the President of the United States*, vols 1 and 2, Pergamon Press, Oxford.

Barrett, S. (1993) *The Theory of Property Rights: Transboundary Resources*, Beijer International Institute of Ecological Economics, Beijer Discussion Paper Series no. 44, Stockholm.

Bartelmus, P. (1990) *Sustainable Development: A Conceptual Framework*, Department of International Economic and Social Affairs, United Nations, October, New York.

Bartelmus, P., Lutz, E. and Scheinfest, S. (1992) *Integrated Environmental and Economic Accounting: A Case Study for Papua New Guinea*, Environment Working Paper No. 54, World Bank, Washington, DC.

Bartelmus, P., Stahmer, C. and Tongeren, J. van (1991) Integrated environmental accounting framework for an SNA satellite system, *The Review of Income and Wealth*, **37**(2), 111–48.

Baumol, W.J. (1968) On the social rate of discount, *American Economic Review*, **58**.

Baumol, W.J. (1972) On taxation and the control of externalities, *American Economic Review*, **62**(3): 307–22.

Baumol, W.J. and Oates, W.E. (1971) The use of standards and prices for protection of the environment, *Swedish Journal of Economics*, **73**(1): 42–54.

Baumol, W.J. and Oates, W.E. (1975) *The Theory of Environmental Policy*, Englewood Cliffs, New Jersey.

Baumol, W.J. and Oates, W.E. (1979) *Economics, Environmental Policy, and the Quality of Life*, Prentice-Hall, New Jersey.

Beckerman, W. (1972) Economists, scientists, and economic catastrophe, *Oxford Economic Papers*, **24**: 237–44.

Beckerman, W. (1974) *In Defence of Economic Growth*, Jonathan Cape, London.

Beder, S. (1993) *The Nature of Sustainable Development*, Scribe Publications, Sydney.

Bi, J. (1993) A study of the ecological impacts of the Three-Gorges project, in *Development and the Environment*, Proceedings of the 13th Annual Meeting of the International Association for Impact Assessment, 12–15 June, Shanghai, China, IAIA.

Bishop, R. (1982) Option value: an exposition and extension, *Land Economics*, **58**: 1–15.

Bishop, R. and Heberlein, T.A. (1979) Measuring values of extramarket goods: are indirect measures biased? *American Journal of Agricultural Economics*, **61**.

Bojo, J., Maler, K.-G. and Unemo, L. (1992) *Environment and Development: An Economic Approach* (2nd edn), Kluwer, Dordrecht.

Bolin, B. (1989) Changing climates, in *The Fragile Environment*, The Darwin College Lectures, Cambridge University Press, Cambridge.

Bos, E., Vu, M.T., Levin, A. and Bolatao, R. (1992) *World Population Projections 1992–93*, World Bank, Johns Hopkins University Press, Baltimore, MD.

Boulding, K.E. (1966) The economics of the coming spaceship Earth, in *Environmental Quality in a Growing Economy (ed. H. Jarrett)*, Johns Hopkins Press, Baltimore.

Boulding, K.E. (1985) *The World as a Total System*. Sage, Beverly Hills, CA.

Brady, G. and Bower, B. (1981) *Air Quality Management: Quantifying Benefits*, Research Report 7, East–West Centre, Honolulu, Hawaii.

Brandt Commission Report (1980) *North–South: A Program for Survival*, Report of the Independent Commission on International Development Issues, Pan Books, London.

Bresser, J. (1988) A comparison of the effectiveness of incentives and directives: the case of Dutch water quality policy, *Policy Studies Review*, **7**(3): 500–18.

Bromley, D.W. (ed.) (1995) *The Handbook of Environmental Economics*, Blackwell, Oxford.

Brookshire, D.S., Eubanks, L.S. and Randall, A. (1983) Estimating option prices and existence values for wildlife resources, *Land Economics*, **59**: 1–15.

Brookshire, D.S., Thayer, M.A., Schulze, W. and D'Arge, R. (1982) Valuing public goods: a comparison of survey and hedonic approaches, *American Economic Review*, **79**: 165–72.
Brown, L.R., Flavin, C. and French, H. (1998) *The State of the World*, Worldwatch Institute, Washington, DC.
Brown, R. and Singer, P. (1996) *The Greens*, Text Publishing, Melbourne.
Brown, S., Donovan, D., Fisher, B. *et al.* (1997) *The Economic Impact of International Climate Change Policy*, Research Policy 97.4, Australian Bureau of Agricultural and Resource Economics, Canberra.
Brown, T. (1984) The concept of value in resource allocation, *Land Economics*, Fall 1984.
Brundtland, G.H. (1993) Population, environment and development, in *The Rafael M. Salas Memorial Lecture*, UN Population Fund, New York. See also World Commission on Environment and Development (1987, 1992).
Brydges, T.G. and Wilson, R.B. (1991) Acid rain since 1985 – times are changing, *Proceedings of the Royal Society of Edinburgh*, **97b**: 1–16.
Budiansky, S. (1997) *Nature's Keepers*, Weidenfeld and Nicholson, New York.
Bureau of Industry Economics (1990) *Environmental Assessment: Impact on Major Projects*, Research Report No. 35, Australian Government Publishing Service, Canberra.
Butterwick, L., Harrison, R. and Merritt, M. (1992) *Handbook for Urban Air Improvement*, The European Union, Brussels.
Cairncross, F. (1992) *Costing the Earth: The Challenge for Governments: The Opportunities for Business*, Harvard Business School, Boston.
Cairncross, F. (1995) *Green Inc: A Guide to Business and the Environment*, Earthscan, London.
Calow, P. (ed.) (1998) *Handbook of Environmental Risk Assessment and Management*, Blackwell, Oxford.
Calow, P. (ed.) (1998) *Encyclopedia of Ecology and Environmental Management*, Blackwell, Oxford.
Cameron, J. and Aboucher, J. (1991) The precautionary principle: a fundamental principle of law and policy for the protection of the global environment, *Boston College International and Comparative Law Review*, **14**: 1–27.
Carlin, A. (1992) *The US Experience with Economic Incentives to Control Environmental Pollution*, US EPA Economic and Innovations Division, Washington, DC.
Carpenter, R.A. and Dixon, J.A. (1985) Ecology meets economics: a guide to sustainable development, *Environment*, **27** (5).
Carson, R. (1951) *The Sea Around US*, MacGibbon and Kee, London.
Carson, R. (1955) *The Edge of the Sea*, MacGibbon and Kee, London.
Carson, R. (1962) *Silent Spring*, Houghton Mifflin, Boston.
Chen, G. (1993) Impacts of the Three-Gorges project on the eco-environment and countermeasures, in *Development and the Environment*, Proceedings of the 13th Annual Meeting of the International Association for Impact Assessment, 12–15 June, Shanghai, China, IAIA, pp. 18–19.
Cipolla, C.M. (1967) *The Economic History of World Population*, Penguin, Harmondsworth.
Ciriacy-Wantrup, S.V. (1968) *Resource Conservation: Economics and Policies (3rd edn)*, University of California Press, Berkeley, CA.
Clark, C. (1977) *Population, Growth and Land Use (2nd edn)*, Macmillan, London.
Clark, R. and Canter, L.W. (eds) (1996) *Environmental Policy and NEPA: Past, Present and Future*, St Lucie Press, New York.

Clawson, M. (1957) *Federal Lands: Their Use and Management,* Resources For the Future, Washington DC.
Clawson, M. (1966) *Economics of Outdoor Recreation,* Resources For the Future, Washington DC.
Clawson, M. (1975) *Forests for Whom and for What?* Resources For the Future, Washington DC.
Clawson, M. (1983) *Federal Lands Revisited,* Resources For the Future, Washington DC.
Clercq, M. de (1996) The implementation of green product taxes: the Belgian experience, in *The Dublin Workshop, National Experiences of Environmental Taxes and Charges,* The European Foundation for the Improvement of Living and Working Conditions, Dublin.
Coase, R.H. (1937) The nature of the firm, *Economica.*
Coase, R.H. (1960) The problem of social cost, *Journal of Law and Economics,* **3** (October): 1–44.
Cohen, J. (1997) *How Many People can the World Support?* Norton, New York.
Committee on Air Pollution (1954) *Report* (The Beaver Report), Cmnd 9322, London.
Common, M.S. (1988) *Environmental and Resource Economics: An Introduction,* Longman, Harlow.
Common, M.S. (1995) *Sustainability and Policy: Limits to Economics,* Cambridge University Press, Cambridge.
Commoner, B. (1963) *Science and Survival,* The Viking Press, New York.
Commoner, B. (1972) *The Closing Circle: Nature, Man and Technology,* Knopf, New York.
Commonwealth Department of Home Affairs and Environment (1981) Costs and benefits of environment protection, *Proceedings of a National Conference,* October, Melbourne, AGPS.
Coombs, H.C. (1990) *The Return of Scarcity: Strategies for an Economic Future,* Cambridge University Press, Cambridge.
Cornière, P. (1986) Natural resource accounts in France, in *Information and Natural Resources,* OECD, Paris.
Costanza, R. (1997) *Frontiers in Ecological Economics: Transdisciplinary Essays,* Edward Elgar, Cheltenham.
Costanza, R. and Perrings, C. (1990) A flexible assurance bonding system for improved environmental management, *Ecological Economics,* **2** (1): 57–76.
Council on Environmental Quality, US (annual) *Environmental Quality,* CEQ, Washington, DC.
Cropper, M.L. and Oates, W.E. (1992) Environmental economics: a survey, *Journal of Economic Literature,* **XXX**: 675–740.
Cummings, R.G., Brookshire, D.S. and Schulze, W.D. (1986) *Valuing Environmental Goods: An Assessment of the Contingent Valuation Method,* Rowman and Allenheld, New Jersey.
Daily, G.C., Ehrlich, P.R., Mooney, H.A. and Ehrlich, A.H. (1991) Greenhouse economics: learn before you leap, *Ecological Economics,* **4**: 1–10.
Dalkey, N.C. and Helmer, O. (1963) An experimental application of the Delphi method to the use of experts, *Management Science,* **9**: 458–67.
Daly, H.E. (1991) Towards an environmental macroeconomics, *Land Economics,* **67** (2): 2.
Daly, H.E. (1991) *Steady State Economics* (2nd edn), Island Press, Washington, DC.
Daly, H.E. and Cobb, J.B. (1989) *For the Common Good: Redirecting the Economy towards Community, the Environment, and a Sustainable Future,* Beacon Press, Boston.

Dasgupta, P. (1989) Exhaustible resources, in *The Fragile Environment,* the Darwin College Lectures, Cambridge University Press, Cambridge, pp. 107–23.
Dasgupta, P. (1995) The population problem: theory and evidence, *Journal of Economic Literature,* **XXXIII** (December): 1879–902.
Dasgupta, P. (1996) *The Economics of the Environment,* Beijer International Institute of Ecological Economics, Beijer Discussion Paper Series No. 80, Stockholm.
Dasgupta, P. and Heal, G.M. (1992) *The Economic Theory of Exhaustible Resources,* Cambridge University Press, Cambridge.
Dasgupta, P. and Maler, K.-G. (eds) (1997) *The Environment and Emerging Development Issues,* vols 1 and 2, Oxford University Press, Oxford.
Dasgupta, P., Kristrom, B. and Maler, K.-G. (1995) Current issues in resource accounting, in *Current Issues in Environmental Economics,* Manchester University Press, Manchester, pp. 117–52.
Demeny, P. and McNicoll, G. (eds) (1996) *The Earthscan Reader in Population and Development,* Earthscan, London.
Department of Finance (1991) *Handbook of Cost–Benefit Analysis,* Australian Government Publishing Service, Canberra, p. 6.
Diesendorf, M. and Hamilton, C. (1997) *Human Ecology: Human Economy,* Allen and Unwin, Sydney.
Dixon, J.A. and Hufschmidt, M.M. (1986) *Economic Valuation Techniques for the Environment: A Case Study Workbook,* Johns Hopkins Press, Baltimore.
Dixon, J.A., Scura, L.F., Carpenter, R. and Sherman, P.B. (1998) *Economic Analysis of Environmental Impacts* (2nd edn), Earthscan, London.
Dollar, D. and Pritchett, L. (1998) *Assessing Aid: What Works, What Doesn't and Why,* World Bank, Washington, DC.
Dorfman, R. (ed.) (1965) *Measuring Benefits of Government Investments,* Brookings Institution, Washington, DC.
Dorfman, R. and Dorfman, N.S. (eds) (1993) *Economics of the Environment: Selected Readings* (3rd edn), Norton, New York.
Dovers, S. (1994) *Sustainable Energy Systems,* Cambridge University Press, Cambridge.
Doxiados, C.A. (1977) *Ecology and Ekistics,* Elek Books, London; University of Queensland Press, Australia.
Dubos, R.B. (1968) *Man, Medicine and the Environment,* Penguin, Harmondsworth.
Dupuis, J. (1844) On the measurement of the utility of public works (translated), in *International Economic Papers,* **2** (1952): 83–110.
Eckersley, R. (ed.) (1998) *Measuring Progress: Is Life Getting Better?* Commonwealth Scientific and Industrial Research Organization, Canberra.
Eckstein, O. (1958) *Water Resources Development: The Economics of Project Evaluation,* Harvard University Press, Cambridge, MA.
Economic and Social Commission for Asia and the Pacific (1990) *State of the Environment in Asia and the Pacific,* ESCAP, New York.
Economist, The (1989) A survey of the environment: costing the Earth, *The Economist,* London, 2 September.
Economist, The (1995) A survey of cities: turn up the lights, *The Economist,* London, 29 July.
Economist, The (1996) A survey on living with the car: taming the beast, *The Economist,* London, 22 June.
Economist, The (1997) A warming world; Taxes for a cleaner planet, *The Economist,* London, 28 June.
Economist, The (1998) A survey of women and work: for better, for worse, *The Economist,* London, 18 July.

Eggertsson, T. (1993) *Economic Perspectives on Property Rights and the Economics of Institutions*, The Beijer International Institute of Ecological Economics, Beijer Discussion Paper Series No. 40, Stockholm.
Ehrlich, P.R. and Ehrlich, A.H. (1968) *The Population Bomb*, Ballantine, New York.
Ehrlich, P.R. and Ehrlich, A.H. (1970) *Population, Resources, Environment: Issues in Human Ecology*, Freeman, San Francisco.
Ehrlich, P.R. and Ehrlich, A.H. (1996) *The Betrayal of Science and Reason*, Island Press, New York.
El Serafy, S. and Lutz, E. (1989) Environmental and natural resource accounting, in *Environmental Management and Economic Development* (eds G. Schramm and J.J. Warford), The World Bank, Washington, DC, pp. 23–38.
Ellerman, A. and Montero, J. (1996) *Why are Allowance Prices so Low? Analysis of the Sulphur Dioxide Emissions Trading Program*, Massachusetts Institute of Technology, Cambridge.
Elton, C.S. (1927) *Animal Ecology*.
Elton, C.S. (1930) *Animal Ecology and Evolution*.
Environmental Law Network International (1997) *International Environmental Impact Assessment: European and Comparative Law*, Cameron May, London.
Evelyn, J. (1661) *Fumifugium: Or the Smoake of London Dissipated*, National Smoke Abatement Society Reprint, 1933.
Faber, M., Manstetton, R. and Proops, J. (1996) *Ecological Economics: Concepts and Methods*, Edward Elgar, Cheltenham.
Fearnside, P.M. (1993) The Canadian feasibility study for the Three-Gorges dam proposal for China's Yangtze River: a grave embarrassment to the impact assessment profession, in *Impact Assessment*, **12** (1), Spring 1994. A paper presented at the 13th Annual Meeting of the IAIA, 12–15 June, Shanghai, China, IAIA.
Field, B.C. (1997) *Environmental Economics: An Introduction* (2nd edn), McGraw-Hill, New York.
Fisher, A.C. and Krutilla, J.V. (1975) Resource conservation, environmental preservation, and the rate of interest, *Quarterly Journal of Economics*, **89**: 358–70.
Fisher, A.C. and Peterson, F.M. (1976) The environment in economics: a survey, *The Journal of Economic Literature*, **14** (1): 1–33.
Fisher, B. (1998) An economic assessment of the Kyoto Protocol using the ABARE global trade and environmental model, *Proceedings of the 27th Conference of Economists*, University of Sydney.
Fisher, J.L. and Potter, N. (1964) *World Prospects for Natural Resources: Some Projections of Demand and Indicators of Supply to the Year 2000*, Resources for the Future, Washington, DC.
Folmer, H., Gabel, H.L. and Opschoor, H. (eds) (1995) *Principles of Environmental and Resource Economics: A Guide for Students and Decision-makers*, Edward Elgar, Cheltenham.
Food and Agriculture Organization (1995) *Forest Resources Assessment: Global Synthesis*, FAO, Rome.
Food and Agriculture Organization (1996) *The Sixth World Food Survey*, FAO, Rome.
Food and Agriculture Organization (1997) *The Seventh World Food Survey*, FAO, Rome.
Freeman, A.M. (1979) *The Benefits of Environmental Improvements: Theory and Practice*, Johns Hopkins University Press, Baltimore, MD.
Friday, L. and Laskey, R. (1989) *The Fragile Environment: The Darling College Lectures*, Cambridge University Press, Cambridge.

Friend, A.M. (1986) *Natural Resource Accounting and its Relationship with Economic and Environmental Accounting,* Statistics Canada, Ottawa.
Friend, A.M. (1991) *Towards a Pluralistic Approach in National Accounting,* Occasional Paper 3, University of Ottawa, Ottawa.
Galbraith, J.K. (1974) *Economics and the Public Purpose,* Andre Deutsch, London.
Garnasjordet, P.A. and Saebo, H.V. (1986) A system of natural resource accounts in Norway, in *Information and Natural Resources,* OECD, Paris, pp. 15–39.
Garrod, G. and Willis, K.G. (1998) *Economic Valuation of the Environment,* Edward Elgar, Cheltenham.
Gilbert, A.J. and James, D.E. (1988) *Natural Resource Accounting: A Review of Current Activity and its Application in Australia,* AGPS, Canberra.
Gilpin, A. (1963) *Control of Air Pollution,* Butterworths, London.
Gilpin, A. (1986) *Dictionary of Economics and Financial Markets* (5th edn), Butterworths, London.
Gilpin, A. (1986) *Environmental Planning: A Condensed Encyclopedia,* Noyes Publications, New Jersey.
Gilpin, A. (1990) *An Australian Dictionary of Environment and Planning,* Oxford University Press, Melbourne, Australia.
Gilpin, A. (1995) *Environmental Impact Assessment: Cutting Edge for the Twenty-first century,* Cambridge University Press, Cambridge.
Gilpin, A. (1996) *Dictionary of Environment and Sustainable Development,* John Wiley, Chichester.
Gilpin, A. and Lewis, S.J.L. (1990) *Environmental Impact Assessment in Taiwan and Australia: A Comparative Study,* Working Paper No. 20, US East–West Centre, Honolulu, Hawaii.
Global Environment Facility (1995) *Capacity Building Requirements for Global Environmental Protection,* Working Paper No. 12, UNDP, UNEP, World Bank.
Goldsmith, E., Allen, R., Allaby, M., Davoll, J. and Lawrence, S. (1972) *Blueprint for Survival,* Penguin, Harmondsworth, UK.
Goodstein, E. (1995) *Economics and the Environment,* Prentice Hall, London.
Gore, A. (1991) *Earth in Balance,* Routledge, London.
Gowdy, J.M. and O'Hare, S. (1995) *Economic Theory for Environmentalists,* St Lucie Press, Florida.
Gren, I.-M. and Tomasz, Z. (1993) *Cost-effectiveness of the Baltic Sea Clean-up: Will the Wetlands Reconcile Efficiency with Biodiversity?* Beijer International Institute of Ecological Economics, Beijer Discussion Paper Series 24, Stockholm.
Hammond, R.J. (1958) *Benefit–Cost Analysis and Water Pollution Control,* Stanford University Press, Stanford, CA.
Hanna, S. and Munasinghe, M. (eds) (1995) *Property Rights and the Environment: Social and Ecological Issues,* The Beijer International Institute of Ecological Economics and the World Bank, Washington, DC.
Hanna, S., Folke, C. and Mahler, K.-G. (eds) (1996) *Rights to Nature: Ecological, Economic, Cultural and Political Principles,* Island Press, Washington, DC.
Hardin, G. (1968) The tragedy of the commons, *Science,* **162**: 1243–8.
Hardy, A. (1968) Charles Elton's influence on ecology, *Journal of Animal Ecology,* **37**: 1–8.
Hartwick, J.M. (1977) Intergenerational equity and the investing of revenues from exhaustible resources, *American Economic Review,* **66**: 972–4.
Hartwick, J.M. (1990) Natural resources, national accounting, and economic depreciation, *Journal of Public Economics,* **43**: 291–304.

Herfindahl, O.C. and Kneese, A.V. (1965) *Quality of the Environment: An Economic Approach to Some Problems in using Land, Water and Air*, Resources for the Future, Washington, DC.
Herfindahl, O.C. and Kneese, A.V. (1974) *Economic Theory of Natural Resources*, Merrill, Columbus.
Hicks, J.R. (1939) Foundations of welfare economics, *Economic Journal*, **49** (December): 696–712.
Hirsch, F. (1977) *The Social Limits to Growth*, Routledge and Kegan Paul, London.
Hotelling, H. (1931) The economics of exhaustible resources, *Journal of Political Economy*, **39**: 137–75.
House of Commons (1996) *World Trade and the Environment*, vols 1 and 2, HMSO, London.
Hufschmidt, M.M. (ed.) (1980) *New Approaches to the Economic Analysis of Natural Resources and Environmental Quality*, Proceedings of a conference on extended benefit–cost analysis held at the East–West Center, 17–26 September 1979, Environment and Policy Institute, Hawaii.
Hufschmidt, M.M. and Hyman, E.L. (eds) (1982) *Economic Approaches to Natural Resource and Environmental Quality Analysis*, Bray Icooly, Wicklow.
Hufschmidt, M.M., Bower, B.T., James, D., Meister, A.D., Pearson, C.S. and Yamauchi, H. (1981) *The Role of Benefit–Cost Analysis in Environmental Quality and Natural Resource Management*, Environment and Policy Institute, East-West Center, Hawaii.
Hufschmidt, M.M., James, D.E., Meister, A.D., Bower, B.T. and Dixon, J.A. (1983) *Environment, Natural Systems, and Development: An Economic Valuation Guide*, Johns Hopkins Press, Baltimore.
Hulme, M. and Kelly, M. (1993) Exploring the links between desertification and climate change, *Environment*, **35**(6): 4–11, 39–46.
Hyman, E.L. (1981) The valuation of extramarket benefits and costs in environmental impact assessment, *Environmental Impact Assessment Review*, **2**(3), Plenum, New York, pp. 227–58.
Imber, D., Stevenson, G. and Wilks, L. (1991) *A Contingent Valuation Survey of the Kakadu Conservation Zone*, Research Paper No. 3, Resource Assessment Commission, AGPS, Canberra.
Independent Commission on Disarmament and Security (1982) *Common Security*.
Independent Commission on Population and Quality of Life (1996) *Caring for the Future: Making the Next Decades provide a Life Worth Living: The Report of the ICPQL*, Oxford University Press, Oxford.
Intergovernmental Panel on Climate Change (1995) *The Second Assessment Report*, comprising three volumes: *Vol. 1 The Science of Climate Change* (the contribution of Working Group I); *Vol. 2 Impacts, Adaptations and Mitigation of Climate Change; Scientific and Technical Analyses* (the contribution of Working Group II); *Vol. 3 Economic and Social Dimensions of Climate Change* (the contribution of Working Group III), Cambridge University Press, Cambridge.
International Atomic Energy Agency (1991) *Electricity and the Environment*, IAEE-Tecdoc-624, background papers of a senior expert symposium held in Helsinki, 13–17 May, IAEE, Vienna.
International Institute of Environment and Development (1989) *Blueprint for a Green Economy*, Earthscan, London.
International Institute of Environment and Development (1991) *Blueprint 2: Greening the World Economy*, Earthscan, London.

International Institute of Environment and Development/World Resources Institute (1998) *World Resources 1998*, Basic Books, New York.

International Oil Pollution Compensation Fund (1998) *Annual Report*, IOPCF, London.

International Union for the Conservation of Nature and Natural Resources, in partnership with UNEP and the World Wide Fund for Nature (1991) *Caring for the Earth: A Strategy for Sustainable Living*, Gland, Switzerland.

Jacobs, M. (1991) *The Green Economy: Environment, Sustainable Development, and the Politics of the Future*, Pluto Press, London.

Jacobs, M. (1993) *Economic Instruments, Objectives or Tools?* Paper delivered at the Environmental Economics Conference, Canberra.

Jakobsonn, K. (1996) *Contingent Valuation and Endangered Species: Methodological Issues and Applications*, Elgar, London.

James, D. (1993) *Economic Instruments for Meeting Environmental Objectives: Australian Experience*, Ecoservices, Canberra.

James, D. (1997) *Environmental Incentives: Australian Experience with Economic Instruments for Environmental Management*, Environment Australia, Canberra.

Jevons, W.S. (1865) *The Coal Question.*

Jodha, N.S. (1993) *Property Rights and Development*, Beijer International Institute of Ecological Economics, Beijer Discussion Paper Series No. 41, Stockholm.

Jones, P.D., Wigley, T.M.L. and Wright, P.B. (1986) Global temperature variations between 1861 and 1984, *Nature* **322** (6078): 430–4.

Joskow, P. and Schmalensee, R. (1996) *The Political Economy of Market-based Environmental Policy: The US Acid Rain Program*, Centre for Energy and Environmental Policy Research, Massachusetts Institute of Technology, Cambridge.

Karadeloglou, P., Ikwue, T. and Skea, J. (1995) Environmental policy in the European Union, in *Principles of Environmental and Resource Economics: A Guide for Students and Decision-makers*, Elgar, Aldershot and Brookfield.

Kirkby, J., O'Keefe, P. and Timberlake, L. (eds) (1995) *The Earthscan Reader in Sustainable Development*, Earthscan, London.

Kneese, A.V. (1962) *Water Pollution Control, Economic Aspects, and Research Needs*, Resources for the Future, Washington, DC.

Kneese, A.V. and Bower, B.T. (1968) *Managing Water Quality: Economics, Technology, Institutions*, Resources for the Future and the Johns Hopkins Press, Baltimore.

Kneese, A.V., Ayres, R.U. and d'Arge, R.C. (1970) *Economics and the Environment: A Material Balance Approach*, Resources for the Future and the Johns Hopkins Press, Baltimore.

Kormondy, E.J. (ed.) (1989) *International Handbook of Pollution Control*, Greenwood Press, Westport, Connecticut.

Kraemer, R.A. (1995) *The Effectiveness and Efficiency of Water Effluent Charge Systems: A Case Study on Germany*, OECD, Paris.

Krutilla, J.V. (1967) Conservation reconsidered, *American Economic Review*, **57**: 777–86.

Krutilla, J. and Echstein, O. (1958) *Multiple Purpose River Development: Studies in Applied Economic Analysis*, Johns Hopkins Press, Baltimore, MD.

Krutilla, J.V. and Fisher, A.C. (1975) *The Economics of Natural Environments*, Johns Hopkins Press, Baltimore.

Larsen, B. and Shah, A. (1995) Global climate change, energy subsidies, and national carbon taxes, in *Public Economics and the Environment in an Imperfect World* (eds Bovenberg and Cnossen), Kluwer, Dordrecht, pp. 113–32.

Laszlo, E. (1977) *Goals for Mankind*, The Club of Rome, Hutchinson, London.

Lave, L.B. and Seskin, E.P. (1977) *Air Pollution and Human Health,* Johns Hopkins Press, Baltimore.

Leonard, H.J. (1988) *Pollution and the Struggle for the World Product,* Cambridge University Press, Cambridge.

Leontief, W. (1970) Environmental repercussions and the economic structure: an input–output approach, *Review of Economics and Statistics,* **52**: 262–71.

Lewis, J., Webb, R. and Kapur, D. (1997) *The World Bank: Its First Half-century,* Brookings Institution, Washington, DC.

Linstone, H.A. and Turoff, M. (1975) *The Delphi Method: Techniques and Applications,* Addison-Wesley, Reading, MA.

Low, P. (ed.) (1992) *International Trade and the Environment,* World Bank Discussion Paper 159, World Bank, Washington, DC.

Luccarelli, M. (1996) *Lewis Mumford and the Ecological Region, the Politics of Planning,* UCL Press, London.

Lutz, W. (ed.) (1994) *The Future of World Population,* Earthscan, London.

Lutz, W. (1996) *The Future Population of the World: What Can We Assume Today?* Earthscan, London.

Lutz, E. and El Sarafy, S. (1988) *Environmental and Resource Accounting: An Overview,* Working Paper No. 6, World Bank, Washington, DC.

Maass, A.A., Hufschmidt, M.M., Dorfman, R., Thomas, H.A., Marglin, S.A. and Fair, G.M. (1962) *Design of Water Resource Systems,* Harvard University Press, Cambridge, MA.

MacKean, R. (1958) *Efficiency in Government through Systems Analysis,* Wiley, New York.

Mahar, D. (1989) *Government Policy and Deforestation in Brazil's Amazon Basin,* World Bank in co-operation with WWF and the Conservation Foundation, Washington, DC.

Mäler, K.-G., Dasgupta, P. and Vercello, A. (eds) (1996) *The Economics of Transnational Commons,* Clarendon Press, Oxford.

Malthus, T.R. (1798) *Essay on the Principle of Population,* Penguin, Harmondsworth.

Mankiw, G. (1997) *Principles of Economics,* Dryden Press.

Marglin, A. (1963) The social rate of discount and the optimal rate of investment, *Quarterly Journal of Economics,* **77**: 95–111.

Markandya, A. (1988) The value of the environment: a state of the art survey, *Journal of Environmental Economics and Management,* **15**: 142–66.

Markandya, A. and Perrings, C. (1991) *Resource Accounting for Sustainable Development: A Review of Basic Concepts, Recent Debate, and Future Needs,* London Environmental Economics Centre, London.

Markandya, A. and Richardson, J. (eds) (1992) *The Earthscan Reader in Environmental Economics,* Earthscan, London.

Marshall, A. (1890) *Principles of Economics,* Macmillan, London.

Marshall, A. (1920) *Principles of Economics: An Introductory Volume* (8th edn), Macmillan, London.

Marx, K. and Engels, F. (1848) *The Communist Manifesto,* Penguin, Harmondsworth.

Mather, A.S. (1990) *Global Forest Resources,* Belhaven Press, London; Timber Press, Portland, Oregon.

McCarthy, P. (1993) *Problems and Prospects for Green GDP in the Australian National Accounts,* Environmental Economics Conference, Department of the Environment, Sport and Territories, Canberra.

McDonald, J. (1995) *The World Trade Organization and Environment Protection Law and Policy,* paper presented to the 2nd Environmental Outlook Conference, March,

Australian Centre for Environmental Law and the Department of the Environment, Sport and Territories, Sydney.

McHarg, I.L. and Steiner, F.R. (eds) (1998) *To Heal the Earth: Selected Writings of Ian L. McHarg,* Island Press, New York.

McLean, B. (1998) *The US Sulphur Permit Trading Experience: Reducing Emissions from 8.7 to 4.5 Million Tons in Four Years Time,* paper presented to the Emissions Trading Conference, Sydney, June 1998.

Meade, J.E. (1952) External economies and diseconomies in a competitive situation, *Economic Journal,* **62**: 54–67.

Meade, J.E. (1973) *The Theory of Economic Externalities: The Control of Environmental Pollution and Similar Social Costs,* Sijthoff-Leiden, Geneva.

Meadows, D.H., Meadows, D.L., Randers, J. and Behrens, W.W. (1972) *The Limits to Growth: A Report to the Club of Rome's Project on the Predicament of Mankind,* Pan Books, London.

Meadows, D.H., Meadows, D.L., Randers, J. and Behrens, W.W. (1974) *The Limits to Growth* (2nd edn), New American Library, New York.

Meadows, D.H., Meadows, D.L. and Randers, J. (1992) *Beyond the Limits: Global Collapse or a Sustainable Future?* Earthscan, London.

Merrett, S. (1997) *Introduction to the Economics of Water Resources: An International Perspective on Supply and Use,* UCL Press, London.

Ministry of Housing and Local Government (1961) *Pollution of the Tidal Thames: The Report of a Departmental Committee,* HMSO, London.

Mishan, E.J. (1967) *The Costs of Economic Growth,* Pelican, Harmondsworth.

Mishan, E.J. (1976) *Cost–Benefit Analysis,* Praeger, New York.

Mishan, E.J. (1982) *Cost–Benefit Analysis: An Informal Introduction* (3rd edn), Allen and Unwin, London.

Morgenstern, R.D. (ed.) (1997) *Economic Analyses at the US EPA: Assessing Regulatory Impact,* Resources for the Future, Washington, DC.

Morris, P. and Therivel, R. (eds) (1999) *Methods of Environmental Impact Assessment,* UCL Press, London.

Mumford, L. (1961) *The City in History: Its Origins, Its Transformations, and Its Prospects,* Penguin, Harmondsworth.

Mumford, L. (1961) *Renewing of Life.*

Munasinghe, M. (1993) *Environmental Economics and Sustainable Development,* World Bank, Washington, DC.

Munasinghe, M. and Cruz, W. (1995) *Economy Wide Policies and the Environment: Lessons from Experience,* World Bank, Washington, DC.

Munasinghe, M. and McNeely, J. (1994) *Protected Area Economics and Policy: Linking Conservation and Sustainable Development,* World Bank, Washington, DC.

Munasinghe, M. and Shearer, W. (eds) (1995) *Defining and Measuring Sustainability: The Biogeophysical Foundations,* UN University and the World Bank, Washington, DC.

Myrdal, G. (1970) *The Challenge of World Poverty: A World Anti-poverty Program in Outline,* Pantheon, New York.

Myrdal, G. (1973) The economics of an improved environment, in *Who speaks for Earth?* Norton, New York, pp. 67–105.

Ness, G. and Gholay, M. (1996) *Population and Strategies for National Sustainable Development: A Handbook on Population Appraisal and Planning Processes* (for IUCN and UNFPA), Earthscan, London.

Nichols, R. and Hyman, E. (1982) Evaluation of environmental assessment methods, *Journal of the Water Resources, Planning and Management Division*, **108** (WR1), March 1982.
Nicolaisen, J., Dean, A. and Hoeller, P. (1991) *Economics and the Environment: A Survey of Issues and Policy Options*, OECD Economic Studies No. 16, Spring.
Nordhaus, W.O. (1973) World dynamics: measurement without data, *Economic Journal*, 83 (December): 1182.
Nordhaus, W.O. (1994) *Managing the Global Commons: The Economics of Climate Change*, MIT Press, Cambridge.
Nordhaus, W.O. (1997) *The Swedish Nuclear Dilemma: Energy and the Environment*, Resources for the Future, Washington, DC.
Norton, G.A. (1984) *Resource Economics*, Arnold, London.
Norwegian Central Bureau of Statistics (1987) Natural resource accounting and analysis: the Norwegian experience, *Social and Economic Studies* (Oslo), **65**.
Norwegian Central Bureau of Statistics (1990) *Natural Resources and the Environment*, Reports NCDS, Oslo.
Norwegian Green Tax Commission (1996) *Policies for a Better Environment and High Employment*, Oslo, Norway.
O'Connor, M. and Spash, C.L. (eds) (1998) *Valuation and the Environment: Theory, Method and Practice*, Edward Elgar, Cheltenham.
OECD (1975) *The Polluter Pays Principle: Definition, Analysis, Implementation*, OECD, Paris.
OECD (1989) *Economic Instruments for Environmental Protection*, OECD, Paris.
OECD (1997a) *Reforming Environmental Regulation in OECD Countries*, OECD, Paris.
OECD (1997b) *Environmental Taxes and Green Tax Reform*, OECD, Paris.
OECD (1997c) *Evaluating Economic Instruments for Environmental Policy*, OECD, Paris.
OECD (1997d) *Sustainable Development: OECD Policy Approaches for the 21st Century*, OECD, Paris.
OECD (1997e) *Economic Globalization and the Environment*, OECD, Paris.
OECD (1997f) *Sustainable Production and Consumption*, OECD, Paris.
Opschoor, J.B. and Turner, R.K. (eds) (1994) *Economic Incentives and Environmental Policies: Principles and Practice*, Kluwer, Dordrecht.
O'Riordan, T. (1994) *Environmental Science for Environmental Management*, Longman, Harlow.
Ortolano, L. (1997) *Environmental Regulation and Impact Assessment*, John Wiley, New York.
Owen, L. and Unwin, T. (eds) (1997) *Environmental Management: Readings and Case Studies*, Blackwell, Oxford.
Palmer, K.L. and Krupnick, A.J. (1991) Environmental costing and electric utilities' planning and investment, *Resources 105*, Resources for the Future, Washington, DC.
Panayotou, T. (1994) *Economic Instruments for Environmental Management and Sustainable Development*, UNEP Paper No. 16, UNEP, Nairobi.
Parry, I.W.H. (1997) *Reducing Carbon Emissions: Interactions with the Tax System*, Resources for the Future, Washington, DC.
Pearce, D.W. (1976) *Environmental Economics*, Longman, London.
Pearce, D.W. (1978) *The Valuation of Social Cost*, Macmillan, London.
Pearce, D.W. (1983) *Cost–Benefit Analysis* (2nd edn), Macmillan, London.
Pearce, D.W. (1993) *Economic Values and the Natural World*, Earthscan, London.
Pearce, D.W. (1999) *Economics and Environment*, Edward Elgar, Cheltenham.
Pearce, D.W. and Markandya, A. (1989) *Environmental Policy Benefits: Monetary Valuation*, OECD, Paris.

Pearce, D.W. and Moran, D. (1994) *The Economic Value of Biodiversity,* Harvester Wheatsheaf, Hemel Hempstead.
Pearce, D.W. and Turner, R.K. (1990) *Economics of Natural Resources and the Environment,* Harvester Wheatsheaf, Hemel Hempstead; Johns Hopkins University Press, Baltimore.
Pearce, D.W. and Warford, J.J. (1993) *World without End: Economics, Environment, and Sustainable Development,* Oxford University Press, Oxford.
Pearce, D.W., Markandya, A. and Barbier, E.B. (1989) *Blueprint for a Green Economy,* Earthscan, London.
Pearce, D., Barbier, E. and Markandya, A. (1990) *Sustainable Development, Economics and Environment in the Third World Earthscan, London.*
Pearce, F. (1995) The biggest dam in the world, *New Scientist,* January: 25–9.
Perkins, F. (1994) *Practical Cost–Benefit Analysis: Basic Concepts and Applications,* Macmillan, Melbourne.
Perman, R. and McGilvray, J. (1996) *Natural Resource and Environmental Economics,* Longman, London.
Perrings, C.A. (1987) *Economy and Environment,* Cambridge University Press, New York.
Perrings, C.A. (1989) Environmental bonds and environmental research, *Ecological Economics,* **1**(1): 95–115.
Perrings, C.A., Turner, K. and Folke, C. (1995) *Ecological Economics: The Study of Interdependent Economic and Ecological Systems,* Beijer Dicussion Paper, Series No. 55, Stockholm.
Peskin, H. and Lutz, E. (1990) *A Survey of Resource and Environmental Accounting in Industrialized Countries,* World Bank Working Paper No. 37, World Bank, Washington, DC.
Peterson, R.T. (1934) *A Field Guide to the Birds, giving Field Marks of All Species found East of the Rockies,* Boston Houghton Mifflin, New York.
Peterson, R.T. (1954) *A Field Guide to the Birds of Britain and Europe,* Collins, London.
Pezzey, J. (1992) *Sustainable Development Concepts: An Economic Analysis,* World Bank, Washington, DC.
Pickering, K.T. and Owen, L.A. (1997) *An Introduction to Global Environmental Issues* (2nd edn), Routledge, London.
Pigou, A.C. (1946) *The Economics of Welfare* (4th edn), Macmillan, London.
Pimm, S.L., Russell, G.J., Gittleman, J.L. and Brooks, T.M. (1995) The future of biodiversity, *Science,* **269**: 347–50.
Pittevils, I. (1996) *Ecotaxes on Products in Belgium,* paper to the Dublin Workshop of the European Foundation for the Improvement of Living and Working Conditions, Dublin.
Portney, P.R. (ed.) (1990) *Public Policies for Environmental Protection,* Resources for the Future, Washington, DC.
Pratt, E.T. (1995) Intellectual property rights and international trade, *The Economist,* 27 May: 26.
President's Materials Policy Commission (1952) *Resources for Freedom: Foundation for Growth and Security,* Washington, DC.
Prest, A.R. and Turvey, R. (1965) Cost–benefit analysis: a survey, *The Economic Journal,* **75** (December): 683–735.
Qing, Dai (1998) *The River Dragon has come! The Three-Gorges Dam and the Fate of China's Yangtze River and Its People,* Sharpe, New York.
Ramphal, S. and Sinding, S.W. (1996) *Population Growth and Environmental Issues,* Praeger, Westport and London.

Reid, D. (1995) *Sustainable Development: An Introductory Guide*, Earthscan, London.

Repetto, R. (1987) *Natural Resource Accounting for Countries with Natural Resource-based Economies*, a report for the Australian Environment Council, Canberra.

Repetto, R. (1993) *Natural Resource Accounting: Benefits and Prospects*, Environmental Economics Conference, Canberra, 15–17 November.

Repetto, R., Magrath, W., Wells, M., Beer, C. and Rossini, F. (1989) *Wasting Assets: Natural Resources in the National Income Accounts*, World Resources Institute, Washington, DC.

Repetto, R., Magrath, W., Wells, M., Beer, C. and Rossini, F. (1992) Wasting assets: natural resources in the national income accounts, in *The Earthscan Reader in Environmental Economics* (eds A. Markandya and J. Richardson), Earthscan, London.

Resource Assessment Commission (1992) *Multi-criteria Analysis as a Resource Assessment Tool*, Research Paper No. 6. AGPS, Canberra.

Rich, B. (1994) *Mortgaging the Earth: The World Bank, Environmental Impoverishment and the Crisis of Development*, The World Bank, Washington, DC.

Riley, C.J. (1996) *The New UK Landfill Tax*, paper to the *Dublin Workshop*, European Foundation for the improvement of living and working conditions, Dublin.

Roback, J. (1982) Wages, rents and the quality of life, *Journal of Political Economy*, **90**: 1257–78.

Robison, H.D. (1985) Who pays for industry pollution abatement? *Review of Economic Statistics*, **67** (4): 702–6.

Rugman, A.M. and Kirton, J.J. (eds) (1998) *Trade and the Environment: Economic, Legal and Policy Perspectives*, Edward Elgar, Cheltenham.

Salim, E. (1988) *Social Impact Assessment: The Indonesian Experience*, Presentation to the 7th Meeting of the International Association of Impact Assessment, Griffith University, Australia.

Samuelson, P. and Nordhaus, W. (1995) *Economics*, McGrawHill, New York.

Savage, E. and Hart, A. (1993) Economic instruments: balancing equity and efficiency, *Proceedings of the Environmental Economics Conference: Moving to Sustainability*, Canberra.

Scheumann, W. and Schiffler, M. (eds) (1998) *Water in the Middle East: Potential for Conflicts and Prospects for Cooperation*, Springer-Verlag, Germany.

Schofield, J.A. (1987) *Cost–Benefit Analysis in Urban and Regional Planning*, Allen and Unwin, London.

Serageldin, I. (1996) *Sustainability and the Wealth of Nations: First Steps in an Ongoing Journey*, Monograph Series No. 5, World Bank, Washington, DC.

Serageldin, I. and Steer, A. (1994) *Making Development Sustainable: From Concepts to Action*, World Bank, Washington, DC.

Shah, A. and Larsen, B. (1992) *Carbon Taxes, the Greenhouse Effect and Developing Countries*, Working Paper WPS 957, World Bank, Washington, DC.

Siebert, H. (1998) *Economics of the Environment: Theory and Policy* (5th edn), Springer Verlag, Germany.

Simon, J.L. (1981) *The Ultimate Resource*, Princeton University Press, Princeton.

Simon, J.L. (ed.) (1995) *The State of Humanity*, Blackwell, Oxford.

Simon, J.L. and Kahn, H. (1984) *The Resourceful Earth: A Response to GLOBAL 2000*, Blackwell, Oxford.

Sinden, J.A. (1990) *Valuation of Unpriced Benefits and Costs of River Management*, Office of Water Resources, Victorian Department of Conservation and Environment, Melbourne.

Sinden, J.A. and Thampapillai, D.J. (1995) *Introduction to Benefit–Cost Analysis*, Longman, Melbourne.

Sinden, J.A. and Worrell, A.C. (1979) *Unpriced Values: Decisions without Market Prices*, Wiley, Chichester and New York.
Smil, V. (1996) *Environmental Problems in China: Estimates of Economic Costs*, US East–West Center, Honolulu, Hawaii.
Smith, A. (1776) *The Wealth of Nations*, reprinted 1961 by Methuen, London.
Smith, K.R. (1988) *The Risk Transition*, Environment and Policy Institute, East–West Center, Honolulu, Hawaii.
Smith, V.K. (ed.) (1996) *Estimating Economic Values for Nature: Methods for Non-market Valuation*, Edward Elgar, Cheltenham.
Smith, V.K. and Desvousges, W.H. (1986) Asymmetries in the valuation of risk and the siting of hazardous waste disposal facilities, *American Economic Review, Proceedings*, **75** (May): 291–4.
Solow, R. (1974a) The economics of resources or the resources of economics, *American Economic Review*, **64**: 1–21.
Solow, R. (1974b) Intergenerational equity and exhaustible resources, *Review of Economic Studies Symposium*, 29–45.
Solow, R. (1986) On the intergenerational allocation of natural resources, *Scandinavian Journal of Economics*, **88**: 141–9.
Solow, R. (1992) *An Almost Practical Step Toward Sustainability*, Resources for the Future, Washington, DC.
Squire, L. and van der Tak, H.G. (1975) *Economic Analysis of Projects*, a World Bank Research Publication, Johns Hopkins University Press, Baltimore.
Stigler, G.J. (1996) *Theory of Price* (3rd edn), Macmillan, London.
Streeting, M. (1990) *A Survey of the Hedonic Price Technique*, Research Paper No. 1, September, Resource Assessment Commission, Canberra.
Suzuki, D. and O'Connell, A. (1997) *The Sacred Balance: Rediscovering Our Place in Nature*, Allen and Unwin, Sydney.
Swanson, T. and Johnston, S. (1999) *Global Environmental Problems and International Environmental Agreements*, Edward Elgar, Cheltenham.
Taylor, G.R. (1970) *The Doomsday Book*, Panther, London.
Thampapillei, D.J. (1991) *Environmental Economics*, Oxford University Press, Oxford.
The White House (1965) *Restoring the Quality of Our Environment: The Report of the President's Science Advisory Committee*, The White House, Washington, DC, November 1965.
The White House (1992) *National Report to the UN Conference on Environment and Development 1992, Rio de Janeiro*, The White House, Washington, DC.
Tietenberg, T. (1985) *Emissions Trading*, Resources for the Future, Washington, DC.
Tietenberg, T. (1990) Economic instruments for environmental regulation, *Oxford Review of Economic Policy*, vol. 16, Oxford University Press, Oxford.
Tietenberg, T. (1994) *Economics and Environmental Policy*, Edward Elgar, Cheltenham.
Tietenberg, T. (1996) *Environmental and Natural Resource Economics* (4th edn), Harper Collins, New York.
Tietenberg, T., Button, K.G. and Nijkamp, P. (eds) (1999) *Environmental Instruments and Institutions*, Edward Elgar, Cheltenham.
Tisdell, C.A. (1990) *Natural Resources: Growth and Development*, Praeger, New York.
Tisdell, C.A. (1991) *Economics of Environmental Conservation*, Elsevier, Amsterdam.
Tisdell, C.A. (1993) *Environmental Economics: Policies for Environmental Management and Sustainable Development*, Edward Elgar, Cheltenham.
Tisdell, C.A. (1998) *Biodiversity, Conservation and Sustainable Development*, Edward Elgar, Cheltenham.

Tobey, J.A. (1989) *The Impact of Domestic Environmental Policies on International Trade,* Dissertation, Department of Economics, University of Maryland, MD.
Tobey, J.A. (1990) *Patterns of World Trade: An Empirical Test,* Kyklos.
Trexler, M. and Haugen, C. (1995) *Keeping it Green: Tropical Forestry Opportunities for Mitigating Climate Change,* World Resources Institute, Washington, DC.
Turner, R.K. (ed.) (1993) *Sustainable Environmental Economics and Management: Principles and Practice,* Belhaven Press, London and New York.
Turner, R.K., Pearce, D.W. and Bateman, I. (1993) *An Introduction to Environmental Economics,* Harvester Wheatsheaf, Hemel Hempstead.
Turvey, R. (1963) On divergencies between social cost and private cost, *Economica,* **30** (August).
UN (1993) *Handbook of National Accounting: Integrated Environmental and Economic Accounting,* Series F, No. 61, UN, New York,
UN (1994) *Proceedings of the UN Expert Group Meeting on Population, Environment, and Development,* January 1992, UN, New York.
UN (1995) *Report of the Conference of the Parties to the Framework Convention on Climate Change,* held in Berlin, 28 March–7 April, UN, New York.
UN Centre for Human Settlements (1994) Cities in the developing world, in *Population, Environment and Development: Proceedings of an Expert Group Meeting,* UN, New York.
UNDP (1992) *Long-range Population Projections: Two Centuries of Population Growth 1950–2150,* UNDP, New York.
UNDP (1998) *Human Development Report,* Oxford University Press, Oxford.
UNEP (1982) *The State of the Environment 1972–1982,* UNEP, Nairobi.
UNEP (1991) *Urban Air Pollution,* UNEP, Nairobi.
UNEP (1992) *Workshop on Environmental and Natural Resource Accounting: Summary Record,* UNEP, Nairobi.
UNEP (1992) *Chemical Pollution: A Global Overview,* UNEP, Nairobi.
UNEP (1993a) *Global Biodiversity,* UNEP, Nairobi.
UNEP (1993b) *Environmental Accounting: A Review of the Current Debate* (eds A. Markandya and R. Costanza), UNEP, Nairobi.
UNEP (1994a) *The Pollution of Lakes and Reservoirs,* UNEP, Nairobi.
UNEP (1994b) *Air Pollution in the World's Megacities,* UNEP, Nairobi.
UNEP (1995a) *UNEP's New Way Forward: Environmental Law and Sustainable Development* (eds Lin and Kurukulasuriya), UNEP, Nairobi.
UNEP (1995b) *Energy, Pollution, Environment and Health,* UNEP, Nairobi.
UNEP (1996) *Environmental Accounting* (ed. Abaza), UNEP, Nairobi.
UNICEF (1998) *The Progress of Nations,* UNICEF, New York.
Urban, F. and Trueblood, M. (1990) *World Population by Country and Region 1950–2020,* Economic Research Service, US Department of Agriculture, Washington, DC.
US Department of Commerce (1996) *Pollution Abatement Costs and Expenditures,* Government Printing Office, Washington, DC.
US National Commission on Materials Policy (1973) *Materials Needs and the Environment: Today and Tomorrow,* Government Printing Office, Washington, DC.
US EPA (1985) *Costs and Benefits of Reducing Lead in Gasoline: Final Regulatory Impact Analysis,* EPA, Washington, DC.
US EPA (1990) *Environmental Investments: The Cost of a Clean Environment,* November, EPA, Washington, DC.
Van Den Bergh, J.C.J.M. (ed.) (1999) *Handbook of Environmental and Resource Economics,* Edward Elgar, Cheltenham.

Ward, B. (1966) *Spaceship Earth*, Hamish Hamilton, London.
Ward, B. (1976) *The Home of Man*, Penguin, Harmondsworth.
Ward, B. and Dubos, R. (1972) *Only One Earth: The Care and Maintenance of a Small Planet*, Penguin, Harmondsworth.
Weber, J.L. (1983) The French natural patrimony accounts, *Statistical Journal of the UNECE*, **1**: 419–44.
Weizsacker, E. von, Lovins, A.B. and Lovins, L.H. (1997) *Factor Four: Doubling Wealth, Halving Resource Use. The New Report to the Club of Rome*, Allen and Unwin, Sydney.
Wells, G. (1989) Observing Earth's environment from space, *The Fragile Environment*, the Darwin College Lectures (eds L. Friday and R. Laskey), Cambridge University Press, Cambridge.
Wills, I. (1997) *Economics and the Environment*, Allen and Unwin, Sydney.
Wills, I. (1998) Information exchange in pollution regulation, *Proceedings of the 27th Conference of Economists*, University of Sydney, Sydney.
Winpenny, J.T. (1991) *Values for the Environment: A Guide to Economic Appraisal*, HMSO, London.
Wolozin, H. (ed.) (1966) *The Economics of Air Pollution: A Symposium*, Norton, New York.
World Bank (1994) *Making Development Sustainable*, World Bank, Washington, DC.
World Bank (1995a) *Monitoring Environmental Progress*, World Bank, Washington, DC.
World Bank (1995b) *Mainstreaming the Environment*, World Bank, Washington, DC.
World Bank (1996) *Environment Matters and the World Bank*, World Bank, Washington, DC.
World Bank (1997) *Defining and Measuring Sustainability: the Biogeophysical Foundations* (eds M. Munasinghe and W. Shearer), World Bank, Washington, DC.
World Business Council for Sustainable Development/UNEP (1998) *Industry, Fresh Water, and Sustainable Development*, Nairobi.
World Commission on Environment and Development (1987) *Our Common Future*, Oxford University Press, Oxford.
World Commission on Environment and Development (1992) *Our Common Future Reconvened*, WCED, London.
World Conservation Union, UNEP and WWF for Nature (1991), *Caring for the Earth: A Strategy for Sustainable Living*, WCU, Gland, Switzerland.
World Health Organization (1990) *Potential Health Effects of Climatic Change*, WHO, Geneva.
World Resources Institute (1996) *World Resources 1996–97*, World Resources Institute, UNEP, UDEP and the World Bank, Oxford University Press, Oxford.
World Resources Institute (1998) *World Resources 1998–99*, Oxford University Press, Oxford.
World Trade Organisation (1996) *Report of the Committee on Trade and the Environment*, WTO, New York.
Yang, H. (1993) Analysis and assessment of the Three-Gorges project on the natural scenery of the Three-Gorges, in *Development and the Environment*, Proceedings of the 13th Annual Meeting of the International Association for Impact Assessment, 12–15 June, Shanghai, China.
Young, R. (1988) *Environmental Degradation in the Murray–Darling Basin: An Evaluation of Some Options*, paper presented to the 33rd Annual Conference of the Australian Agricultural Economics Society, La Trobe University, Melbourne.
Zylicz, T. (1992) *Debt-for-Environment Swaps: The Institutional Dimension*, Beijer International Institute of Ecological Economics Discussion Paper No. 18, Stockholm.

Name Index

Index compiled by Annette Musker

Subject Index

Note: Page references in *italics* refer to Figures; those in **bold** refer to Tables

Index compiled by Annette Musker